Molecular Design and Bioorganic Catalysis

NATO ASI Series

Advanced Science Institutes Series

A Series presenting the results of activities sponsored by the NATO Science Committee, which aims at the dissemination of advanced scientific and technological knowledge, with a view to strengthening links between scientific communities.

The Series is published by an international board of publishers in conjunction with the NATO Scientific Affairs Division

A Life Sciences **B Physics**	Plenum Publishing Corporation London and New York
C Mathematical and Physical Sciences **D Behavioural and Social Sciences** **E Applied Sciences**	Kluwer Academic Publishers Dordrecht, Boston and London
F Computer and Systems Sciences **G Ecological Sciences** **H Cell Biology** **I Global Environmental Change**	Springer-Verlag Berlin, Heidelberg, New York, London, Paris and Tokyo

PARTNERSHIP SUB-SERIES

1. Disarmament Technologies	Kluwer Academic Publishers
2. Environment	Springer-Verlag / Kluwer Academic Publishers
3. High Technology	Kluwer Academic Publishers
4. Science and Technology Policy	Kluwer Academic Publishers
5. Computer Networking	Kluwer Academic Publishers

The Partnership Sub-Series incorporates activities undertaken in collaboration with NATO's Cooperation Partners, the countries of the CIS and Central and Eastern Europe, in Priority Areas of concern to those countries.

NATO-PCO-DATA BASE

The electronic index to the NATO ASI Series provides full bibliographical references (with keywords and/or abstracts) to more than 50000 contributions from international scientists published in all sections of the NATO ASI Series.
Access to the NATO-PCO-DATA BASE is possible in two ways:

– via online FILE 128 (NATO-PCO-DATA BASE) hosted by ESRIN, Via Galileo Galilei, I-00044 Frascati, Italy.

– via CD-ROM "NATO-PCO-DATA BASE" with user-friendly retrieval software in English, French and German (© WTV GmbH and DATAWARE Technologies Inc. 1989).

The CD-ROM can be ordered through any member of the Board of Publishers or through NATO-PCO, Overijse, Belgium.

Series C: Mathematical and Physical Sciences – Vol. 478

Molecular Design and Bioorganic Catalysis

edited by

Craig S. Wilcox

and

Andrew D. Hamilton

Department of Chemistry,
University of Pittsburgh,
Pittsburgh, PA, U.S.A.

Kluwer Academic Publishers

Dordrecht / Boston / London

Published in cooperation with NATO Scientific Affairs Division

Proceedings of the NATO Advanced Research Workshop on
Bioorganic Catalysis
Johnstown, Pennsylvania, U.S.A.
May 17–21, 1995

A C.I.P. Catalogue record for this book is available from the Library of Congress

ISBN-13:978-94-010-7252-6 e-ISBN-13:978-94-009-1679-1
DOI:10.1007/978-94-009-1679-1

Published by Kluwer Academic Publishers,
P.O. Box 17, 3300 AA Dordrecht, The Netherlands.

Kluwer Academic Publishers incorporates the publishing programmes of
D. Reidel, Martinus Nijhoff, Dr W. Junk and MTP Press.

Sold and distributed in the U.S.A. and Canada
by Kluwer Academic Publishers,
101 Philip Drive, Norwell, MA 02061, U.S.A.

In all other countries, sold and distributed
by Kluwer Academic Publishers Group,
P.O. Box 322, 3300 AH Dordrecht, The Netherlands.

Printed on acid-free paper

This book contains the proceedings of a NATO Advanced Research Workshop held within the programme of activities of the NATO Special Programme on Supramolecular Chemistry as part of the activities of the NATO Science Committee.

Other books previously published as a result of the activities of the Special Programme are:

WIPFF, G. (Ed.), *Computational Approaches in Supramolecular Chemistry.* (ASIC 426) 1994. ISBN 0-7923-2767-5

FLEISCHAKER, G.R., COLONNA, S. and LUISI, P.L. (Eds.), *Self-Production of Supramolecular Structures.* From Synthetic Structures to Models of Minimal Living Systems. (ASIC 446) 1994. ISBN 0-7923-3163-X

FABBRIZZI, L., POGGI, A. (Eds.), *Transition Metals in Supramolecular Chemistry.* (ASIC 448) 1994. ISBN 0-7923-3196-6

BECHER, J. and SCHAUMBURG, K. (Eds.), *Molecular Engineering for Advanced Materials.* (ASIC 456) 1995. ISBN 0-7923-3347-0

LA MAR, G.N. (Ed.), *Nuclear Magnetic Resonance of Paramagnetic Macromolecules.* (ASIC 457) 1995. ISBN 0-7923-3348-9

SIEGEL, JAY S. (Ed.), *Supramolecular Stereochemistry.* (ASIC 473) 1995. ISBN 0-7923-3702-6

Table of Contents

Preface ix

List of Participants and Observers xi

J. L. Liras and Eric V. Anslyn 1
The Site of Cleavage of Pyranoside Acetals: Endo *vs*. Exocyclic Cleavage

P. Dowd, Z. B. Zheng, R. Hershline, S. W. Ham, S. Naganathan and J. Kerns 15
Vitamin K Mechanism of Action and Its Bearing on Vitamin E

B. Imperali, R. S. Roy, G. K. Walkup and L. Wang 35
Unnatural Amino Acids for the Design of Functional Proteins: Biomimetic Catalysis Using Coenzyme-Amino Acids

H.-D. Jakubke, U. Eichhorn, S. Gerisch, M. Hänsler, K. Salchert, G. Ullmann and D. Ullmann 53
Nonconventional Approaches to Enzyme-Catalyzed Peptide Synthesis

R. Cacciapaglia and L. Mandolini 71
Alkaline-Earth Metal Ion Catalysis of Ester Cleavage

C. Raposo, M. Crego, M. Almaraz, M. L. Mussons, M. Martin, V. Weinrich, M. C. Caballero and J. R. Morán 87
A Study of the Catalytic Properties of Some Hydrogen Bonding Receptors

J. S. Nowick, I. Q. Lee, G. Mackin, M. Pairish, A. J. Shaka, E. M. Smith and J. W. Ziller 111
Design, Synthesis, and Study of Artificial β-Sheets

D. M. Rudkevich, J. Scheerder and D. N. Reinhoudt 137
Anion Recognition by Neutral Receptors

F. Yang and A. J. Russell 163
Enzymes in the Continuous Gas Phase Bioreactor

F. P. Schmidtchen, M. Berger, A. Metzger, K. Gloe and H. Stephan 191
Foldable Anion Hosts Based on Bicyclic Guanidinium Anchor Groups

P. Scrimin, P. Tecilla, U. Tonellato and A. Veronese 211
Transition Metal Ion-Based Supramolecular Catalysts

B. D. Smith, J. T. Bien and M. P. Hughes 227
Neutral Organoboron Receptors with Oriented Dipoles

P. L. Toogood 243
Regulation of Protein Synthesis by Antitumor Peptides

Author Index 263

Subject Index 264

Preface

One of the most active areas of contemporary organic chemistry involves the search for new catalysts that borrow concepts, strategies and even components from enzymes but yet are not found in nature. Such artificial enzymes not only give enormous insights into the mechanisms of enzyme catalysis but also offer the potential for catalyzing a wide range of chemical reactions with no counterpart in nature. Several approaches have been taken in the development of new catalysts, some based on biological methods and others on synthetic techniques.

Site directed mutagenesis has allowed the direct replacement of amino acids in an enzyme with resulting changes in stability, selectivity and mechanism. Recent developments have shown that even non-natural amino acids can be incorporated into proteins and also that enzymes can function effectively in organic solvents. A different biological route to artificial enzymes has exploited the immune system and its ability to generate millions of antibodies to a given antigen. Novel antigens have been designed to mimic the transition states of chemical reactions. Antibodies elicited against these antigens thus contain an active site that is complementary to transition state structure and can potentially catalyze target reactions. A broad range of reactions can now be catalyzed using the method with rate accelerations reaching 10^6 compared to the control reactions. Protein engineering and catalytic antibodies represent complex solutions to the problem of artificial enzymes. Their complexity is however their principal limitation. Little is currently known about either the structure or the mechanism of catalytic antibodies yet their potential is enormous. At the opposite end of the scale, synthetic hosts offer a simple approach to artificial enzymes. It is possible to synthesize small molecule receptors that marshal several catalytic groups and that show shape and binding complementarity to the transition state of a reaction. Such small receptors provide structural, thermodynamic and kinetic information often in a detail not possible with large biomolecules. Hence, a greater insight may be obtained on the nature of the catalytic effect. However, synthetic receptors are limited by their simplicity and the difficulty of building in both the number and precise positioning of reactive groups. As a result the rate accelerations that can be achieved are often small and almost always the reactions do not turnover, since the catalysts are strongly inhibited by the product. One of the largest unsolved problems in the synthetic area is the design of catalytic receptors that bind more strongly to reaction transition states than to products.

In order to discuss possible solutions to these and other problems, a NATO Advanced Research Workshop on Bioorganic Catalysis was organized. The meeting was held in May 1995 on the campus of the University of Pittsburgh, Johnstown, nestled in the foothills of the Laurel Highlands. Leaders in the field of bioorganic chemistry and catalysis were brought together to discuss the state-of-the-art in artificial enzymes and possible future directions of the field. Participants were drawn from the many areas of the field including protein engineering, catalytic antibodies, RNA catalysis, enzymes in organic solvents, micellar catalysis, and general supramolecular catalysis. Particular focus was placed on discussing the strengths and weaknesses of each strategy and what could be learned from a candid assessment of the present achievements of the different approaches. The meeting led not only to a cross-fertilization of ideas among proponents of the different approaches but also to a sober (and occasionally less sober) reflection on the current goals and missed opportunities in the area.

The papers making up this volume are drawn from those presented at the meeting. They provide an excellent perspective on the advantages and disadvantages of the different approaches to understanding and mimicking bioorganic catalysis. There is little doubt that the next ten years will see continued success in the development of highly active catalysts. What is equally certain is that this progress will be inspired by and derived from enzymatic mechanisms of catalytic behavior.

We are indebted to Ms. Fran Nagy for her professional and friendly assistance in preconference arrangements, and in manuscript preparation. On behalf of all those who were promptly reimbursed for travel and lodging, we thank Ms. Cynthia Funtal for her excellent help with account management. Many of our students and research associates were involved in making the meeting run smoothly, and for this we thank them. Finally, we wish to gratefully acknowledge the administrators and scientific panel of the NATO Special Programme on Supramolecular Chemistry for the financial support which made the meeting and this book possible.

Andrew D. Hamilton and Craig S. Wilcox
Department of Chemistry
University of Pittsburgh
Pittsburgh, PA 15260 USA

List of Participants and Observers

Dr. James C. Adrian, Jr.
Chemistry Department
Union College
Schenectady, NY 12308
518/388-6406 *phone*
518/388-6795 *fax*

Dr. Eric Anslyn
Department of Chemistry
The University of Texas
Austin, TX 78712
512/471-0068
512/471-8696
anslyn@ccwf.cc.utexas.edu

Mr. Philippe Barthelemy
Emory University
Department of Chemistry
P. O. Box 127
1515 Pierce Drive
Atlanta, GA 30322
404/727-0027
404/727-6586

Mr. Claus Beckman
Chemistry Department
Lensfield Road
Cambridge CB2 1EW
UNITED KINGDOM

Professor G. M. Blackburn FRSC
Department of Chemistry
Sheffield S3 7HF
UNITED KINGDOM
44 1142 824474
44 1142 738673
g.m.blackburn@sheffield.ac.uk

Dr. Andreas Bommarius
Degussa AG
P. O. Box 1345
Hanau 63403
GERMANY
49 6181 593 219
49 6181 594 312

Ms. Lynn Bretscher
Department of Chemistry
University of Pittsburgh
Pittsburgh, PA 15260
412/624-8275
412/624-8552

Dr. Roberta Cacciapaglia
Dipartimento di Chimica
Universita La Sapienza
P. le A. Moro 5 Roma 00185
ITALY
39 6 495 7808
39 6 490 421

Ms. Hilal Aksoy Cam
Department of Chemistry
Middle East Technical University
Ankara 06531
TURKEY
90 312 210 00/3250
90 312 210 12 80

Dr. James Canary
Department of Chemistry
New York University
31 Washington Place
New York, NY 10003-5180
212/998-8422
212/260-7905

Ms. Heather Carlson
Yale University
P.O. Box 208107
New Haven, CT 06520-8107
203/432-6288
203/432-6144

Dr. Barbara J. Carter
2801 West Bancroft Street
Department of Chemistry
University of Toledo
Toledo, OH 43606-3390
419/537-7867
419/537-4940

Dr. Shelley D. Copley
University of Colorado at Boulder
Dept. of Chemistry & Biochem.
Campus Box 215
Boulder, CO 80309
303/492-6328
303/492-5894
copley@terra.colorado.edu

Dr. Jeffrey T. Davis
Dept. of Chemistry & Biochemistry
University of Maryland
College Park, MD 20742
301/405-1845
301/314-9121

Dr. Ulf Diederichsen
Inst. of Org. Chemie & Biochemie II
TU - Muenchen, Lichtenbergstr. 4
Garching 85747
GERMANY
49 89 3209 3333
49 89 170 8294

Mr. Cesar A. Raposo Funcia
D. Quimica Organica
U. Salamanca
Plaza de los Caidios 1-5 Salamanca
37008 SPAIN
34 23 294 481

Mr. Carsten Gittel
Lehrstuhl für Organische Chemie und Biochemie
Technische Universität München
Lichtenbergstrasse 4
D-85747 Garching
FEDERAL REPUBLIC OF GERMANY
49 89 3209 3338
49 89 3209 3345

Mr. Neil Glagovich
University of Pittsburgh
Department of Chemistry
Pittsburgh, PA 15260
412/624-8269
412/624-8552

Ms. Gülin Güler
Facettepe University
Department of Biology
06532, Beytepe, Ankara
TURKEY
90 312 235 2341
90 312 235 2341

Professor Andrew D. Hamilton
University of Pittsburgh
Department of Chemistry
Pittsburgh, PA 15260
412/624-8340
412/624-8552

Ms. Mu He
University of Pittsburgh
Department of Chemistry
Pittsburgh, PA 15260
412/624-8275
412/624-8552

Dr. Chris Hunter
Department of Chemistry
University of Sheffield
Sheffield S3 7HF, UNITED KINGDOM
44 114 282 4502
44 114 273 8673
c.hunter@sheffield.ac.uk

Prof. Dr. Hans-Dieter Jakubke
Leipzig University
Faculty of Biosciences, Pharmacy and Psychology
Department of Biochemistry
Talstr. 33, D-04103, Leipzig
GERMANY
49 341 97 36 901
49 341 96 03 099
jakubke@iz.uni-leipzig.de

Mr. Eun-il Kim
University of Pittsburgh
Chemistry Department
Pittsburgh, PA 15260
412/624-8269
412/624-8552

Dr. Nikos Kyriakidis
University of Athens
Iera Odos 75
Votanikos
11855 Athens
GREECE
30 1 529 4691

Mr. Stephen Lee
Emory University
Chemistry Department
1515 Pierce Drive
Atlanta, GA 30322
404/727-0027
404/727-6586

Dr. Mark Lipton
Department of Chemistry
Purdue University
West Lafayette, IN 47907-1393
317/494-0132
317/494-0239

Dr. Helen S. M. Lu
DuPont Merck Pharmaceutical Company
P. O. Box 80328
Wilmington, DE 19880-0328
302/695-2051
302/695-1128

Mr. David McElfresh
Department of Chemistry
University of Pittsburgh
Pittsburgh, PA 15260
412/624-8275
412/624-8552

Ms. Corinne Mounier
Georgia Institute of Technology
School of Chemistry and Biochemistry
Atlanta, GA 30332
404/894-4042
404/894-7452

Dr. James S. Nowick
University of California, Irvine
Department of Chemistry
Irvine, CA 92717-2025
714/824-6091
714/824-8571
jsnowick@uci.edu

Mr. Sunil Paliwal
University of Pittsburgh
Department of Chemistry
Pittsburgh, PA 15260
412/624-8269
412/624-8552
Dr. Dmitry Rudkevich
Laboratories of Organic Chemistry
University of Twente
P.O. Box 217
7500 AE Enschede
THE NETHERLANDS
xx xxx xxxxx
31 533 12738 *fax*

Mr. Christian Seitz
Institut fur Org. Chemie
University of Munchen
Garching 85747
GERMANY
49 89 3209 3044
49 89 3209 3345

Mr. Shu-ou Shan
Department of Biochemistry
B400 Beckman Center
Stanford University
Stanford, CA 94305-5307
415/725-6376
415/723-6783

Dr. Jumi Shin
Department of Chemistry
University of Pittsburgh
Pittsburgh, PA 15260
412/624-4927
412/624-8552

Dr. Bradley Smith
Department of Chemistry &
Biochem.
University of Notre Dame
Notre Dame, IN 46556
219/631-8632
219/631-6652
bradley.d.smith.115@nd.edu

Mr. Lance Steward
Department of Chemistry
University of California
Irvine, CA 92717-2025
714/824-6787
714/824-7089

Dr. Peter Toogood
Department of Chemistry
University of Michigan
930 N. University Avenue
Ann Arbor, MI 48109
313/763-4922
313/747-4865
peter.toogood@chem.1sa.umich.edu

Dr. Dirk Ullmann
University of Leipzig
Department of Biochemistry
Talstrasse 33 Leipzig
GERMANY
49 341 7165 315
49 341 960 3099

Dr. Scott Van Arman
Department of Chemistry
Franklin and Marshall College
Lancaster, PA 17603
717/291-4355

Dr. Andrea Veronese
Centro Neccanismi Reazioni Organiche
Via Marzolo
Padova 35128
ITALY
39 49 831 353
39 49 831 222

Professor Normand Voyer
Unviersite De Sherbrooke
Departement de Chimie
Faculte des Sciences
Sherbrooke (Quebec)
CANADA J1K 2R1
819/821-7005
819/821-8017
nvoyer@structure.chimie.usherb.ca

Ms. Helma Wennemers
Department of Chemistry
Columbia University
Box 768 Havemeyer Hall
New York, NY 10027-6948
212/854-8702
212/854-5429

Professor Craig S. Wilcox
Department of Chemistry
University of Pittsburgh
Pittsburgh, PA 15260
412/624-8270
412/624-8552

Dr. Nicholas Williams
Department of Chemistry
McGill University
801 Sherbrooke Street West
Montreal, Quebec H3A 2K6
CANADA
514/398-6229
514/398-3797

Ms. Zhizhen Zheng
Department of Chemistry
University of Pittsburgh
Pittsburgh, PA 15260
412/624-8275
412/624-8552

Mr. Zhijian Zhao
University of Pittsburgh
Department of Chemistry
Pittsburgh, PA 15260
412/624-8275
412/624-8552

Mr. Zhaohui Zhou
The Scripps Research Institute
10666 N. Torrey Pines Road
La Jolla, CA 92037
619/554-9704
619/554-6725

THE SITE OF CLEAVAGE OF PYRANOSIDE ACETALS: ENDO- vs EXOCYCLIC CLEAVAGE

A Review of Some Pertinent Literature

Jennifer L. Liras and Eric V. Anslyn*
Department of Chemistry and Biochemistry
The University of Texas at Austin
Austin, Texas 78712

Manipulation of carbohydrates via glycosyl transfer and hydrolysis represents a fundamental biological process. Investigating these reactions is an important goal of chemistry and biochemistry, because any attempt to alter enzymatic transformations or to mimic enzymes with a synthetic catalyst requires an understanding of the mechanistic pathways involved. Additionally, the study of the mechanism of glycosyl transfer provides a system for testing a broad theory of reactivity: the theory of stereoelectronic control. This paper will review the chemistry of the hydrolysis of glycosides and related pyranose acetals by enzymatic and acid catalysis. We specifically discuss the evidence relating to the site of pyranoside cleavage and its relevance to the theory of stereoelectronic control.

The Mechanistic Pathways

Acetal hydrolysis in acidic aqueous media is a fundamental organic reaction which has been studied extensively.[1] The cyclic nature of the pyranoside acetals complicates the mechanism by providing two possible sites of protonation and cleavage. As shown in Scheme **1** using maltose as an example, the first possibility involves the protonation of the exocyclic acetal oxygen, followed by expulsion of this leaving group to yield a cyclic oxocarbenium ion. The second involves protonation of the pyranose ring oxygen, with expulsion of the endocyclic oxygen to yield an acyclic oxocarbenium ion. Both pathways ultimately lead to the same hemiacetal, although the endocyclic pathway involves more intermediates.

Early literature discusses both pathways as mechanistically viable.[2] Subsequently, it was found that the ^{18}O kinetic isotope effect for the oxygen of the methoxyl group of methyl α-D-glucopyranoside was $k_{16}/k_{18} = 1.03$.[3] This isotope effect was interpreted to support exocyclic bond fission. However, the ring opening mechanism had not been excluded for the equatorial anomers. More recently, a detailed study of the ^{18}O kinetic isotope effects for both α- and β-methyl glucopyranosides showed an inverse isotope effect for the pyranose oxygen of both anomers, but these effects were much smaller than predicted.[4] A complication to the isotope effects has recently been discussed by Kirby.[5] He proposes that acetal hydrolysis involves a loose

C. S. Wilcox and A. D. Hamilton (eds.), Molecular Design and Bioorganic Catalysis, 1–14.

transition state which may not yield isotope effect data that can be easily interpreted and compared from one system to another. Therefore, the current isotope effect studies involving equatorial anomers of pyranoside acetals are not able to clearly delineate exocyclic from endocyclic cleavage.

Scheme 1

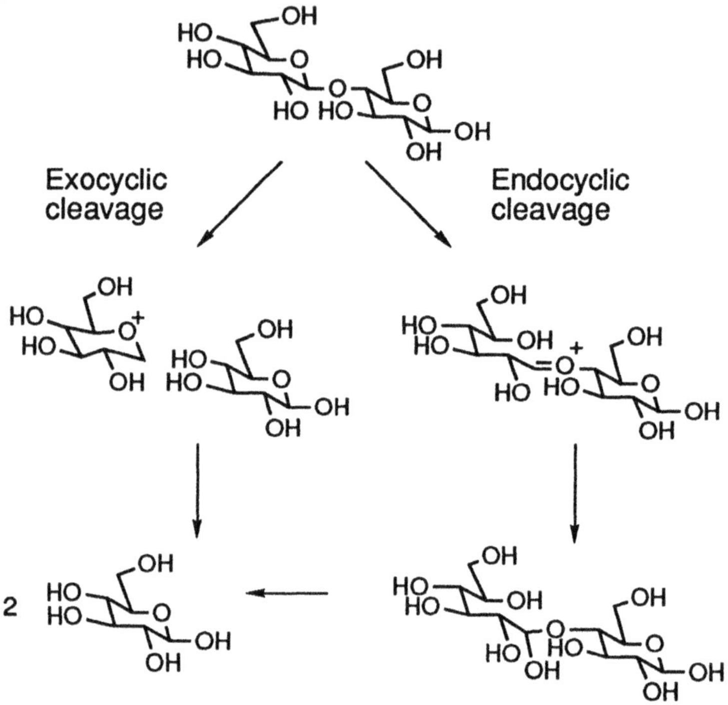

Glycoside hydrolysis has received renewed interest as the result of a calculation by the Karplus group.[6] Based upon a 50 ps molecular dynamics calculation, Karplus and Post have proposed an alternative mechanism (path A, Scheme 2) involving endocyclic cleavage for the hydrolysis of β-glycosides by lysozyme. In this mechanism Glu35 protonates the pyranose oxygen instead of the aglycon oxygen as postulated in the Phillips mechanism (path B, Scheme 2).

Scheme 2: A) Karplus Mechanism B) Phillips Mechanism

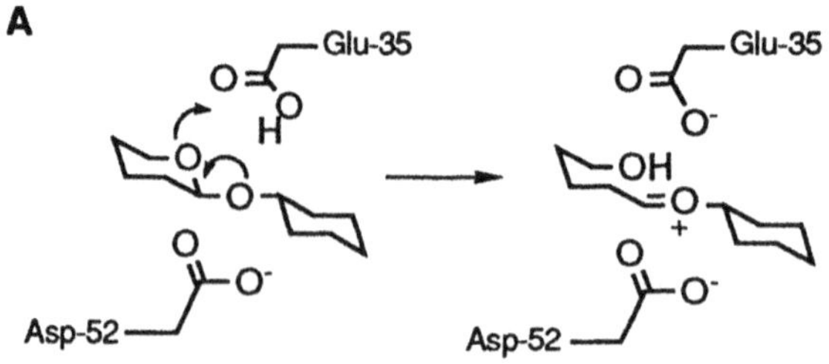

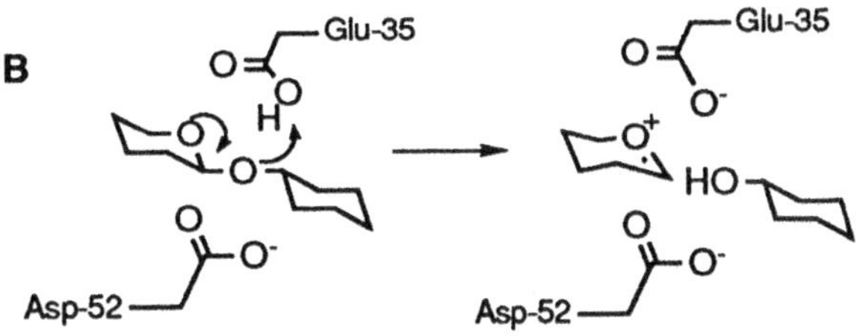

Glycosidase Mechanistic Investigations

LYSOZYME INHIBITION STUDIES

Studies of the mechanism of hydrolysis of β-glycosides by the class of enzymes known as β-glycosidases have led to the general consensus that the site of cleavage is between the exocyclic oxygen and the anomeric carbon. In a recent review Franck contends that many of the observations are equally consistent with the endocyclic cleavage mechanism.[7] He discusses an important paper by Leinhard and coworkers on the mechanism of lysozyme which helped to reaffirm the exocyclic cleavage mechanism. This study examined the inhibition of lysozyme by N,N',N",N'"-tetraacetylchitotetraose lactone (TACL). The authors proposed that this cyclic lactone, which bound more tightly than the unmodified tetrasaccharide, was a good mimic of the postulated sofa transition state structure (see the following sections) for exocyclic cleavage (Scheme 3A). Franck notes that the observation of TACL binding does not rule out the endocyclic transition state which has the ring oxygen protonated and a shortened exocyclic C-O bond. He also points out that the corresponding acyclic acid (TACA) binds lysozyme below pH 4.5 (likely when it is unionized) with an affinity comparable to that of the lactone. Moreover, lysozyme catalyzes the hydrolysis of this lactone to the acyclic acid (Scheme 3B). Generation of this acyclic product requires protonation of the ring oxygen and raises the question whether protonation of the same site is involved in the first step for the hydrolysis of natural saccharide substrates. This would be consistent with the Karplus endocyclic cleavage mechanism for lysozyme.

Scheme 3

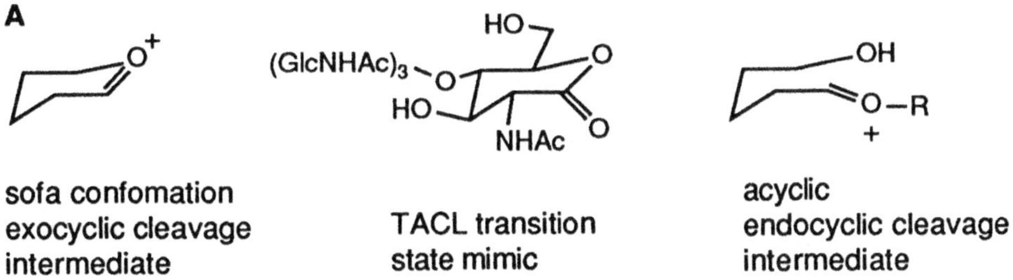

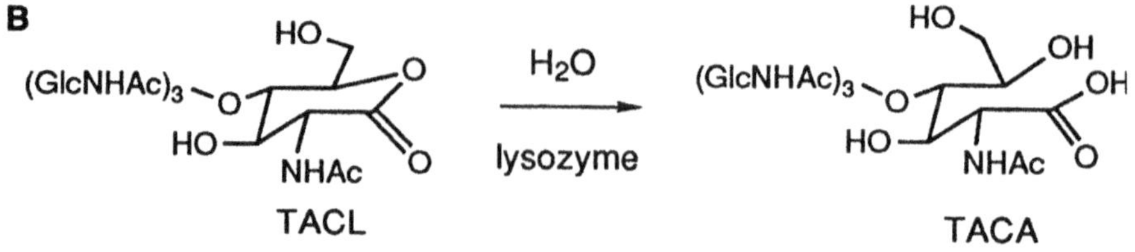

Other glycosidase inhibition studies have been used to support endocyclic cleavage. In contrast to Leinhard's conclusions, Fleet's interpretation of lysozyme inhibition studies with polyhydroxylated piperidine compounds (Scheme **4**) prompted him to suggest the endocyclic cleavage mechanism.[8] It was noted that these compounds were the most potent, competitive reversible inhibitors of glycosidase activity. Given the well accepted postulate that the strongest inhibitors most resemble the transition state structures along the enzymatic pathway, Fleet concludes that these inhibitors are better analogs of the cationic transition state that leads to endocyclic cleavage rather than exocyclic cleavage.

Scheme **4**

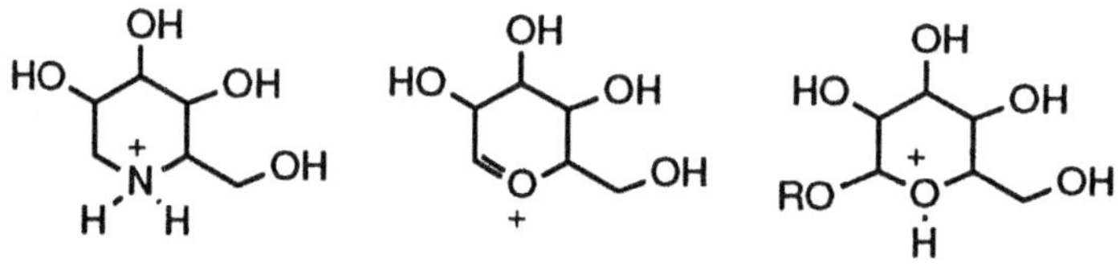

In conclusion, the inhibitor studies can be interpreted to support either endocyclic or exocyclic cleavage, since both possible oxocarbenium ions are quite similar in shape and in charge distribution.

Crystallographic Evidence

The first mechanistic information derived from structural elucidation of hen egg-white lysozyme (HEWL) in 1965 was the identification of its active-site residues.[9] Subsequent studies led to the proposed general acid catalysis by Glu35 with exocyclic cleavage.[10] Further resolution (1.5 Å) of the crystal structure with the trisaccharide N-acetyl muramic acid–N-acetyl glucosamine–N-acetyl muramic acid (NAM-NAG-NAM) afforded the coordinates of the saccharide occupying the catalytic subsite. This residue is distorted from its chair conformation into a sofa conformation where ring atoms C-1, C-2, O-5, and C-5 are nearly coplanar, and the hydroxymethyl group is directed axially. This distortion is used to support exocyclic cleavage via its putative sofa oxocarbenium ion intermediate. In addition, the lack of catalysis of N-acetylxylosamine substrates has been interpreted as evidence of the importance of the hydroxymethyl group in invoking the distortion of the substrate.[11]

A crystal structure of a covalently bound enzyme-substrate intermediate was obtained from a T4 lysozyme mutant, providing further insight into the structure of the wild-type enzyme intermediate.[12] An active site threonine residue was modified by site-directed mutagenesis to glutamic acid (T26E). The result was an enzyme that cleaved the cell wall of *E. coli*, but now afforded a cyclic acylal enzyme intermediate. The saccharide in the catalytic subsite was also found to be distorted as discussed above.

The crystal structures are the best evidence for exocyclic cleavage at the active site of lysozyme. Yet, they do not definitively rule out endocyclic cleavage. Although the residue of the saccharide of NAM-NAG-NAM in the catalytic subsite was found in a puckered conformation thus supporting exocyclic cleavage, the fact that the sugar is in its cyclic form does not preclude the possibility of an acyclic structure during hydrolysis. Furthermore, the structure of the T4 lysozyme-bound intermediate is that of a mutant, and therefore the wild-type mechanism may have been altered.

Stereoelectronic Considerations

ANTIPERIPLANAR LONE PAIR HYPOTHESIS

Examination of the stereoelectronic factors present in tetrahydropyranyl acetals also suggests the probability of a difference in mechanism for the equatorial and axial anomers. The traditionally accepted mechanism of exocyclic cleavage follows a stereoelectronically favored path for the hydrolysis of the axial anomers. The lone pair of the pyranose oxygen can readily achieve antiperiplanarity with the exocyclic leaving group to generate a cyclic oxocarbenium ion (Scheme 5A).

Scheme 5

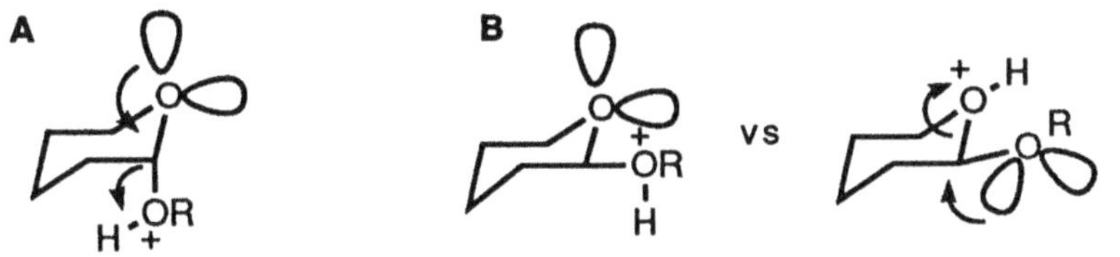

However, when this mechanism is invoked for the equatorial anomer, there can be no electronic assistance to the leaving group departure without distortion of the chair ring conformation. On the other hand, the endocyclic cleavage mechanism follows a stereoelectronically favored pathway, where the exocyclic oxygen lone pair can achieve a conformation antiperiplanar to the endocyclic C-O bond as in the exo-anomeric effect (Scheme 5B).

THE EXO-ANOMERIC EFFECT

In 1969 Lemieux and coworkers suggested that the same effect responsible for the preferred axial orientation of alkoxy groups on the anomeric center (anomeric effect) was also responsible for the preferred torsion angles of the aglycon groups in β-acetals (exo-anomeric effect).[13] Certain observations, such as the conformation and C-O bond lengths of aglycon groups of saccharides as determined by crystal structures, have been postulated to support the exo-anomeric effect .[14] Furthermore the exo-anomeric effect has been demonstrated in solution by NMR studies of the conformational behavior of pyranosides.[15] The 3-bond coupling constants and NOE resonances of 2-methoxy- and 2-methylamino-tetrahydropyrans indicate that there is a preference for conformations in which an exocyclic heteroatom lone pair is antiperiplanar to the endocyclic C2-O bond.[16]

Another piece of evidence that the exo-anomeric effect is a strong influence in the cleavage pathway of pyranosides is the C-O bond length of the acetal linkage. Crystal structures of saccharides indicate that the exocyclic C-O bond is shorter than the typical C-O ether bond, and that the endocyclic C-O bond is longer.[14] Thus the ground state of β-pyranosides with alkyl aglycon groups are apparently pre-activated toward endocyclic cleavage.

BASICITY CONSIDERATIONS

Another consequence of the exo-anomeric effect for the equatorial anomers is the increased basicity of the pyranose ring oxygen. As discussed by Deslongchamps, stereoelectronic effects influence the basicity of oxygens of an acetal.[17] Structural evidence suggests that the exo-anomeric effect is stronger for the equatorial anomers. Unlike the axial anomer, where the delocalization of electrons from the exocyclic oxygen toward the anomeric carbon is opposed by the similar effect from the ring oxygen, the equatorial anomer has an unopposed exo-anomeric effect.[18] This enhances the basicity of the pyranose ring oxygen for the equatorial anomers compared to the axial anomers, and thereby supports endocyclic cleavage for the equatorial anomers. This predicted basicity is in agreement with *ab initio* calculations by Fraser-Reid.[19]

In summary, the theory of stereoelectronic control would predict endocyclic cleavage of equatorial pyranoside acetals. In order for an exocyclic pathway to be viable, the theory states that the reaction would proceed via a higher energy sofa conformation, as has been postulated for lysozyme. Endocyclic cleavage for lysozyme would not require this substrate destabilization. The theory, however, has come under strong criticism,[22] and even one of its original proponents has stated "It is not necessary to take literally the earliest formulation of the stereoelectronic theory, ... but it is important to apply the general principle involved."[5]

Kinetic Data

The rates of the acid-catalyzed hydrolysis of glycosides and rigid pyranoside acetals have been measured. However the factors governing these rates are not well understood. The ratio of the rates of hydrolysis for axial and equatorial acetals was found to be 0.52 ± 0.12.[20] This observation has provoked several explanations to describe the reactivity of glycosides.[24] One theory is that the hydrolysis may occur via a synperiplanar rather than an antiperiplanar conformation.[21] Another is that the hydrolysis mechanism follows the principle of least motions and that stereoelectronic effects are not important factors governing the reactivity.[22] A third ascribes the difference in rates to differences in energy for the chair (α) and twist boat (β) late transition states proceeding toward the half-chair oxocarbenium ion intermediate.[23]

Van Eikeren has measured rates of hydrolysis of rigid acetals derived from *trans*-decalin.[24] He reported the axial to equatorial hydrolysis rate ratio as 1.51 ± 0.22, and found that both the entropy and enthalpy of activation of the axial anomer were greater than the equatorial. From these data, it was concluded that the hydrolysis of the different anomers must proceed through different transition states such that there is more bond breaking in the axial anomer's structure.

Scheme **8**

$\Delta H^{\ddagger} = 24.6$ kcal/mol	$\Delta H^{\ddagger} = 26.1$ kcal/mol
$\Delta S^{\ddagger} = 16$ cal/mol	$\Delta S^{\ddagger} = 21$ cal/mol

It is interesting to note that the activation parameters are in agreement with the alternative endocyclic cleavage pathway for the equatorial anomer. The cleavage into two molecules, as in the case for the exocyclic cleavage of the axial anomer, would be expected to afford an greater positive entropy term than that from the endocyclic pathway for the axial anomer. This expectation is based upon the concept that creating two molecules as in exocyclic cleavage would be more entropically favored than ring opening in the endocyclic pathway.

In summary, the relative rates of axial and equatorial anomers are complex functions of the relative energies of the ground and transition states, and therefore cannot be used reliably to support either endo- or exocyclic cleavage.

Direct Evidence of Endocyclic Cleavage

LEWIS ACID CATALYSIS

Guidon has found that the reaction of tetrahydropyranyl and tetrahydrofuranyl acetals with dimethylboron bromide at -78 °C yields acyclic α-bromo intermediates.[25] The reaction was also studied employing fully protected O-methyl-β-D-2-deoxyglucopyranosides and a variety of nucleophiles to displace the bromide.[26] In most cases the endocyclic cleavage product predominated. The authors rationalize their results on the stereoelectronic grounds discussed previously. They liken the Lewis acid-catalyzed reaction in non-polar solvents to the lysozyme reaction, and suggest that lysozyme may be "exploiting the same preference for endocyclic bond cleavage."

Scheme 9

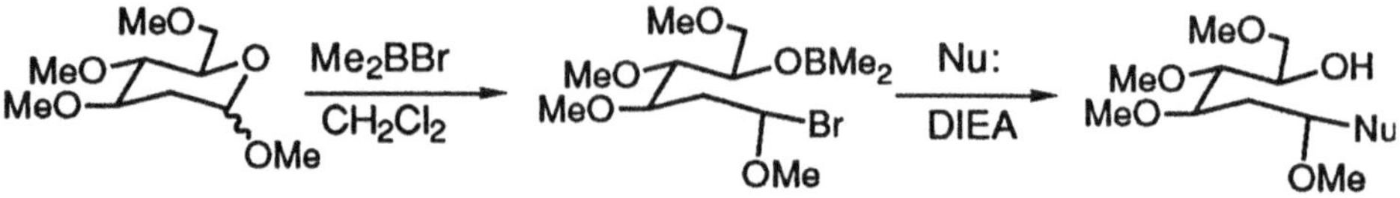

ACETOLYSIS OF METHYL-D-GLUCOPYRANOSIDES

Endocyclic and exocyclic cleavage have been quantitated by Fraser-Reid and coworkers in a non-polar aprotic system.[27] The experiment involved trapping the oxocarbenium ions resulting from acetolysis of protected α- and β-methyl-D-glucopyranosides. The solvolysis was effected with acetic anhydride, and the products obtained were peracetylated pyranoses, furanoses, and acyclic derivatives. The percent of endo- and exocyclic cleavage from each anomer was quantitated by product analysis. The acid in this medium was acetylium ion and the nucleophile was acetate. The results clearly show that the α and β anomers react through different intermediates. The α anomer afforded almost exclusively products generated from exocyclic cleavage, whereas the β gave mostly products arising from endocyclic cleavage.

Scheme 10

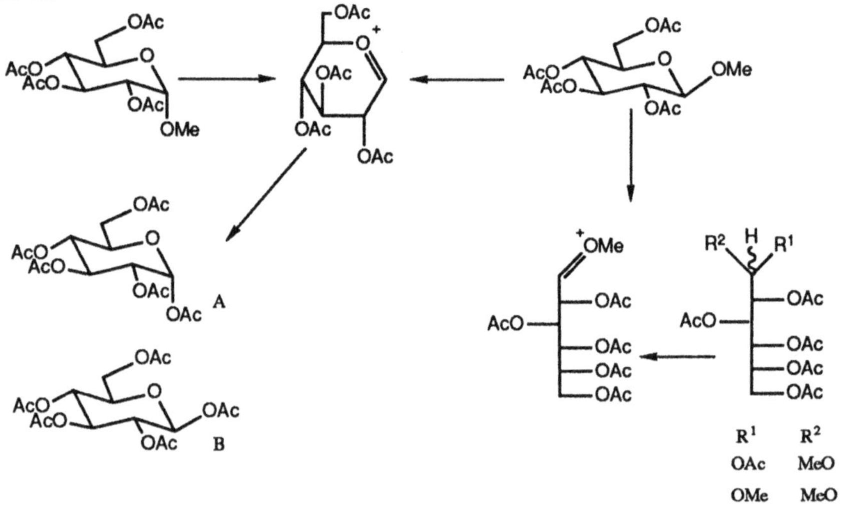

Starting Material	A	B	C	D
methyl α-D-glucopyranoside	73	18	8	tr
methyl β-D-glucopyranoside	19	5	48	23
acyclic acyl acetal (epimeric)	2	tr	46	52
acyclic acyl acetal (1S)	2	tr	46	52
acyclic acyl acetal (1R)	2	tr	47	51
acyclic dimethyl acetal			66	34

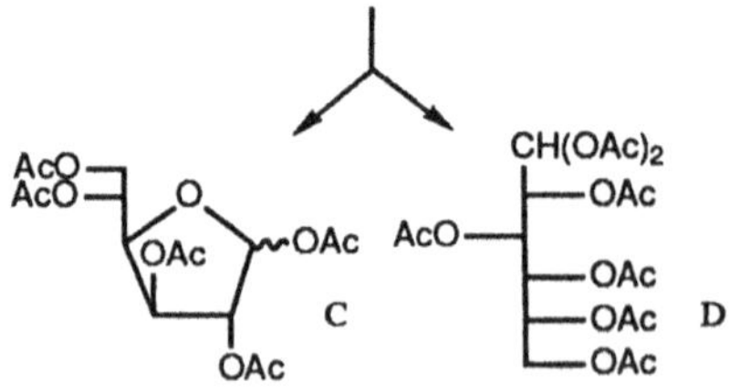

ENAMINE TRAPPING OF METHANOLYSIS INTERMEDIATES

Franck and Gupta have examined the solvolysis of equatorial anomers of tetrahydropyranyl (THP) acetals in a polar solvent containing a Brønsted acid.[28] The system contained an intramolecular enamine designed to trap the intervening oxocarbenium ions. After treatment with methanol and Dowex-H^+ resin, the ratio of endo- to exocyclic cleavage was determined by product analysis. A problem arose since the rate of trapping by methanol is competitive with enamine. Nevertheless, the product ratio indicated between 45-54 percent endo- and 38-45 percent exocyclic cleavage with approximately 10 percent of the products resulting from competitive solvent trapping of the intermediates (Scheme **11**). No axial anomers were studied and the authors suggest a reexamination of the previous conclusion about the site of cleavage for THP acetal hydrolysis.

The past three examples definitively show endocyclic cleavage and therefore demonstrate the viability of both endo- and exocyclic reaction pathways. However, the first two do not involve general or specific acid catalysis as in enzymes or protic media. The Guidon work involves a strong Lewis acid rather than a Brønsted acid. Similarly, the Fraser-Reid work involves acetylium as a Lewis acid, and the low dielectric solvent acetic anhydride. The Franck work in contrast uses an acidic resin in the protic solvent methanol, but some of the products can only be rationalized by intermediates that are trapped by solvent.

DETECTION OF ENDOCYCLIC CLEAVAGE IN METHANOL AND WATER WITH AN ISOTOPICALLY LABELED PROBE

In an attempt to distinguish between the endo- and exocyclic cleavage pathways with specific acid in high dielectric-protic media, we have designed a pyranoside probe (**1**) that contains an intramolecular nucleophile.[29] The effective molarity of the nucleophile was 30,000 to 50,000 M.[30] The key to distinguishing between exo- and endocyclic cleavage resides in the differing symmetry of the oxocarbenium ions generated by each cleavage pathway (Scheme **12**). During solvolysis of **1**, the asymmetric oxocarbenium ion (**2**) from exocyclic cleavage cannot be trapped because of the insufficient length of the appended nucleophile. Only the α- and β-methyl acetals **3** and **4** can be formed. However, the symmetric oxocarbenium ion **5** generated from endocyclic cleavage yields not only **3** and **4**, but also their deuterium-scrambled counterparts **6** and **7**. Furthermore, by reversal of the ring opening, deuterium can be incorporated into the pyran ring of the starting material (**8**). Therefore exocyclic cleavage yields no deuterium scrambling, but endocyclic cleavage results in a portion of deuterium-scrambled products and starting material. Endocyclic cleavage is thereby indicated by the appearance of ^{1}H NMR resonances for the C$\underline{H}_2$OH appendage during solvolysis.

The methanolysis (CD_3OD) of β-**1** was performed with a catalytic amount of DCl and monitored by ^{1}H NMR spectroscopy. Due to the fact that the products are also acetals subject to further solvolysis, the reaction was terminated at thirty percent completion so that the primary solvolysis reaction would predominate. After quenching the reaction with triethylamine, the α-anomer products (**3** and **6**), the β-anomer products (**4** and **7**), and the recovered starting material (**1** and **8**) were isolated, and the C$\underline{H}_2$OH ^{1}H NMR resonances analyzed. In the formation of α-methyl acetal products (**3** and **6**, Scheme **12**), 14 percent endocyclic cleavage was observed. For the mixture of β-methyl acetal products (**4** and **7**), approximately 30 percent endocyclic cleavage was detected. Scrambling was detected but not quantifiable in the recovered starting material.

Scheme **11**

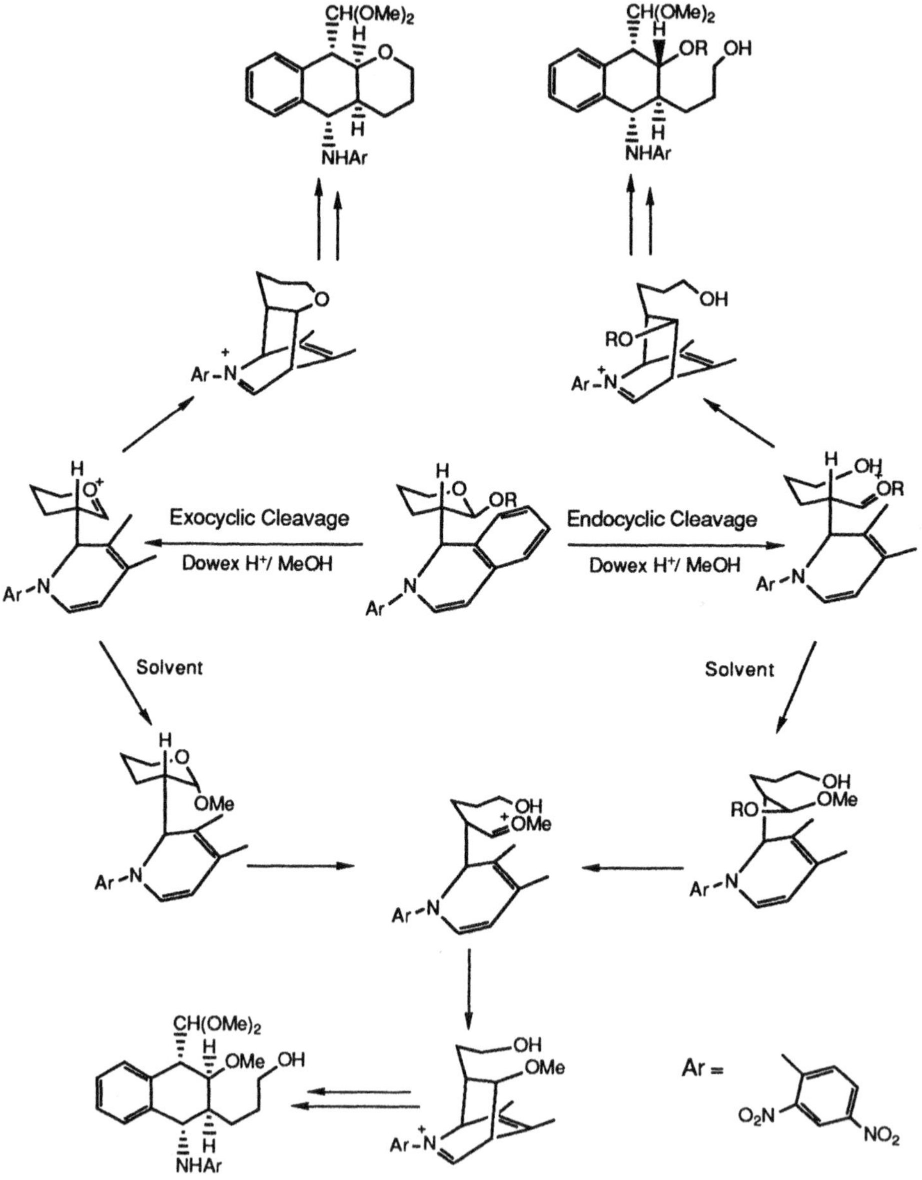

Scheme 12

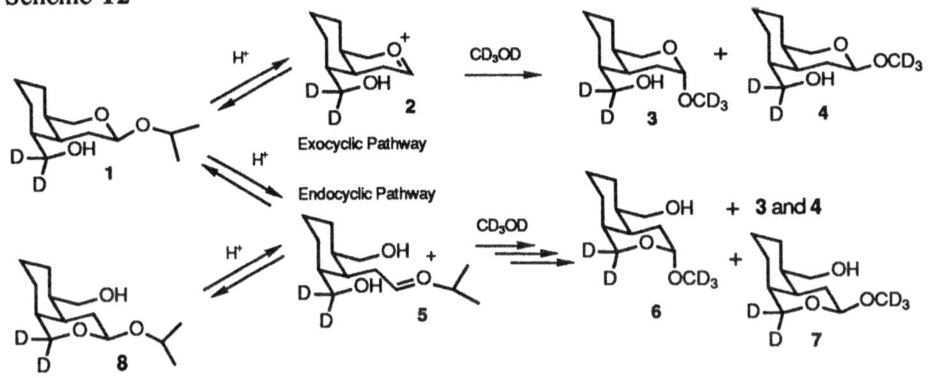

In order to measure the importance of stereoelectronic considerations for the proposed pathways of solvolysis of acetals, the methanolysis of an α-isopropyl probe (α-**1**) was conducted. As discussed above, the theory of stereoelectronic control predicts that α-anomers would proceed via the exocyclic mechanism. Indeed, the products from α-**1** showed negligible scrambling even after one hundred percent completion of primary solvolysis.

Endocyclic cleavage was also detected in water using probe β-**1**. However, analysis of the products in the same manner as that in the methanolysis studies provides no valuable information because mutarotation generates scrambling in the hemiacetal product at a significantly greater rate than hydrolysis. In contrast, the starting material can only scramble via endocyclic cleavage. Therefore, the DCl-catalyzed hydrolysis of β-**1** in D_2O[31] was allowed to proceed to 55 percent completion, and the non-hydrolyzed material (**1** and **8**) was isolated and analyzed (Scheme **12**). It indicated incorporation of the deuterium-labeled appendage into the pyran ring corresponding to eight percent endocyclic cleavage. The eight percent represents only a minimum for endocyclic cleavage, whereas the actual value can be detected only as the reaction approaches completion.

Conclusions

We have shown that much of the current evidence in the literature can be used to support either endocyclic or exocyclic cleavage for equatorial THP acetals. Isotope effects, crystal structures, enzyme inhibitors, kinetics, and stereoelectronics, contain certain questionable aspects in their interpretation. In constrast, direct experimental observation of products that can only arise from endocyclic cleavage is solid evidence for such a pathway. Four such solution experiments, all of which show either complete endocyclic cleavage, or mixtures of endocyclic and exocyclic cleavage, have recently been published. It is now left to a clever experiment with lysozyme to finally settle this issue at the enzyme active site.

References

1. (a) Cordes, E.H. (1967) Mechanism and Catalysis for the Hydrolysis of Acetals, Ketals, and Ortho Esters, *Prog. Phys. Org. Chem.* **4**, 1-44. (b) Cordes, E.H. and Bull, H.G. (1974) Mechanism and Catalysis for the Hydrolysis of Acetals, Ketals, and Ortho Esters, *Chem. Rev.* **74**, 5, 581-603. (c) Fife, T.H. (1972) General Acid Catalysis of Acetal, Ketal, and Ortho Ester Hydrolysis, *Acc. Chem Res.* **5**, 264-272.

2. Bunton, C.A., Lewis, T.A., Llewellyn, D.R. and Vernon, C.A. (1955) Mechanisms of Reactions in the Sugar Series. Part I. The Acid-catalyzed Hydrolysis of α- and β-Methyl and α- and β-Phenyl D-Glucopyranosides, *J. Chem. Soc.* 4419-4423.

3. Banks, B.E.C., Meinwald, Y., Rhind-Tutt, A.J., Sheft, I. and Vernon, C.A. (1961) Mechanism of Reactions in the Sugar Series. Part IV. The Structure of the Carbonium Ions formed in the Acid-catalyzed Solvolysis of Glucopyranosides, *J. Chem. Soc.* 3240-3246.

4. Bennet, A.J. and Sinnott, M.L. (1986) Complete Kinetic Isotope Effect Description of Transition States for Acid-Catalyzed Hydrolyses of Methyl α- and β-Glucopyranosides, *J. Am. Chem. Soc.* **108**, 7287-7294.

5. Kirby, A.J. (1987) Mechanism and Stereoelectronic Effects in the Lysozyme Reaction, *CRC Crit. Rev. Biochem.* **22,** 283-314.

6. Post, C.B. and Karplus, M. (1986) Does Lysozyme Follow the Lysozyme Pathway? An Alternative Based on Dynamic, Structural, and Stereoelectronic Considerations, *J. Am. Chem. Soc.* **108**, 1317-1319.

7. Franck, R.W. (1992) The Mechanism of β-Glycosidases: A Reassessment of Some Seminal Papers, *Bioorg. Chem.* **20**, 77-88.

8. Fleet, G.W.J. (1985) An Alternative Proposal for the Mode of Inhibition of Glycosidase Activity By Polyhydroxylated Piperidines, Pyrrolidines and Indolizidines: Implications for the Mechanism of Action of Some Glycosidases, *Tetrahedron Lett.* **26**, 5073-5076.

9. Blake, C.C.F., Koenig, D.F., Mair, G.A., North, A.C.T., Phillips, D.C. and Sarma, V.R. (1965) Structure of Hen Egg-white Lysozyme, *Nature* **206**, 757-763.

10. (a) Phillips, D.C. (1966) The Three-Dimensional Structure of an Enzyme Molecule, *Sci. Am.* **215** (5), 78-90. (b) Blake, C.C.F., Mair, G.A., North, A.C.T., Phillips, D.C. and Sarma, V.R. (1967) On the Conformation of the Hen Egg-White Lysozyme Molecule, *Proc. Roy. Soc. ser. B* **167**, 365-377. (c) Blake, C.C.F., Johnson, L.N., Mair, G.A., North, A.C.T., Phillips, D.C. and

Sarma, V.R. (1967) Crystallographic Studies of the Activity of Hen Egg-White Lysozyme, *Proc. Roy. Soc. ser. B* **167**, 378-388.

11. Ballardie, F.W., Capon, B., Cuthbert, M.W. and Derie, W.M. (1977) Some Studies on Catalysis by Lysozyme, *Bioorg. Chem.* **6**, 483-509.

12. Kuroki, R., Weaver, L.H. and Matthews, B.W. (1993) A Covalent Enzyme-Substrate Intermediate with Saccharide Distortion in a Mutant T4 Lysozyme, *Science* **262**, 2030-2033.

13. Lemiux, R.U., Pavia, A.A., Martin, J.C. and Wantanabe, K.A. (1969) Solvation effects on conformational equilibria. Studies related to the conformational properties of 2-methoxytetrahydropyran and related methyl glycopyranosides, *Can. J. Chem.* **47**, 4427-4439.

14. (a) Allen, F.H. and Kirby, A.J. (1984) Bond Length and Reactivity. Variable Length of the C-O Single Bond, *J. Am. Chem. Soc.* **106**, 6197-6200. (b) Briggs, A.J., Glenn, R., Jones, P.G., Kirby, A.J. and Ramaswamy, P. (1984) Bond Length and Reactivity. Stereoelectronic Effects on Bonding in Acetals and Glucosides, *J. Am. Chem. Soc.* **106**, 6200-6207. (c) Jones, P.G. and Kirby, A.J. (1984) Simple Correlation between Bond Length and Reactivity. Combined Use of Crystallographic and Kinetic Data to Explore a Reaction Coordinate, *J. Am. Chem. Soc.* **106**, 6207-6212.

15. Lemieux, R.U. and Koto, S. (1974) The Conformational Properties of Glycoside Linkages, *Tetrahedron* **30**, 1933-1944.

16. Booth, H., Dixon, J.M., Khedhair, K.A. and Redshaw, S.A. (1990) Experimental Studies of the Anomeric Effect. Part III. Rotameric Preferences about the Exocyclic C2-X Bond in Equatorial and Axial 2-Methoxy- and 2-Methylamino-Tetrahydropyrans, *Tetrahedron* **46**, 1625-1652.

17. Deslongchamps, P. (1983) *Stereoelectronic Effects in Organic Chemistry*, Pergamon Press, New York.

18. Praly, J.-P. and Lemieux, R.U. (1987) Influence of solvent on the magnitude of the anomeric effect, *Can. J. Chem.* **65**, 213-223.

19. Andrews, C.W., Bowen, J.P. and Fraser-Reid, B. (1989) An *Ab initio* Study of Dimethoxymethane Protonation and its Relevance to Glycoside Hydrolysis, *J. Chem. Soc., Chem. Commun.* 1913-1916.

20. Feather, M.S. and Harris, F. (1965) The Acid-Catalyzed Hydrolysis of Glycopyranosides, *J. Org. Chem.* **30**, 153-157.

21. Ratcliffe, A.J., Mootoo, D.R., Andrews, C.W. and Fraser-Reid, B. (1989) Concerning the Antiperiplanar Lone Pair Hypothesis: Oxidative Hydrolysis of Conformationally Restrained 4-Pentenyl Glycosides, *J. Am. Chem. Soc.* **111**, 7661-7662.

22. Sinnot, M.L. (1990) Catalytic Mechanisms of Enzymic Glycosyl Transfer, *Chem. Rev.* **90**, 1171-1202.

23. Deslongchamps, P. (1993) Intramolecular Strategies and Stereoelectronic Effects. Glycoside Hydrolysis Revisited, *Pure & Appl. Chem.* **65**, 1161-1178.

24. van Eikeren, P. (1980) Models for Glycoside Hydrolysis. Synthesis and Hydrolytic Studies of the Anomers of a Conformationally Rigid Acetal, *J. Org. Chem.* **45**, 4641-4645.

25. Guidon, Y., Bernstein, M.A. and Anderson, P.C. (1987) Ring Cleavage of THP and THF Ethers Using Dimethylboron Bromide, *Tetrahedron Lett.* **28**, 2225-2228.

26. Guidon, Y. and Anderson, P.C. (1987) Stereoelectronic Effects in the Ring Cleavage of Methyl Glycopyranosides Using Dimethylboron Bromide, *Tetrahedron Lett.* **28**, 2485-2488.

27. McPhail, D.R., Lee, J.R. and Fraser-Reid, B. (1992) Exo and Endo Activation in Glycoside Cleavage: Acetolysis of Methyl α- and b-Glucopyranosides, *J. Am. Chem. Soc.* **114**, 1905-1906.

28. Gupta, R.B. and Franck, R.W. (1987) Direct Experimental Evidence for Cleavage of Both Exo- and Endocyclic C-O Bonds in the Acid-Catalyzed Reaction of Alkyl β-Tetrahydropyranyl Acetals, *J. Am. Chem. Soc.* **109**, 6554-6556.

29. Liras, J.L. and Anslyn, E.V. (1994) Exocyclic and Endocyclic Cleavage of Pyranosides in Methanol and in Water Detected by a Novel Probe, *J. Am. Chem. Soc.* **116**, 2645-2646.

30. Mega, T.L., Cortes, S. and Van Etten, R.L. (1990) The ^{18}O Isotope Shift in the ^{13}C Nuclear Magnetic Resonance Spectroscopy. Oxygen Exchange at the Anomeric Carbon of D-Glucose, D-Mannose, and D-Fructose, *J. Org. Chem.* **55**, 522-528.

31. The D_2O contained ten percent CD_3CN to solubilize **1**.

VITAMIN K MECHANISM OF ACTION AND ITS BEARING ON VITAMIN E

PAUL DOWD, ZHIZHEN BARBARA ZHENG, ROGER HERSHLINE,
SEUNG WOOK HAM, SRIRAM NAGANATHAN
AND JEFFREY KERNS
University of Pittsburgh
Department of Chemistry
Pittsburgh, PA 15260

1. Introduction

Vitamin K and vitamin E are relatively simple molecules (in comparison to other complex natural products), but they have a profound effect on human health and well being. Likewise, the chemistry surrounding the mechanism of action of vitamin K is far more complex than might be inferred on the basis of structure alone.

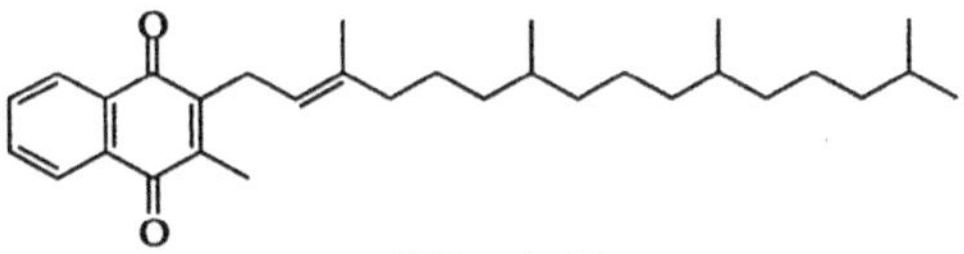

Vitamin K_1

Vitamin K is the blood-clotting vitamin, first adumbrated by the work of Henrik Dam in his investigation into the metabolism of cholesterol in chicks [1]. For almost forty years vitamin K was assigned a non-specific role as a redox electron carrier [2], because it is a quinone. This state of affairs underwent a dramatic change in 1974, when it was discovered that vitamin K, in the form of

C. S. Wilcox and A. D. Hamilton (eds.), Molecular Design and Bioorganic Catalysis, 15–34.

its hydroquinone, is an obligatory cofactor for an enzyme that effects carboxylation of the blood clotting proteins [3] (**Scheme 1**). For the blood

Scheme 1

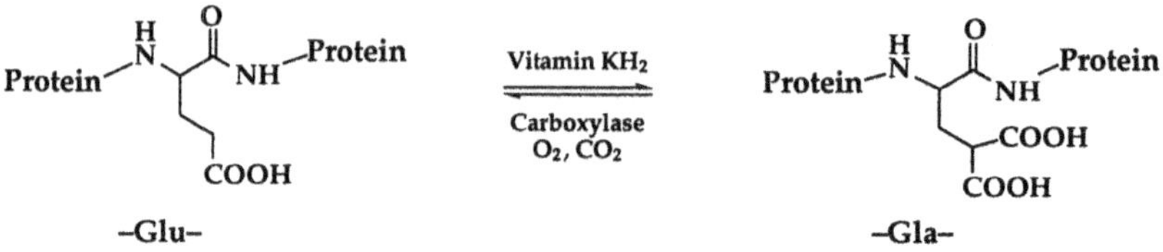

clotting factors II, VII, IX and X, to be activated and carry out their function, nine to twelve glutamates at the N-termini of each protein must be carboxylated. The result of the carboxylation is a new amino acid, γ-carboxyglutamate (Gla) [3]. The vitamin K-dependent carboxylase (**Scheme 1**) requires vitamin K hydroquinone (vitamin KH_2), molecular oxygen and carbon dioxide. On this basis, and that of many subsequent experiments, the general outline in **Scheme 2**

Scheme 2

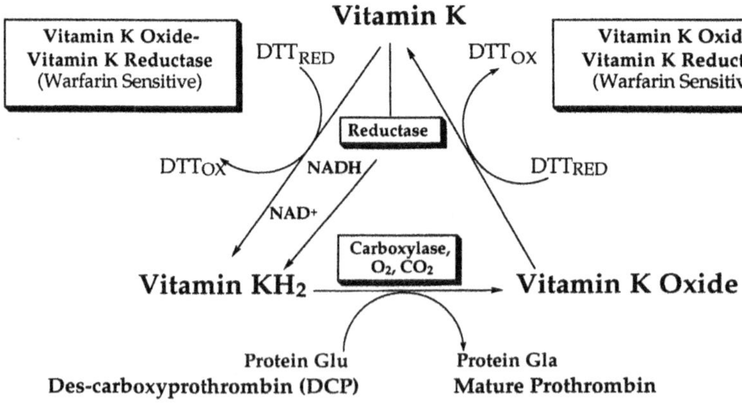

has been developed [3]. Thus, vitamin K is itself not catalytically active. It is reduced to the active cofactor vitamin KH_2 by either of two enzymes. Following reduction, vitamin KH_2 is transformed to vitamin K oxide by the carboxylase using molecular oxygen. At the same time, in a step that is coupled to the oxygenation of vitamin KH_2 to vitamin K oxide, the protein bound glutamate residues are carboxylated yielding the γ-carboxyglutamates essential for blood-clotting. The problem posed by this system is to discern the mechanistic connection between the oxygenation and the carboxylation [4].

We have demonstrated by means of calorimetry and the resulting thermochemical cycle that a large amount of energy (-58 kcal/mole) is made available as a consequence of the oxygenation of vitamin KH_2 [4g]. It is our contention that this energy is manifested in the form of a very strong base that removes a proton from the γ–position of glutamate to facilitate, indeed, make possible, the carboxylation leading to γ-carboxyglutamate (**Scheme 3**). The

Scheme 3

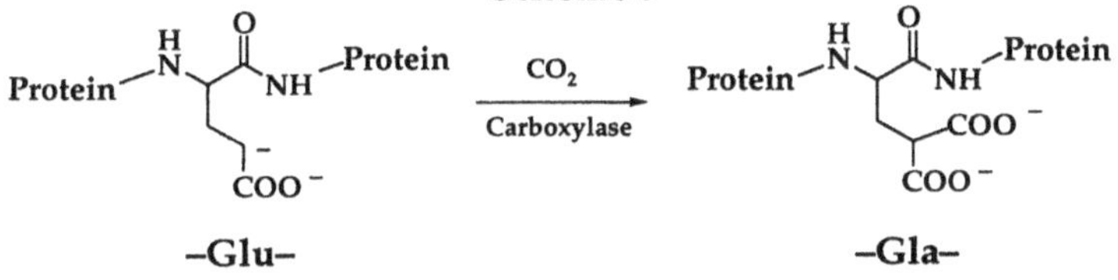

factual basis of this hypothesis stems from model experiments (4a,b] and from oxygen-18 labeling studies [4f].

1.1. MODEL EXPERIMENTS

Model studies were initiated to illuminate the hypothetical mechanism in **Scheme 4** in which vitamin K hydroquinone is oxidized to vitamin K oxide

Scheme 4

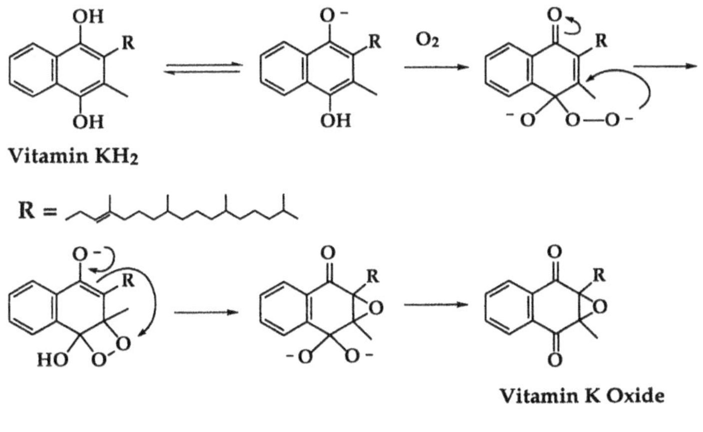

[4a]. One of the key features of **Scheme 4** is the formation of the carbon-oxygen bond leading to the peroxide. This is followed by dioxetane formation that makes possible the subsequent production of vitamin K oxide.

Potassium α-naphthoxide was used for the model studies because it allowed mapping of the course of the oxygenation. The essence of the model is to couple the oxygenation step with carbon-carbon bond formation, thereby mimicking the carbon-carbon bond forming carboxylation effected by the enzyme [4b]. The enzyme enjoys the luxury of placing carbon dioxide in the right position at the right time. The Dieckmann reaction of diethyl adipate was used as an "intramolecular" carboxylation model [4b]. When the α-naphthoxide was treated with diethyl adipate under an inert atmosphere, no reaction took place. The naphthoxide is too weak a base to carry out the Dieckmann reaction under the mild conditions employed in this series of experiments (THF, 25 °C, 18-crown-6). However, injection of 1.7 eq of O_2 into the reaction mixture resulted in rapid transformation of the naphthoxide to the corresponding keto epoxy alcohol in quantitative yield (**Scheme 5**) while the diethyl adipate

Scheme 5

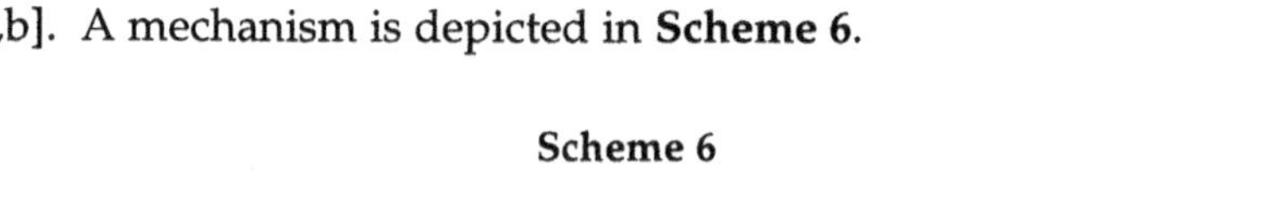

gave the cyclic product, cyclopentanonecarboxylate, in approximately 40% yield [4a,b]. A mechanism is depicted in **Scheme 6**.

Scheme 6

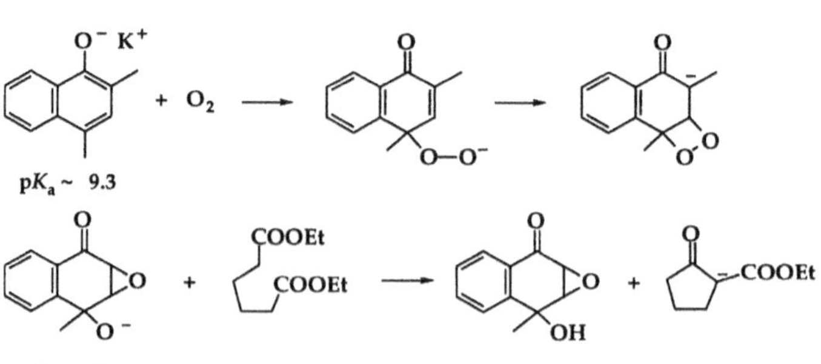

The crucial point in **Scheme 6** is that by action of molecular oxygen, a weak base (the naphthoxide) is transformed to a strong base (the tertiary alkoxide) capable of promoting the carbon-carbon bond forming Dieckmann condensation. Thus, addition of oxygen yields the first peroxide adduct that then forms the dioxetane enolate by addition to the enone. The dioxetane is now in a favorable orientation to interact with the enolate to form the epoxide and the tertiary alkoxide. The latter effects the Dieckmann reaction. We term this a *base strength amplification* reaction.

A central intermediate in this transformation is the dioxetane. What is the evidence garnered in support of the dioxetane? There are three key features. 2,3,4-Trimethylnaphthol leads to an isolable hydroperoxide when treated with molecular oxygen in chloroform. X-ray analysis confirms the hydroperoxide structure shown in **Scheme 7** [4b]. Treatment of the hydroperoxp[pide

Scheme 7

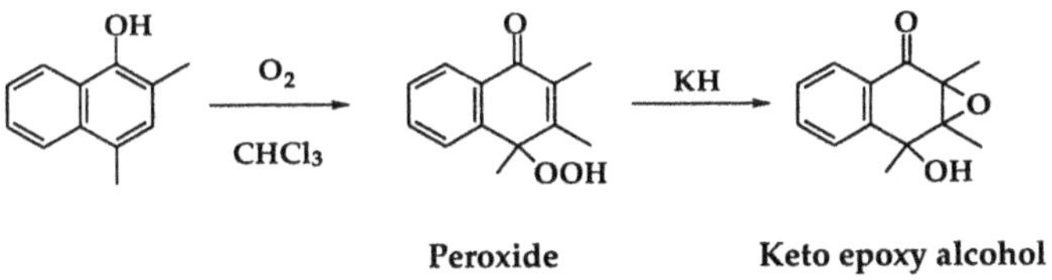

with potassium hydride then yields the corresponding keto epoxy alcohol (**Scheme 7**) in quantitative yield. This experiment speaks to the direct involvement of the hydroperoxide leading to the dioxetane intermediate. Second, an x-ray structure of the keto epoxy alcohol (**Scheme 8**) shows that the

Scheme 8

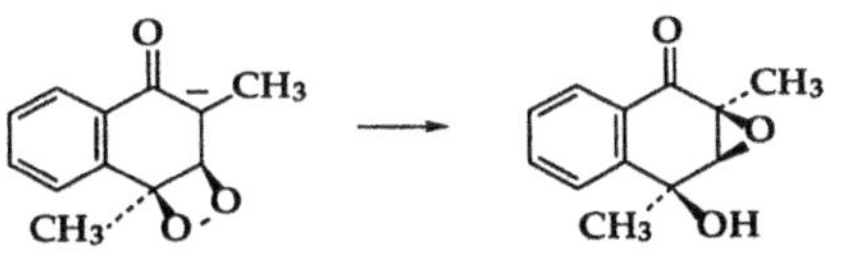

Keto epoxy alcohol

epoxide and alcohol groups are cis to one another as demanded by the intermediacy of the dioxetane [4b]. Finally, there exists the possibility, in addition to intramolecular reaction through the dioxetane intermediate, that epoxide formation is an intermolecular reaction - one hydroperoxide serving to epoxidize another enone. These two possibilities can be distinguished by carrying out the oxygenation under an atmosphere consisting of an approximately 50 : 50 mixture of $^{16}O_2$ and $^{18}O_2$ (**Scheme 9**). In the event, the

Scheme 9

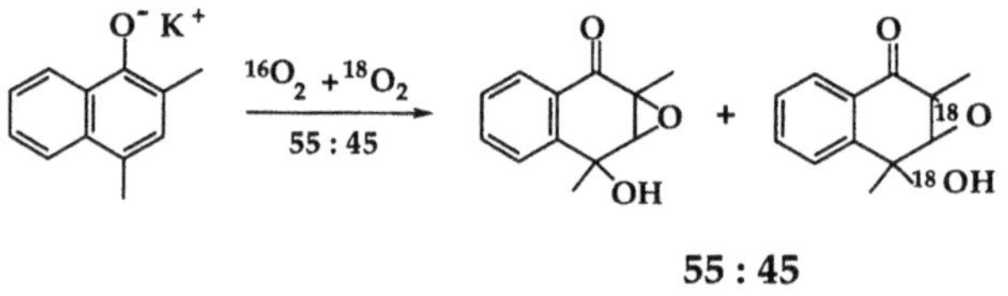

product consisted only of unlabeled and doubly labeled product [4c] (**Scheme 9**). The complete absence of crossover product indicates that the reaction is strictly intramolecular and must proceed through the dioxetane intermediate.

1.2. OXYGEN-18 INCORPORATION INTO VITAMIN K OXIDE

The successful application of ^{18}O labeling studies to the vitamin K model suggests that this might be a useful device to be used in the study of the enzyme catalyzed production of vitamin K oxide. Such an experiment was carried out some years ago by Sadowski, Schnoes and Suttie who were interested at that time in the origin of the epoxide oxygen in vitamin K oxide [5]. They observed the incorporation of an atom of ^{18}O at the epoxide position. In addition, they thoughtfully published the full mass spectra of the results from both the $^{16}O_2$ and $^{18}O_2$ experiments. This was ideal for our purpose because we had a detailed mechanism in mind that might entail the incorporation of a second atom of ^{18}O into the carbonyl groups of vitamin K oxide. Thus, according to the new mechanism a tetrahedral intermediate will form following opening of the dioxetane (**Scheme 10**). The two oxygens in the tetrahedral intermediate are

Scheme 10

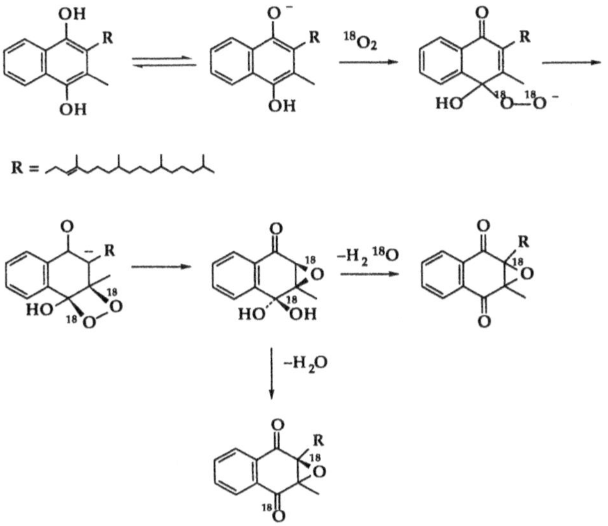

diastereotopic so they will have different chemical properties outside the enzymic milieu, and, presumably, this difference will be accentuated in contact with the enzyme active site. Accordingly, we anticipate that the rates of loss of H_2O and $H_2{}^{18}O$ from the intermediate epoxide will differ. Upon examination of the mass spectra from an experiment, carried out under an atmosphere of $^{18}O_2$ with rat liver microsomes, which form one of the principal sources of the enzyme [3], the rate ratio is found to be approximately 1 : 5. Thus, we observe approximately 18% of a second atom of ^{18}O incorporated into the vitamin K oxide from the experiment This value is thoroughly reproducible in a dozen experiments carried out with the vitamin K-dependent carboxylase in the liver microsomes [4f,j].

With the incorporation of the second ^{18}O into vitamin K oxide, one can now ask which carbonyl group carries the label - that is - which carbon-oxygen bond constitutes the active site of vitamin K hydroquinone. To answer this question, we prepared samples of vitamin K specifically labeled, to a high degree of enrichment, with ^{18}O in each of the carbonyl groups (**Scheme 11**). These

Scheme 11

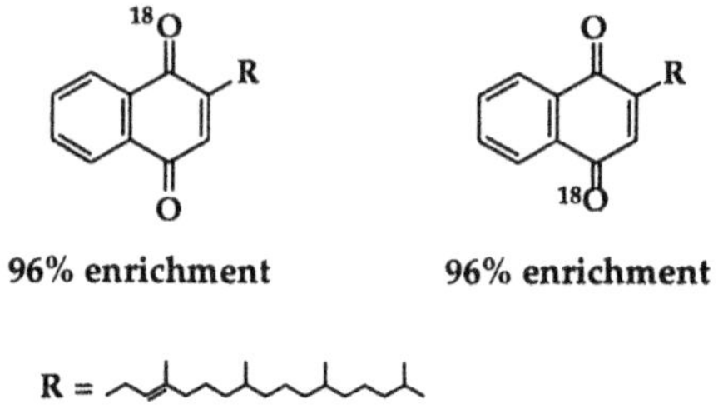

substances were treated individually with the carboxylase in liver microsomes. The product vitamin K oxide was isolated in each case and examined for ^{18}O content by mass spectrometry. The vitamin K-^{18}O with the label next to methyl exchanged 18% of its ^{18}O. Thus, it appeared at this point that all the action occurs adjacent to the methyl group. This conclusion was confirmed when it was found that the other isotopomer, with ^{18}O next to the phytyl sidechain (**Scheme 11**), underwent no exchange of its ^{18}O label. Accordingly, it now becomes possible to designate the active site of vitamin K hydroquinone as shown in **Scheme 12** [4j], in which molecular oxygen attacks the carbon-carbon bond carrying the methyl group.

Scheme 12

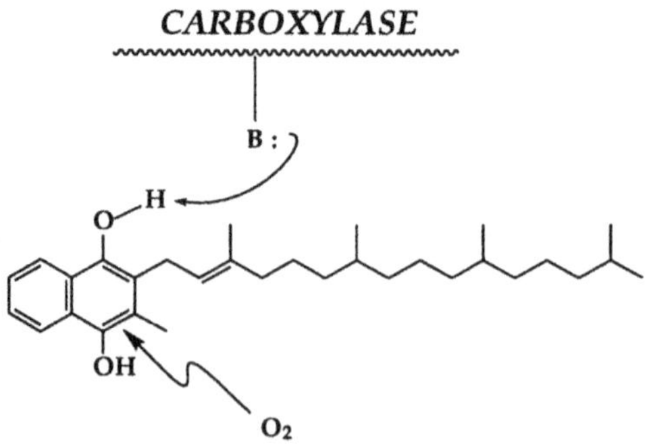

Thus, we envision an enzyme bound base (the weak base) initiating the oxygenation process. A candidate for this role is the thiolate anion (**Scheme 12**). With a pK_a of approximately 8.5, it would be in excellent position to remove a proton from the vitamin K hydroquinone (pK_a 9.3). In work done about fifteen years ago, it was shown that the carboxylase is irreversibly inhibited by the sulfur reagents p-mercuribenzoate and N-ethylmaleimide (NEM) [6]. Furthermore, there appear to be, at least, two active site thiols. One thiol group is protected by prior addition of vitamin K hydroquinone [6a]. The second thiol is not so protected. Its blockage results in inhibition of the carboxylation. Although these observations have been in the literature for a number of years, no mechanistic role has been attached to these active site cysteines. With the advent of the base strength amplification model, the active site cysteines become very important. Thus, we can flesh out the model as diagrammed in **Scheme 13**.

Scheme 13

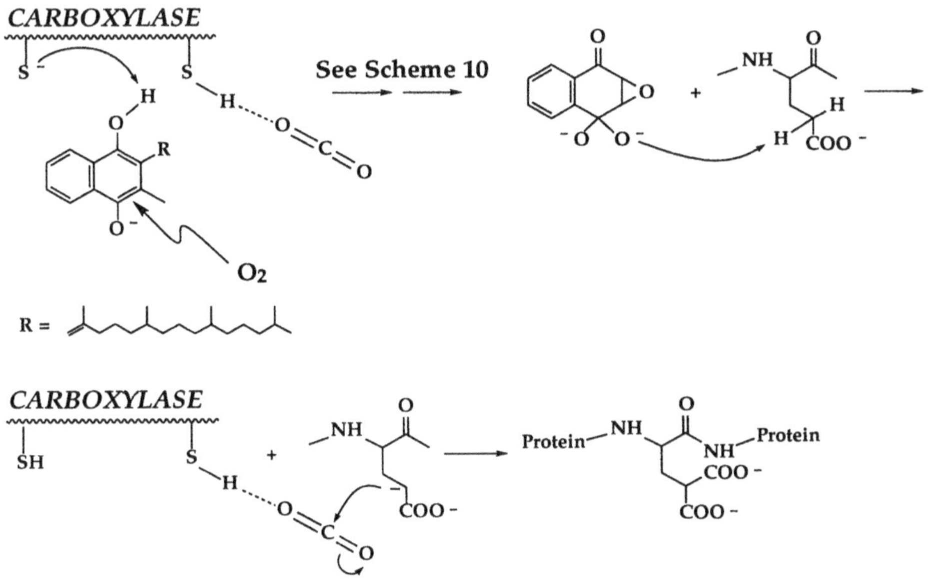

The double thiol at the active site accounts very nicely for the mechanism of action of vitamin K and its associated carboxylase.

1.2.1. *Inhibitors of the Carboxylase*

Of course, this is only an hypothesis, and it is worthwhile to ask whether it can account for other phenomena in the vitamin K-dependent blood-clotting arena. For example, the derivative chloro K in which a chlorine replaces the methyl group in vitamin K is an irreversible inhibitor of the vitamin K-dependent carboxylase [7]. As expected for a vinylogous acid chloride, chloro K reacts instantaneously with alkyl thiols (**Scheme 14**). Therefore, it is entirely

Scheme 14

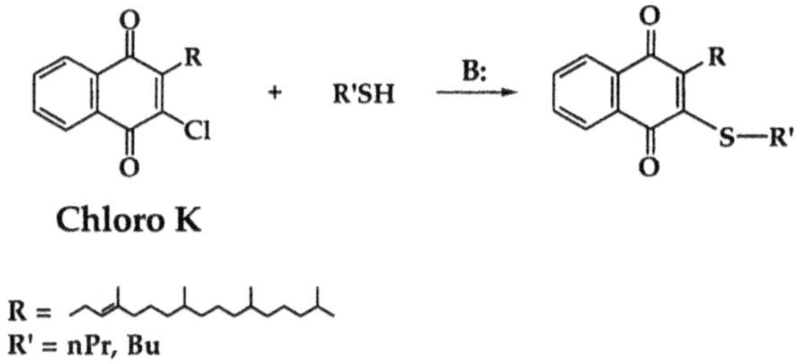

appropriate that the active site thiols of the carboxylase react with chloro K in an irreversible inhibitory fashion (**Scheme 15**).

Scheme 15

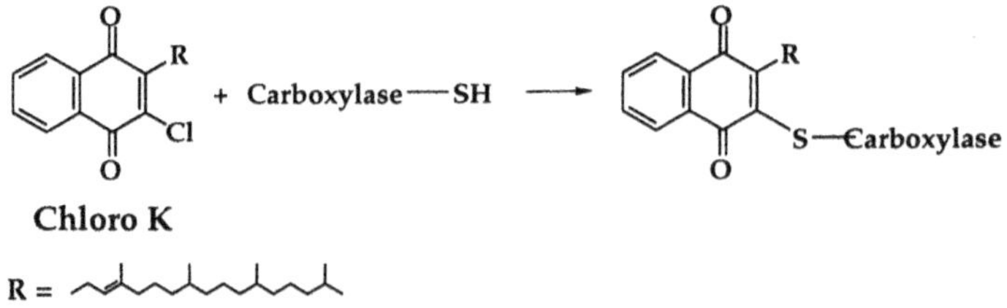

An example that may be somewhat less obvious, but which is also confirmatory of the general hypothesis, is found in desmethyl vitamin K (**Scheme 16**) in

Scheme 16

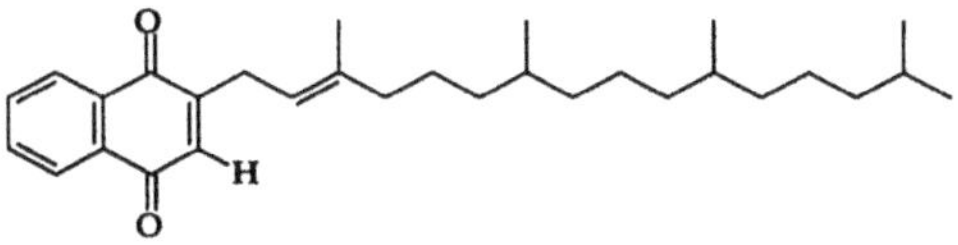

Desmethyl vitamin K

which the quinone methyl group is simply replaced by hydrogen [8]. This substance is also an irreversible inhibitor of the vitamin K-dependent carboxylase. However, the mechanism of inhibition is not known. We find that exposure of des-methyl vitamin K to butanethiol and imidazole at 40 °C leads to the adduct in **Scheme 17** in a few hours [9]. Under the same conditions,

Scheme 17

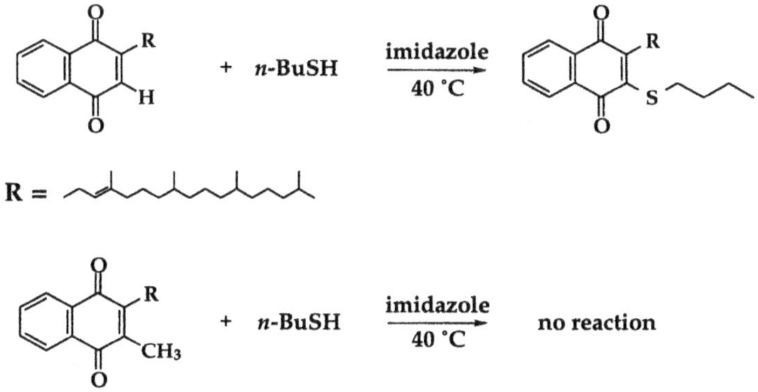

vitamin K yielded no adduct; starting material was recovered. We think it likely that the active site thiols are inhibited by adduct formation to desmethyl vitamin K.

2. VITAMIN E

An even more striking example is provided by consideration of the mechanism of anticlotting action of vitamin E. Although there has been confusion in the literature concerning which oxidation state of vitamin E is responsible for the anticlotting property, we have shown that while α-tocopherol (**Scheme 18**)

Scheme 18

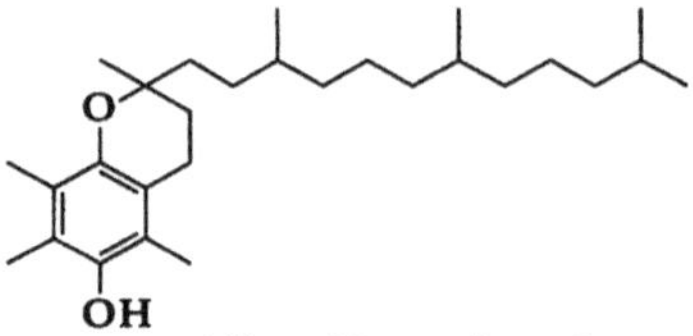

(*d*)–α-**Tocopherol**

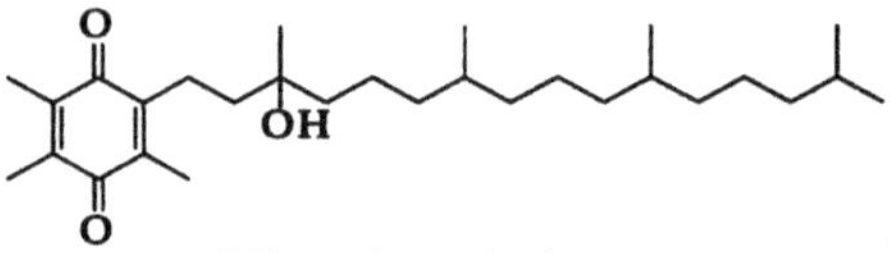

Vitamin E Quinone

has virtually no anticlotting activity (Figure 1), vitamin E quinone (**Scheme 18**) is a potent anticoagulant, as are anhydro vitamin E quinone and the quinone in which an unbranched hydrocarbon chain is substituted for the branched one (Figure 1 and **Scheme 19**) [10].

Scheme 19

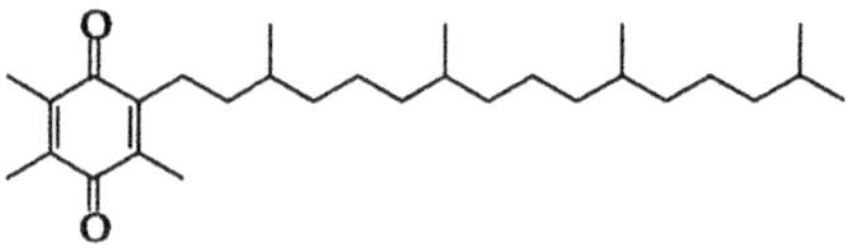

Anhydro Vitamin E Quinone

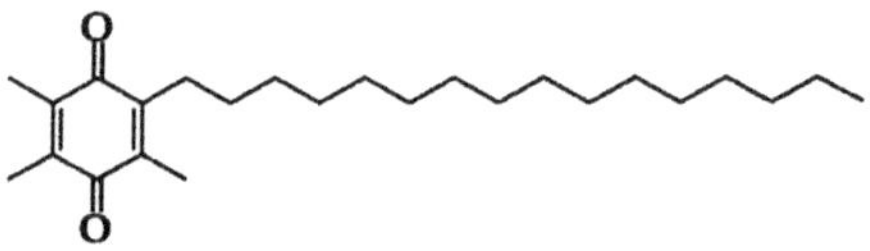

C_{16}-Vitamin E Analogue

The quinones in **Schemes 18** and **19** are approximately equally potent as anticoagulants with the exception that anticoagulant activity drops off as the sidechain of the straight chain synthetic analogues is shortened. For example, the vitamin E quinone analogue with a butyl sidechain shows very little inhibitory activity toward the vitamin K dependent carboxylase.

The anticlotting activity of vitamin E has been known for some time and could have significant impact on health care, if its role is properly understood in mechanistic terms. Oddly, there has been no speculation in the literature regarding the mechanism of vitamin E anticlotting action.

2.1. *Mechanism and Model Studies*

Since the anhydro and straight chain analogues are as potent anticoagulants as is vitamin E quinone, the hydroxyl group in vitamin E quinone plays no role in the inhibition of the carboxylase. That focuses attention squarely on the fully alkylated quinone nucleus. Accordingly, we envisioned that vitamin E quinone might be in equilibrium with various enone tautomers as shown in **Scheme 20**. In

Scheme 20

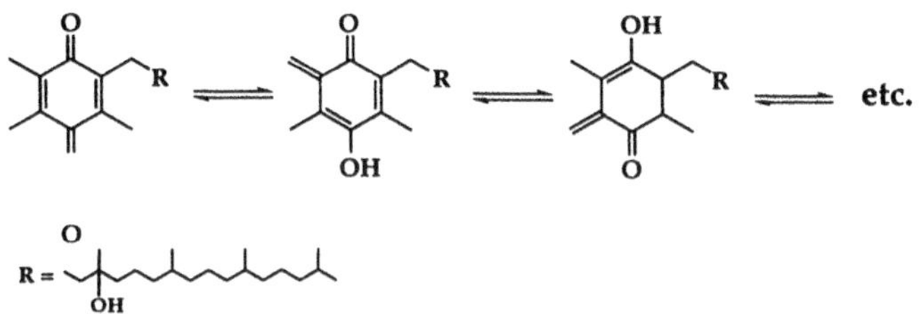

developing this scheme, it is important to be aware that the vitamin K-dependent carboxylase active site contains two thiol groups, as discussed above. The tautomeric enols would then serve as traps for the nucleophilic thiols. Indeed, it is easy to envision the direct involvement of the enzyme thiolate anion in the tautomerization and trapping process (**Scheme 21**). The thiolate anion can first promote the tautomerization by proton removal, and the result-

Scheme 21

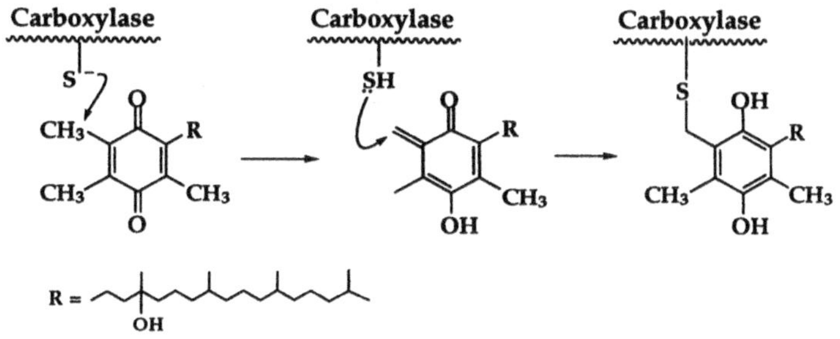

ing thiol can then be trapped by the enone (**Scheme 21**). Moreover, since molecular oxygen is a constituent of the vitamin K-dependent carboxylase active site, the reaction can be made essentially irreversible by oxidation of the first formed hydroquinone by the active site oxygen (**Scheme 22**).

Scheme 22

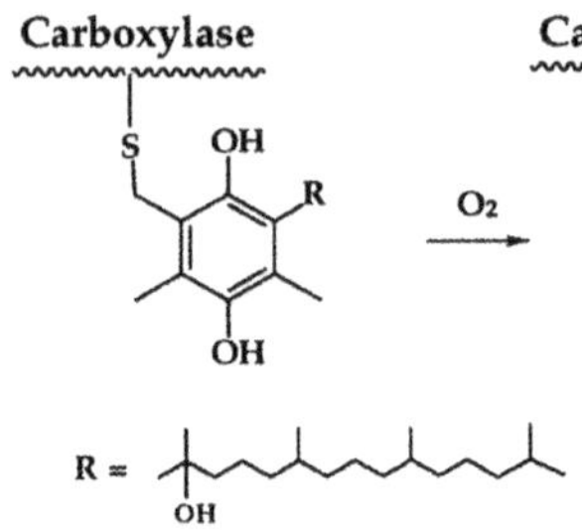

2.2. *Vitamin E Model Studies*

At this juncture, model studies are useful in showing what is feasible and in testing the ideas discussed above. With these considerations in mind, we undertook a series of experiments in which the anticoagulant quinones were exposed to sodium methylthiolate under mild conditions: room temperature, 6 h, methanol/benzene solvent (**Scheme 23**). A mixture of thiomethylated

Scheme 23

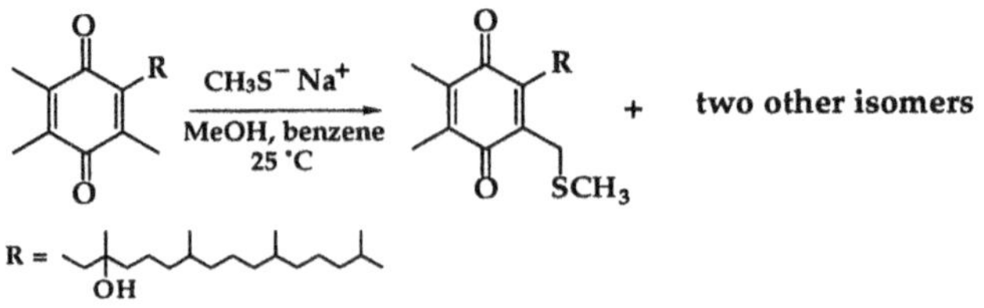

adducts formed that proved difficult to separate using chromatographic techniques. Nonetheless, the probable structures of the thiomethyl derivatives were established by spectroscopic methods as well as by synthesis of an authentic sample of one of the isomers by the path shown in **Scheme 24**. Thus,

Scheme 24

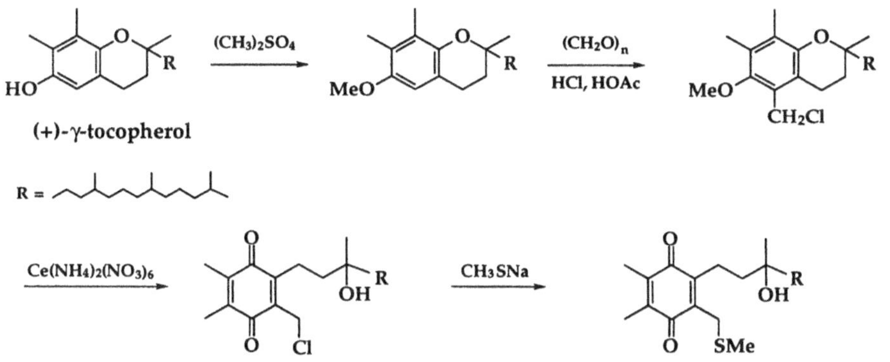

(+)-γ-tocopherol was protected as the methyl ether. Chloromethylation was carried out by treatment with paraformaldehyde and HCl and was followed by oxidation to the quinone with ceric ammonium nitrate. Finally, the thiomethyl group was added by direct displacement. This product compared favorably (HPLC, nmr, ms) with the components of the mixture from treatment of vitamin E quinone with methylthiolate.

A more definitive experiment was carried out on the C_{16} straight chain quinone treated under analogous conditions with sodium methylthiolate. In this instance, one of the thiomethyl adducts was isolated by HPLC (**Scheme 25**). The

Scheme 25

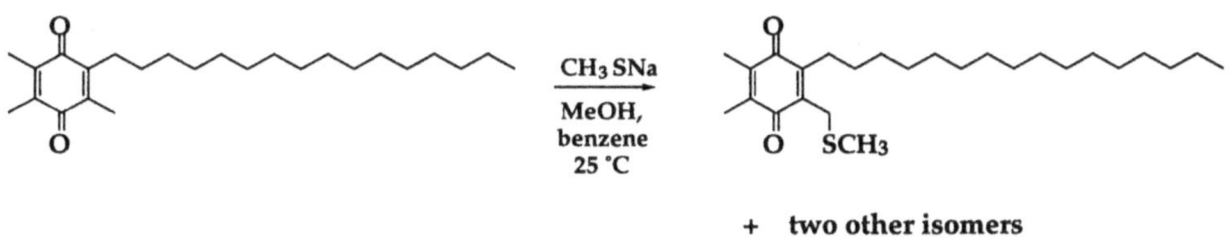

other two isomers formed an inseparable mixture. The structure of the isolated product (**Scheme 25**) was firmly established by spectroscopic means (ir, uv, nmr, ms) and by synthesis of an authentic sample accordingly to the route outlined in **Scheme 26**. Thus, free radical hydroxymethylation of 2,3-dimethylbenzoquin-

Scheme 26

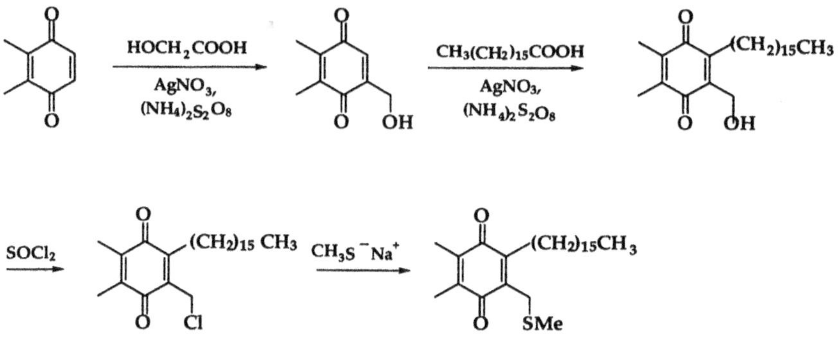

one was followed by a second free radical alkylation to append the sidechain. Thionyl chloride treatment yielded the chloride that was readily transformed to the thiomethyl compound, identical in every respect to the product of exposure of the quinone to sodium methylthiolate.

As noted the other two thiomethyl adducts were not readily separated from one another. Nonetheless, it was useful to prepare authentic samples of these starting from 2,5-dimethylbenzoquinone in one instance and the corresponding 2,6-dimethylbenzoquinone in the other and following the route outlined in **Scheme 26**. These thiomethyl adducts (**Scheme 27**) compared favorably in

Scheme 27

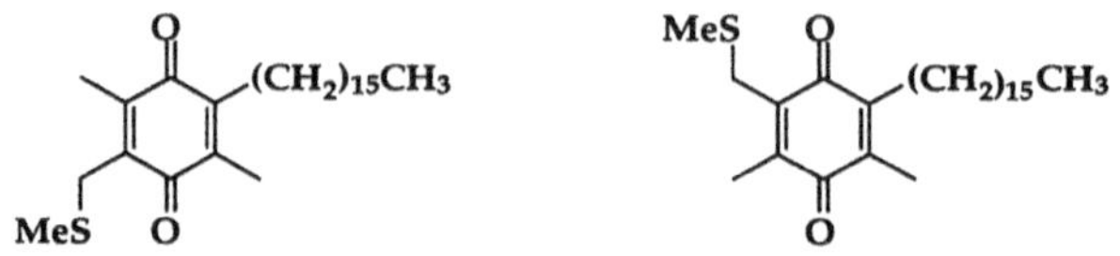

terms of HPLC retention time and nmr spectra with the corresponding components of the mixture.

2.3. *Significance*

The discussion above provides a plausible explanation, in molecular terms, for the beneficent effect of vitamin E on those prone to ischemic heart attack and ischemic stroke. Having demonstrated the potent anticoagulant action of vitamin E quinone, it also becomes clear why vitamin E is detrimental for those liable to hemorrhagic stroke. In our opinion, vitamin E quinone should make a valuable addition to anticoagulant therapy. First, the quinone is probably free of untoward side effects, since commercial samples of vitamin E almost always contain small amounts of the quinone, which are ingested daily by thousands. Second, the mode of anticoagulant action of warfarin and vitamin E quinone are radically different. Warfarin acts to block the reduction of vitamin K oxide to vitamin K (**Scheme 2**). In normal circumstances, the stores of vitamin K are depleted and blood clotting is inhibited. There are several disadvantages to such therapy. It requires time to adjust the warfarin dosage to the desired level of anticlotting activity, often necessitating a hospital stay of 2-3 days or longer. Considerable expense might be spared if this time could be shortened. Second, if the initial stores of vitamin K are high or if dietary intake of vitamin K is higher than normal, then warfarin therapy can fail or be significantly delayed. This delay might have untoward consequences in emergency situations.

Vitamin E quinone functions by a completely different mechanism than does warfarin. The quinone is a direct inhibitor of the carboxylase and depends to a far smaller degree on the levels of dietary intake of vitamin K. Accordingly, we suggest that the efficacy of vitamin E quinone be explored as a clinically useful anticlotting agent.

Finally, because normal samples of vitamin E contain vitamin E quinone and the quinone is a naturally occurring metabolite of α-tocopherol, caution should be exercised in the self-administration of vitamin E.

ACKNOWLEDGMENT

This work was generously supported by the NIHLB of the National Institutes of Health under Grant HL 50667.

REFERENCES

1. Dam, H. (1929) Cholesterinstoffwechsel in Hühnereiern und Hühnchen, *Biochem. Z.* **215,** 475-492. Dam, H. (1930) Cholesterol synthesis in the animal body, *Biochem. Z.* **220**, 158-63. Dam. H. (1934) Hemorrhages in chicks reared on artificial diets. A new deficiency disease, *Nature* **133,** 909-910. McFarlane, W. D., Graham, Jr., W. R. and Hall, G. E. (1931) Studies in protein nutrition of the chick, *J. Nutr.* **4,** 331-349. McFarlane, W. D., Graham, Jr., W. R. and Richardson, F. (1931) The fat-soluble vitamin requirements of the chick, *Biochem. J.* **25**, 358-366. Dam, H. (1935) The antihemmorhagic vitamin of the chick: Occurrence and chemical nature, *Nature* **135,** 652-653. Dam, H. (1935) Antihaemorrhagic vitamin of chick, *J. Biochem.* **29,** 1273-1285.

2. For a review of early work: Wagner, A. F. and Folkers, K. (1964) Vitamins and Coenzymes, Interscience: New York. pp. 407-434.

3. Reviews: Suttie, J. W. (1988) Vitamin K-dependent carboxylation of glutamyl residues in proteins, *Biofactors* **1,** 55-60. Stenflo, J. and Suttie, J. W. (1977) Vitamin K-dependent formation of γ-carboxyglutamic acid, *Ann. Rev. Biochem.* **46,** 157-172. Suttie, J. W. (1985) Vitamin K-dependent carboxylase, *Ann. Rev. Biochem.* **54,** 459-477. Suttie, J. W. (1984) Handbook of Vitamins: Nutritional, Biochemical and Clinical Aspects, L. J. Machlin (ed), Dekker, New York, pp 147-198. Olson, R. E. (1984) The function and metabolism of vitamin K, *Ann. Rev. Nutr.* **4,** 281-337. Vermeer, C. (1990) γ-Carboxyglutamate-containing proteins and the vitamin K-dependent carboxylase, *Biochem. J.* ***266***, 625-636. See also, Girardot, J.-M. (1982) Vitamin K-dependent carboxylase: Partial purification and properties of the enzyme-substrate complex, *J. Biol. Chem.* **257,** 15008-15011.

4. (a) Ham, S.-W. and Dowd, P. (1990) On the mechanism of action of vitamin K. A new nonenzymic model, *J. Am. Chem. Soc.* **112,** 1660-1661. (b) Dowd, P., Ham. S.-W. and Geib, S. J. (1991) Mechanism of action of vitamin K, *J. Am. Chem. Soc.* **113,** 7734-7743. (c) Dowd, P. and Ham, S.-W. (1991) On the intramolecularity of the vitamin K model oxidation, *J. Am. Chem. Soc.* **113,** 9403-9404. (d) Dowd, P. and Ham, S. W. (1991) Mechanism of cyanide

inhibition of the blood-clotting, vitamin K-dependent carboxylase, *Proc. Natl. Acad. Sci., USA* **88**, 10583-10585. (e) Dowd, P., Ham, S.-W., Marchand, A. P. and Rajapaksa, D. (1992) Mechanism of epoxidation of vitamin K with basic hydrogen peroxide, *J. Org. Chem.* **57**, 3514-3516. (f) Dowd, P., Ham. S.-W. and Hershline, R. (1992) Role of oxygen in the vitamin K-dependent carboxylation reaction: Incorporation of a second atom of ^{18}O from molecular oxygen-$^{18}O_2$ into vitamin K oxide during carboxylase activity, *J. Am. Chem. Soc.* **114**, 7613-7617. (g) Arnett, E. M., Dowd, P., Flowers II, R. A., Ham, S.-W. and Naganathan, S. (1992) Thermochemical confirmation of the mechanism of action of vitamin K. *J. Am. Chem. Soc.* **114**, 9209-9210. (h) Naganathan, S., Hershline, R., Ham, S.-W. and Dowd, P. (1993) The active site of vitamin K. Regiospecific oxygenation of vitamin K hydroquinone in its role as carboxylase cofactor, *J. Am. Chem. Soc.* **115**, 5839-5840. (i) Flowers, R. A., II, Naganathan, S., Dowd, P., Arnett, E. M. and Ham. S. W. (1993) Thermochemical investigation of the oxygenation of vitamin K, *J. Am. Chem. Soc.* **115**, 9409-9416. (j) Naganathan, S., Hershline, R., Ham. S. W. and Dowd, P. (1994) The active site of vitamin K and the role of the vitamin K-dependent carboxylase, *J. Am. Chem. Soc.* **116**, 9831-9839.

5. Sadowski, J. A., Schnoes, H. K. and Suttie, J. W. (1977) Vitamin K epoxidase: Properties and relationship to prothrombin, *Biochemistry* **16**, 3856-3863.

6. (a) Canfield, L. M. (1987) Vitamin K-dependent oxygenase/carboxylase; diffential inactivation by sulfhydryl reagents, *Biochem. Biophys. Res. Commun.* **148**, 184-191. (b) Morris, D. P., Soute, B. A. M., Vermeer, C. and Stafford, D. W. (1993) Characterization of the purified vitamin K-dependent γ-glutamyl carboxylase, *J. Biol. Chem.* **268**, 8735-8787. (c) Suttie, J. W. (1985) Vitamin K-dependent carboxylase, *Annu. Rev. Biochem.* **54**, 459-477. (d) Friedman, P. A., Shia, M. (1976) Some characteristics of a vitamin K-dependent carboxylating system from rat liver microsomes, *Biochem. Biophys. Res. Commun.* **70**, 647-657. (e) Canfield, L. M., Sinsky, T. A. and Suttie, J. W. (1980) Vitamin K-dependent carboxylase: Purification of the rat liver microsomal enzyme, *Arch. Biochem. Biophys.*

202, 515-524. (f) Mack, D. O., Suen, E. T., Girardot, J. M., Miller, J. A., Delaney, R. and Johnson, B. C. (1976) Soluble enzyme system for vitamin K-dependent carboxylation, *J. Biol. Chem.* **251**, 3269-3276. (g) Suttie, J. W., Lehrman, S. R., Geweke, L. O., Hageman, J. M. and Rich, D. H. (1979) Vitamin K-dependent carboxylase: Requirements for carboxylation of soluble peptide substrates and substrate specificity, *Biochem. Biophys. Res. Commun.* **86**, 500-507.

7. Lowenthal, J. and Roychowdhury, M. N. (1970) Synthesis of vitamin K_1 analogs. A new class of vitamin K_1 antagonists, *Can. J. Chem.* **48**, 3957-3958.

8. Cheung, A. Y., Wood, G. M., Funakawa, S., Grossman, C. P. and Suttie, J. W. (1988) Vitamin K-dependent carboxylase: Substrates, products and inhibitors, in J. W. Suttie (ed), *Current Advances in Vitamin K Research,* Elsevier Science Publishing Co., Inc. pp. 3-16. Friedman, P. A. and M. W. Smith (1979) Epoxidation of several vitamins K by rat liver microsomes, *Biochem. Pharm.* **28**, 937-938. Friedman, P. A. and Shia, M. (1976) Some characteristics of a vitamin K-dependent carboxylating system from rat liver microsomes, *Biochem. Biophys. Res. Comm.* **70** 647-654. Cheung, A. and Suttie, J. W. (1988) Synthesis of menaquinone-2 derivatives as substrates for the liver microsomal vitamin K-dependent carboxylase, *Biofactors.* **1** 61-65. Olson, R. E., Houser, R. M., Searcey, M. T., Gardner, E. J., Scheinbuks, J., Rao, G. N. S., Jones, J. P. and Hall, A. L. (1978) Nature of the vitamin K-dependent CO_2 fixation in microsomal membranes, *Fed. Proceed* **37.** , 2610-2614.

9. Dowd, P. and Kerns, J. (1995) unpublished research.

10. Dowd P. and Zheng, Z. B. (1995) On the Mechanism of the anticlotting action of vitamin E quinone, *Proc. Natl Acad. Sci., USA, in press.*

UNNATURAL AMINO ACIDS FOR THE DESIGN OF FUNCTIONAL PROTEINS: BIOMIMETIC CATALYSIS USING COENZYME–AMINO ACIDS

B. IMPERIALI, R. SINHA ROY, G. K. WALKUP AND L. WANG
Division of Chemistry and Chemical Engineering
California Institute of Technology
Pasadena, California 91125, USA

Introduction

The array of functional groups that comprises the catalytic machinery within an enzyme active site is generally restricted to the coded amino acids. In several enzymes, however, this limited array of functionality is augmented by the incorporation of coenzymes. These agents greatly broaden the scope of transformations that may be accomplished by these versatile biopolymers. The reactivity and selectivity of the coenzymes are modulated by covalent or non-covalent interactions in the host enzymes. Our research aims at replicating coenzyme chemistry within synthetic polypeptides, thus laying a foundation for the design of new catalytic biomolecules.

While there has been significant progress in the *de novo* design of structural polypeptides in recent years, the synthesis of functional proteins has been hindered by a limited understanding of the protein folding process. The principal difficulty in eliciting function within designed polypeptides is an inability to replicate the exact three-dimensional disposition of functional groups that is the hallmark of an enzyme active site. It would therefore be advantageous to harness the inherent chemical reactivity of coenzyme within a synthetic peptide framework. A disadvantage of using natural coenzymes directly in this capacity is the prerequisite for a coenzyme binding site, the construction of which would again demand a stringent control over the protein architecture.

The chemical modification of amino acid side chains in existing proteins has been utilized as a strategy for the introduction of coenzyme functionality into organized systems.[1-4] However, this conceptual approach lacks versatility and selectivity and is therefore not ideally suited to the protein design process. As an alternative, we have developed and synthesized the coenzyme–amino acid chimeras.[5] This new class of amino acids incorporates the core functionality of select coenzymes as integral constituents of α-amino acid constructs (Table 1). In this "disguise", coenzyme functionality can be incorporated at variable positions in any primary peptide sequence *via* solid phase synthesis methods. The direct covalent attachment of coenzyme functionality to the polypeptide backbone obviates the need for a coenzyme binding site without compromising design versatility. The coenzyme functionalities that we have focused on to date comprise the reactive cores of pyridoxal phosphate (PLP), thiamine pyrophosphate (TPP), and nicotinamide adenine dinucleotide (NADH).

C. S. Wilcox and A. D. Hamilton (eds.), Molecular Design and Bioorganic Catalysis, 35–52.

Table 1. A summary of the coenzymes investigated, the corresponding amio acid chimeras, and the precursors developed for solid phase peptide synthesis.

ENZYME ASSOCIATED COENZYME	FUNCTION OF COENZYME	COENZYME–AMINO ACID CHIMERA		SOLID PHASE PEPTIDE SYNTHESIS PRECURSOR
Pyridoxal Phosphate 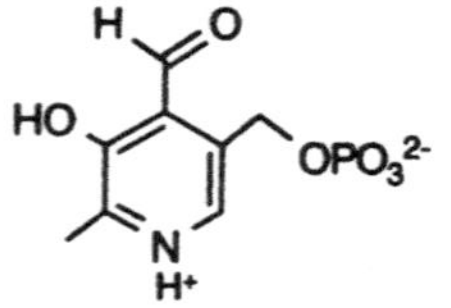	Modification reactions at α-, β-, or γ-position of amino acids. α-Position reactions include racemization,transamination and decarboxylation. β-elimination of serine is key for the biosynthesis of tryptophan.	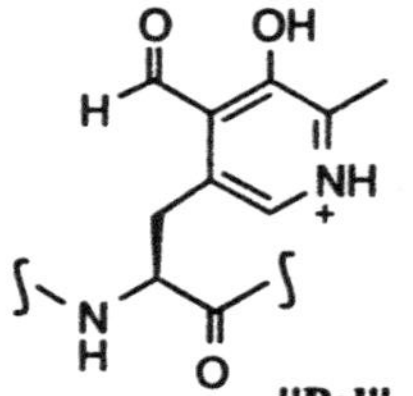 "Pal"	⟹	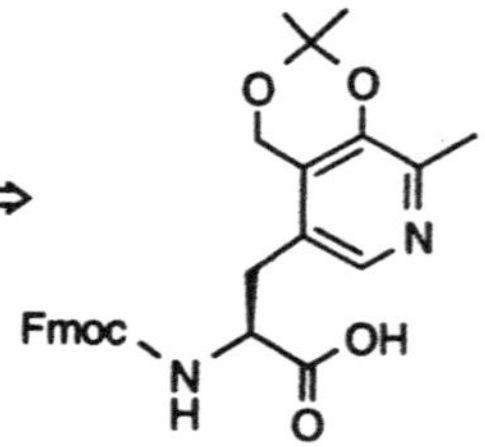
Thiamin Pyrophosphate 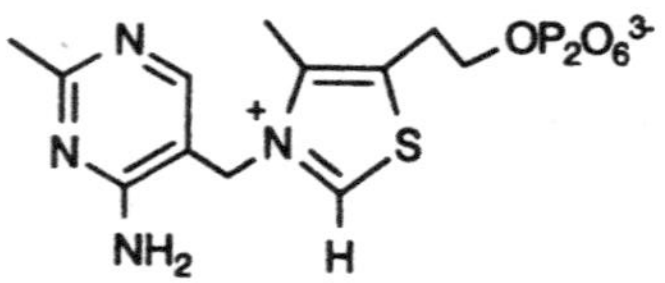	The C-2 methine of the thiazolium heterocycle forms a ylide adduct with carbonyl compounds and provides an electron sink for many transformations. Examples include α-keto acid decarboxylases and transketolases.	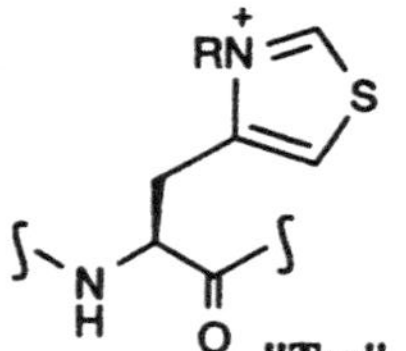 "Taz"	⟹	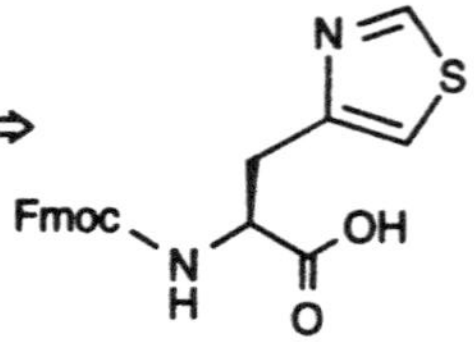
Nicotinamide Adenine Dinucleotide (Red.) 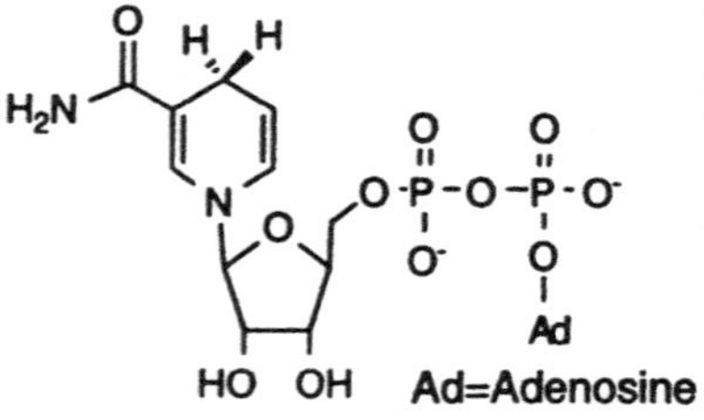	Mediates hydride transfer processes including oxidation and reduction reactions under general acid or Lewis acid catalyzed conditions.	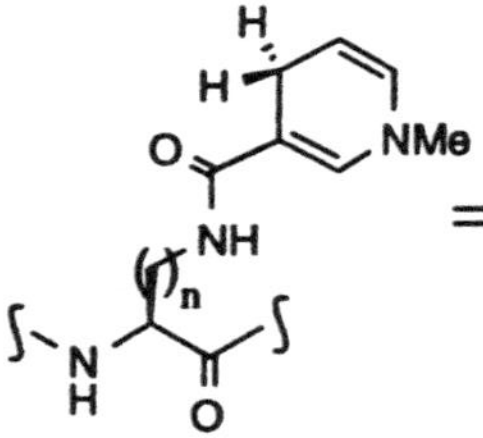 "Nic"	⟹	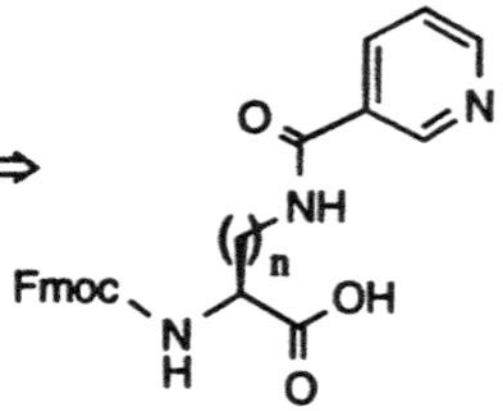

A Pyridoxal-Based Coenzyme Amino Acid Chimera

Our initial investigations into the development of the coenzyme–amino acid chimeras focused on the synthesis of a pyridoxal residue, Pal (Table 1). The highly functionalized character of this coenzyme raised questions regarding the feasibility of handling the residue both during and after polypeptide synthesis. In particular, the 4-formyl functionality of a pyridoxal analog proved difficult to accommodate in the solid phase peptide synthesis protocols. In order to circumvent these problems, a protected pyridoxol amino acid precursor (*S*)-2-amino-N^{α}-9-fluorenylmethyloxycarbonyl-3-(α^4,3-*O*-isopropylidenepyridox-5-yl)propanoic acid (Fmoc-Pol(Iso)-OH, **6**) was synthesized as an alternate target for peptide incorporation.[6] Orthogonal protection of the residue with the acid-labile isopropylidene group ensured compatibility with the base-intensive 9-fluorenylmethyloxycarbonyl (Fmoc) chemistry of peptide synthesis. Transformation of the pyridoxol precursor (Pol) to the catalytically active pyridoxal analog (Pal) was carried out in an oxidation that followed the completion of peptide synthesis.

STEREOSELECTIVE SYNTHESIS OF A PYRIDOXOL AMINO ACID (POL)

A key step in the synthesis of the pyridoxal amino acid coenzyme chimera **6** was the stereoselective alkylation of *N*-(diphenylmethylene)glycine *tert*-butyl ester (**3**) with the bromide **2** using the phase transfer catalyst (PTC) (8*S*, 9*R*)-(-)-*N*-benzylcinchonidinium chloride according to the method of O'Donnell et al[7] (Scheme 1). The alkylation proceeded with modest asymmetric induction (52% ee), affording the desired *S* enantiomer of **4** as the major product. Enantiomerically pure (>99% ee) material was obtained by crystallizing the racemic alkylation product from a solution of the (*S*)-**4** enriched product in hexane. To date, attempts to achieve a more efficient preparation of **6** through an enzymatic resolution have proven unsuccessful. The amino acid was protected as the Fmoc derivative for solid phase peptide synthesis. These optimized procedures allow for the synthesis of multigram quantities of the enantiomerically pure amino acid derivative.

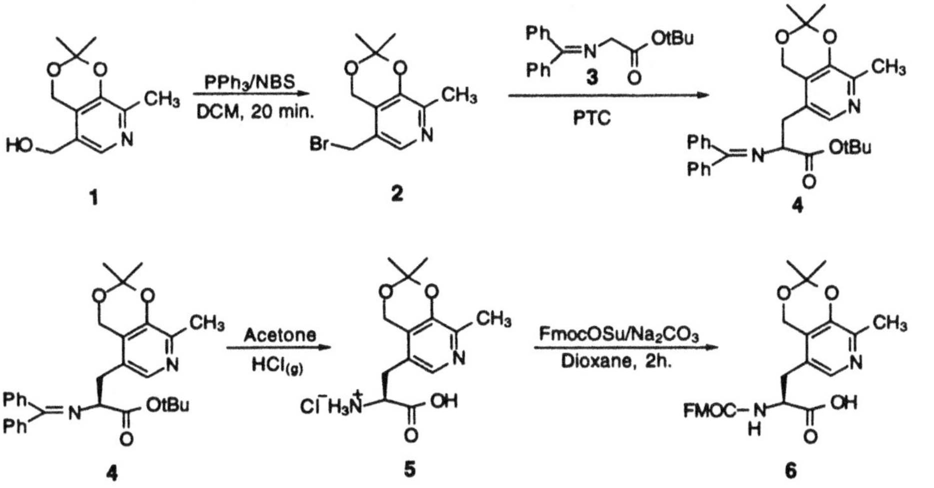

Scheme 1. Synthesis of the protected pyridoxol amino acid.

PEPTIDE INCORPORATION AND OXIDATION TO THE FUNCTIONAL PYRIDOXAL RESIDUE (PAL)

We have demonstrated that the precursor to the pyridoxal coenzyme-amino acid chimera (Pol) is amenable to solid phase peptide synthesis via the preparation of several polypeptides ranging in length from 6-14 residues. In most cases, the Pol residue was incorporated with a coupling efficiency greater than 98%, and the resultant polypeptide products were recovered in good yield and purity. Selective oxidation protocols were developed for conversion of the pyridoxol residue (Pol) into the catalytic pyridoxal analog (Pal). Specifically, acidolytic cleavage of the Pol-peptides from the solid support effected the simultaneous deprotection of the isopropylidene protecting group. The Pol residue was subsequently oxidized to the catalytic pyridoxal analog (Pal) in modest yields, using γ-MnO_2[8] in dimethylsulfoxide (Scheme 2). The oxidation of Pol was selective in the presence of most functionalized residues including serine, threonine, glutamic and aspartic acids, arginine, and histidine. However, it is anticipated that the oxidation process would be incompatible with the sulfur containing amino acids cysteine and methionine.

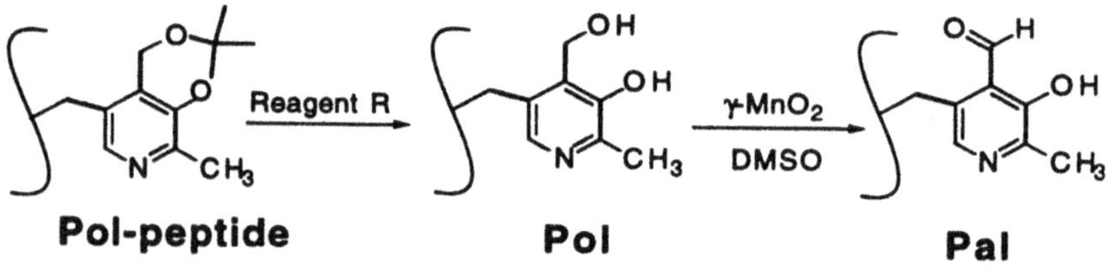

Scheme 2. Post synthesis modification of the Pal peptides to the active pyridoxal analog.

While peptides incorporating the oxidized amino acid are not amenable to rigorous purification, the resulting mixture of reduced and oxidized peptides (composed of 15-40% Pal-peptide) can be quantified by the unique absorption of the pyridoxal moiety in 0.1 N NaOH ($\varepsilon_{390} = 6400\ cm^{-1}M^{-1}$).[9] In general, we chose to assay the products directly to minimize degradation of the reactive Pal functionality. The unoxidized Pol-peptides are catalytically inactive and do not interfere in the kinetic analysis. The identities of the Pal-peptides have been independently confirmed by mass spectral analyses of the trapped ($NaBH_4$) Schiff base intermediates with isoleucine and phenylethylamine.

MODULATION OF VITAMIN B_6 ACTIVITY WITHIN MODEL PAL-PEPTIDES

Hexapeptides **P1-P4** were prepared to establish the viability of the coenzyme amino acid concept in the development of new functional biomolecules (Table 2). The LPro-DAla sequence in peptides **P1-P3** promotes a type-II reverse-turn conformation and results in the flanking residues being proximal to each other in a folded motif (Fig. 1).[10] The reactivity of the Pal chimera (at position 5) was therefore enhanced by the introduction of a general base across the reverse-turn (Xaa = His, **P2**; Xaa = 3-pyridylalanine (Pyr), **P3**).

While pyridoxal-dependent enzymes catalyze a variety of transformations involving α-amino acid substrates,[11] we have elected to focus on assessing the transaminase activity of the Pal-peptides. The distinct absorbances of each intermediate in this process allows for a detailed analysis of the transformation by UV-Vis spectroscopy. The acidic conditions that favor this transformation ($pH_{optimum} \sim 4.0$) also promote a significant reverse-turn population within the heterochiral peptides as evidenced by NMR and CD spectroscopy. For each coenzyme system, the copper-mediated transamination of L-alanine to pyruvic acid was assayed under single turnover conditions (Scheme 3). Although pyridoxal-dependent transaminases *do not* require metal ions for activity, the formation of a metal-Schiff base chelate in simple model systems is required for stabilizing subsequent carbanion intermediates by delocalization of charge into the heteroaromatic ring.[12] Spectra from a typical assay (peptide **P3**) are shown in Fig. 2. The pseudo-first-order rate constants were calculated from the rates of disappearance of Schiff base at 390 nm, or from the rates of appearance of pyridoxamine-peptide (Pmn-peptide) at 324 nm.

P1 Ac-Thr-Val-Pro-DAla-Pal-Gly-NH_2
P2 Ac-Thr-His-Pro-DAla-Pal-Gly-NH_2
P3 Ac-Thr-Pyr-Pro-DAla-Pal-Gly-NH_2
P4 Ac-Thr-His-Gly-Gly-Pal-Gly-NH_2

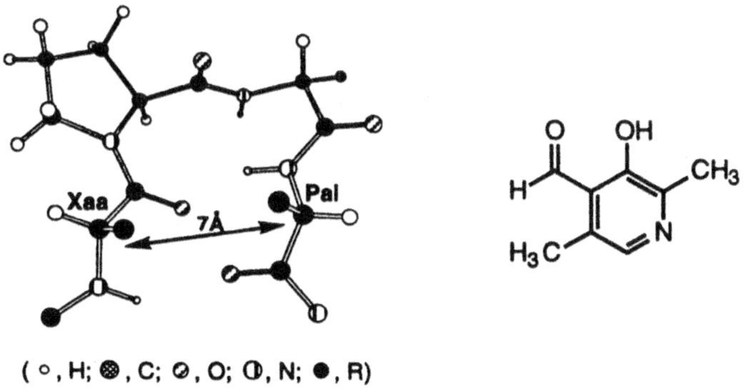

P1-P3 **DPal**

Fig 1. The type-II reverse turn motif in model Pal-hexapeptides brings flanking residues (Xaa_2 and Pal_5) proximal to each other. The transaminase activity of these peptides was compared to 5'-deoxypyridoxal (DPal), an electronically similar model compound.

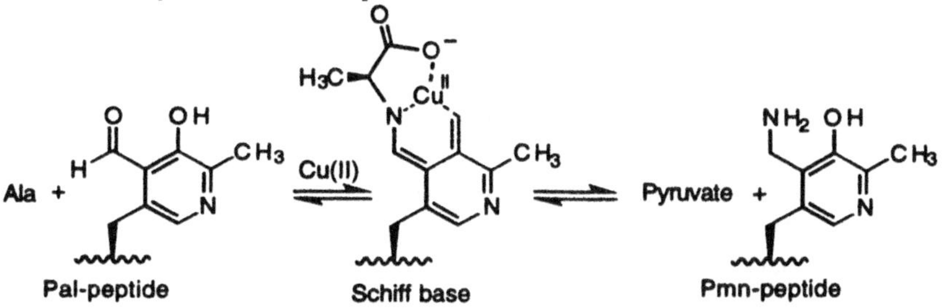

Scheme 3. Copper-mediated transamination of L-alanine to pyruvate, mediated by the Pal-peptides.

Kinetic data for the transamination of L-alanine mediated by the Pal-peptides are illustrated in Table 2. The augmented reactivities of peptides **P2** and **P3** relative to a control peptide (Xaa = Val, **P1**) indicated an enhancement of coenzyme activity by general base catalysis in the former peptides. The increased acidity of the 3-pyridylalanine residue (pKa~5.4, **P3**) relative to histidine (pKa~6.5, **P2**) and valine (**P1**), paralleled the reactivities of these peptides at pH 4.0. The low transamination rate of control peptide **P4** (Xaa = His), which incorporates a flexible glycine spacer instead of the reverse turn motif, argued against an *intermolecular* participation of the general base. The importance of secondary structure in enhancing reactivity was further underscored by the poor reactivity of the model compound, 5'-deoxypyridoxal (DPal).[13]

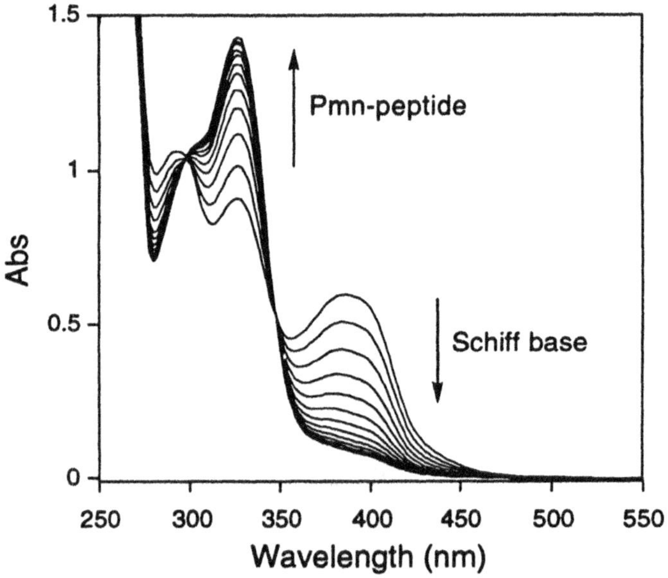

Figure 2. Ultraviolet spectra acquired at 60 min intervals for the transamination of L-Ala to pyruvate by peptide **P3**. The rate constants were derived from the distinct absorbances of the Schiff base intermediate and the pyridoxamine-peptide product.

Table 2. Kinetic data for the Pal hexapeptides.

	Coenzyme System	k_{obs} x 10^5 (min^{-1})	$t_{1/2}$ (h)
P1	Ac-Thr-Val-Pro-DAla-Pal-Gly-NH_2	164.8	7.0
P2	Ac-Thr-His-Pro-DAla-Pal-Gly-NH_2	224.8	5.1
P3	Ac-Thr-Pyr-Pro-DAla-Pal-Gly-NH_2	278.3	4.1
P4	Ac-Thr-His-Gly-Gly-Pal-Gly-NH_2	61.8	18.7
DPal	5'-deoxypyridoxal	58.9	19.6

SEMISYNTHESES OF NEW TRANSAMINASE PROTEINS

The ability to modulate coenzyme function within hexapeptides **P1-P4** reaffirmed the potential utility of coenzyme-amino acid chimeras in the *de novo* design of functional proteins. However, a higher degree of structural complexity was considered necessary for increased efficiency. Towards this end we have introduced the Pol residue into an existing protein template *via* semisynthesis.[14] The Pal residue was incorporated into four ribonuclease-S (RNase-S) complexes in a study aimed at enhancing the reactivity of the coenzyme chimera within an existing protein construct. Ribonuclease-S is a non-covalent protein complex comprising of a 20-residue S-peptide and the larger S-protein (SP).[15] The first 14 residues of S-peptide are sufficient for complex formation.[16] We have attempted to utilize the substrate binding cleft and catalytic residues of this enzyme in the design of a new pyridoxal-dependent biocatalyst.[5]

Synthetic S-peptide derivatives are useful vehicles for the introduction of unnatural residues into the protein complex.[17, 18] A Phe_8-->Pal substitution in S-peptide was made to insert the residue between the general acid-base pair (His_{12}, His_{119})[19] utilized by the native enzyme (Fig. 3). S-peptide$_{1-14}$ derivatives **C0-C3** incorporating the chimera at this position were synthesized by Fmoc-based solid phase methods (Table 3). The Lys residues in native S-peptide$_{1-14}$ were replaced to avoid complications due to intramolecular Schiff base formation. Met_{13} was substituted with a hydrophobic residue to ensure compatibility with the oxidation step that followed peptide synthesis. The Pal-peptides were obtained in 10-29% yield using the oxidation protocols described for hexapeptides **P1-P4**.

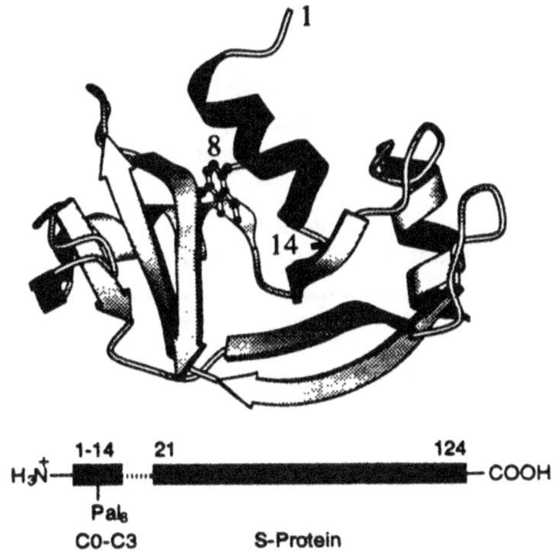

Figure 3. Proposed model of the RNase-S complex incorporating the Pal chimera. The S-peptide fragment has been highlighted,. A linear schematic of the construct is presented for comparison. (Reproduced with permission from *J. Am. Chem. Soc.* **1994**, *116*, 12083.)

For each design, the dissociation constant (K_D) of the complex formed between S-protein (SP) and the Pol-peptide was measured spectroscopically (Fig. 4, Table 3),[20, 21] and used to estimate the K_D of the corresponding Pal system. The

Pal-peptides were subsequently assayed for their ability to catalyze the transamination of L-alanine in the presence and absence of S-protein and Cu(II). Sufficient S-protein was added in appropriate assays to ensure 90% complex formation with the peptides. In addition, for several peptides, the Schiff base intermediate and the pyridoxamine peptide product (Pmn-pep) of the transamination reaction were trapped with sodium borohydride and dansyl chloride respectively, and both species characterized by mass spectroscopy to confirm the nature of the transformation and release of product from the protein complex.

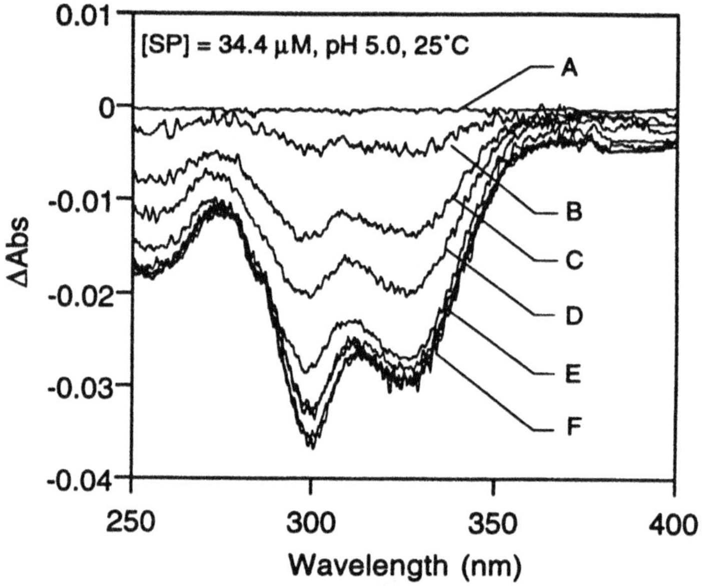

Figure 4. Difference spectra of the Pal residue obtained by the titration of S-protein with peptide **C3**. The absorbance data was used to calculate the dissociation complex of the resulting protein complex. Equivalents of peptide **C3** present: A, 0; B, 0.5; C, 1.0; D, 1.5; E, 2.5; F, 4.0-5.5.

Table 3. Primary sequences of S-peptide$_{1-14}$ and the Pal derivatives **C0**-**C3**. The dissociation constant for the complex of each peptide with S-protein is listed for comparison.

RNase-S Complex: S-peptide$_{1-14}$ sequence	K_D (μM)
SP$_{1-14}$: LysGluThrAlaAlaAlaLysPheGluArgGlnHisMetAspNH$_2$	13
C0-SP: AcGlyGluThrAlaAlaAlaGlyPalGluArgGlnHisIleAspNH$_2$	209
C1-SP: AcNleGluThrAlaAlaAlaGlyPalGluArgGlnHisNleAspNH$_2$	16
C2-SP: AcNleGluThrAlaAlaAlaNlePalGluArgGlnHisNleAspNH$_2$	32
C3-SP: AcNleGluThrAlaAlaAlaPalPheGluArgGlnHisNleAspNH$_2$	11

COPPER-MEDIATED TRANSAMINATION BY THE PAL-RNASE COMPLEXES

First-order kinetic fits to the absorbance data collected for a typical transamination assay are presented in Figure 5. The results are summarized in Table 4. Under similar conditions, DPal exhibited lower transamination rates than either the Pal peptides alone or the semisynthetic complexes ($k_{obs} < 91 \times 10^{-5}$ min^{-1} for all cases). Conversely, the reactivities of all four peptides were affected by S-protein, thus implicating the RNase complex as the prime reactive species under the reaction conditions.

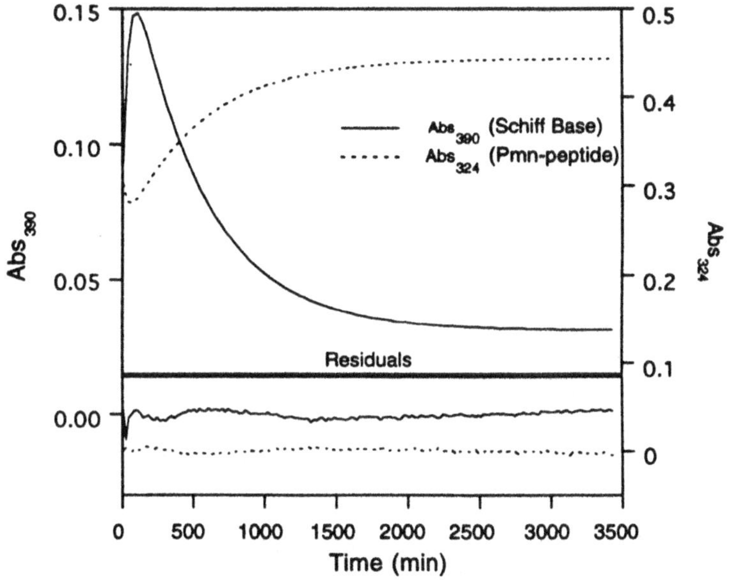

Figure 5. First order kinetic fits to the absorbance data collected for the transamination of L-Ala by complex **C2-SP**. Assay conditions: [SP]=0.6mM; [C2]=0.1mM; [L-Ala]=1M; [Cu(II)]=0.1mM; pH 4.0, 25°C.

The first generation RNase complex (**C0-SP**) exhibited an order of magnitude rate enhancement for transamination, compared to DPal. Unfortunately, the affinity of this peptide for S-protein was 15-fold lower than that of native S-peptide$_{1-14}$. Design efforts were therefore directed at improving the efficiency of complex formation. Specifically, norleucine (Nle) was used as an isosteric replacement for lysine and methionine in the native sequence. The new sequences **C1** and **C2** differed only in the residue preceding the coenzyme chimera. Gly$_7$ was retained in peptide **C1** since the corresponding Nle$_7$ residue in complex **C2-SP** had the potential of occluding the coenzyme chimera (Pal$_8$) from substrate and solvent.

The affinities of peptides **C1** and **C2** for S-protein were comparable to that of the native sequence, and the **C1-SP** complex exhibited modest rates of transamination. In contrast, complex formation in peptide **C2** resulted in a *drop* in activity, consistent with the occlusion of the Pal residue. Further evidence for steric hindrance affecting the reactivity of this system was obtained upon the analysis of peptide **C3**. The Pal residue in this peptide was exposed to solvent by shifting it up one residue to position 7 in the helix. This modification was accompanied by a significant recovery in activity of the complex, consistent with an increased accessibility of the Pal chimera.

Table 4 Copper mediated transamination rates for the various S-peptide derivatives and the resulting complexes with S-protein. *a*. In the absence of S-protein.

Coenzyme System	$k_{obs} \times 10^5$ (min^{-1}) - SP	$k_{obs} \times 10^5$ (min^{-1}) + SP	$t_{1/2}$ (+ SP) (h)
DPal	58.9	81.9	19.6[a]
C0	222	591	2.0
C1	289	510	2.3
C2	282	198	5.8
C3	206	427	2.7

"NATIVE-LIKE" TRANSAMINASE ACTIVITY

We have investigated the ability of the Pal-RNase complexes to mimic native pyridoxal-dependent transaminases, which do not require metal ions for activity.[22] Kinetic analyses had suggested that the transaminase activity of the Pal residue was strongly dependent on local conformational effects. The assays were therefore repeated in the *absence* of Cu(II) to determine whether the peptidyl microenvironment alone was sufficient to activate the Schiff base intermediate towards transamination.

The kinetic data for transamination by the RNase complexes in the presence and absence of Cu(II) are listed for comparison in Table 5. Although two of the three complexes assayed were poor transaminases in the absence of copper, the reactivity of the **C1-SP** complex increased 2.5 fold. Apparently, the restricted conformational freedom of the Schiff base in this system results in the C_{α}-H bond of the substrate being orthogonal to the plane of the pyridoxal ring (satisfying the "Dunathan prerequisite" for transamination, Fig. 6).[23] The slower kinetics of complex **C1-SP** in the presence of Cu(II) indicates the relatively lower activating capacity of the metal ion. In addition, the formation of a copper-alanine chelate decreases the concentration of free amine available for Schiff base formation under the assay conditions (pH 4.0). As discussed above, the coenzyme chimera in complex **C2-SP** is occluded from substrate, and that in complex **C3-SP** experiences no significant steric restraints. Copper-chelation is therefore mandatory for the activation of both these complexes.

Table 5. Transamination kinetics for the Pal-RNase complexes in the presence and absence of Cu(II).

Coenzyme System	$k_{obs} \times 10^5$ (min^{-1}) +Cu(II)	$k_{obs} \times 10^5$ (min^{-1}) -Cu(II)	k_{rel} [-Cu(II)]
C1-SP	510	1288	16.7
C2-SP	198	107	1.4
C3-SP	427	77	1.0

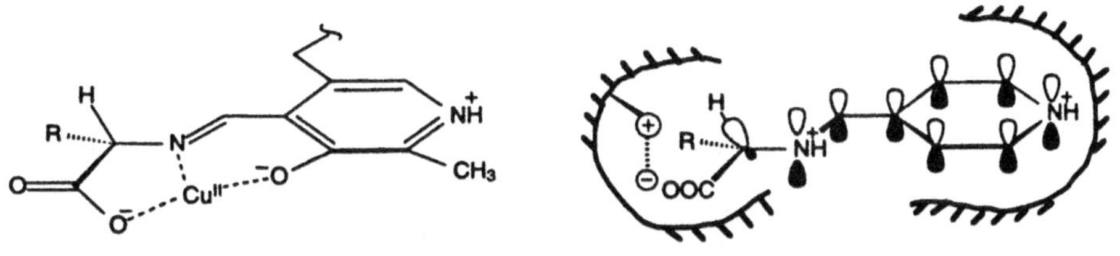

Figure 6. A conformational "lock" on the Schiff base intermediate in complex C1-SP (right) may activate it in a manner analogous to that resulting from Cu(II) chelation (left).

The studies carried out to date demonstrate the feasibility of exploiting the pyridoxal amino acid chimera as a key component in the design of new functional polypeptides. Future efforts will focus on the development of more efficient systems that are capable of catalytic turnover.

A Thiazolium-Based Coenzyme-Amino Acid Chimera

The enzyme catalyzed decarboxylation of α-keto carboxylic acids frequently invokes the assistance of the coenzyme thiamin diphosphate[24, 25] which contains a thiazolium heterocycle as the core functionality. The synthetic strategy that we have adopted for the assembly of peptides incorporating the thiazolium moiety involves the peptide incorporation of a thiazole amino acid (Thz), followed by derivatization to the thiazolium species (Taz, **8**), prior to side chain deprotection and release of the peptide from the solid phase resin. The Fmoc protected thiazole amino acid **7** was synthesized as previously described from L-aspartic acid.[26] This residue has recently become commercially available.

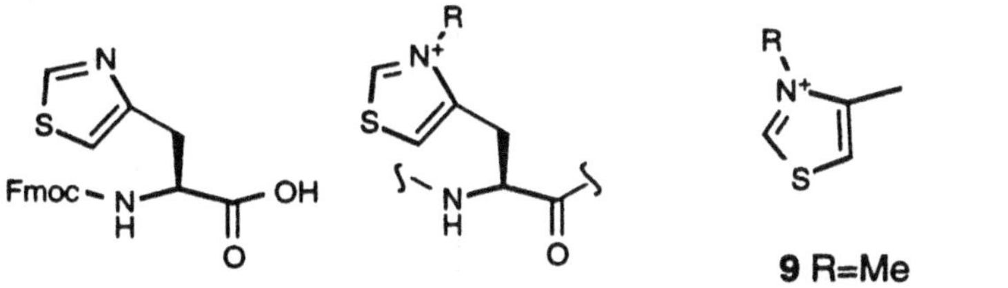

7 **8** **9** R=Me **10** R=PhCH$_2$

The successful syntheses of two thiazole-containing peptides, AcThz-ProDSerPheNH$_2$ and AcSerAlaLeuGluGluThzLeuLysLysLeuAlaGluLeuLeuLysNH$_2$, established the ease and efficiency of solid phase synthesis with the residue. The resultant peptides have been purified by reversed phase HPLC and structurally characterized by mass spectroscopy, amino acid analysis, and NMR.

CHEMICAL MODIFICATION TO THE ACTIVE THIAZOLIUM SPECIES

Activation of the Thz heterocycle to the thiazolium species can be accomplished by alkylation of the thiazole nitrogen after peptide assembly, with the peptide remaining protected on the solid phase support (Scheme 4). This strategy is key to the success of the synthesis, since the thiazolium heterocycle is base-labile,[27] and therefore cannot be introduced until after the final Fmoc deprotection cycle.

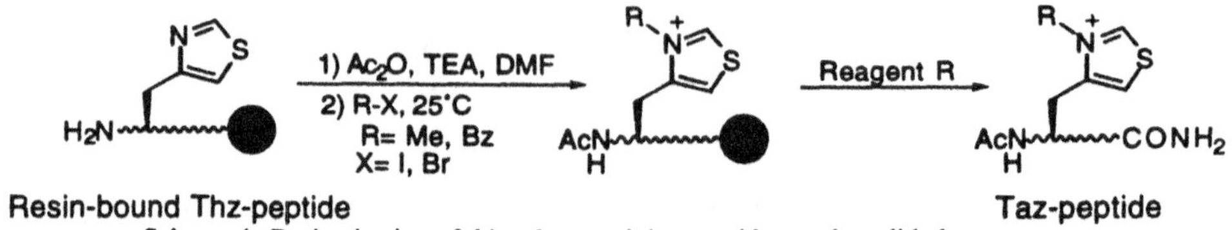

Scheme 4. Derivatization of thiazole-containing peptides on the solid phase support.

The synthetic sequence of *N*-terminal acetylation, thiazole alkylation, peptide cleavage, and HPLC purification was initially established in model studies utilizing 4-methylthiazole and 3,4-dimethylthiazolium bromide (**9**). Effective alkylation was achieved with both methyl iodide and benzyl bromide to afford the corresponding methyl and benzyl derivatives. The alkylation protocol is compatible with many functionalized residues since the side chains remain protected during the alkylation process (methionine would not be compatible with this alkylation sequence). The methyl and benzyl derivatives of the two Thz peptides have been prepared by this method.

DESIGN OF A HELICAL BUNDLE INCORPORATING THE TAZ CHIMERA

We are currently evaluating synthetic constructs that are designed to model the thiamin-dependent enzyme pyruvate decarboxylase (PDC), which carries out the decarboxylation of pyruvic acid to acetaldehyde (Scheme 5). This reaction represents the simplest of the thiamin dependent processes since it is unimolecular in substrate.

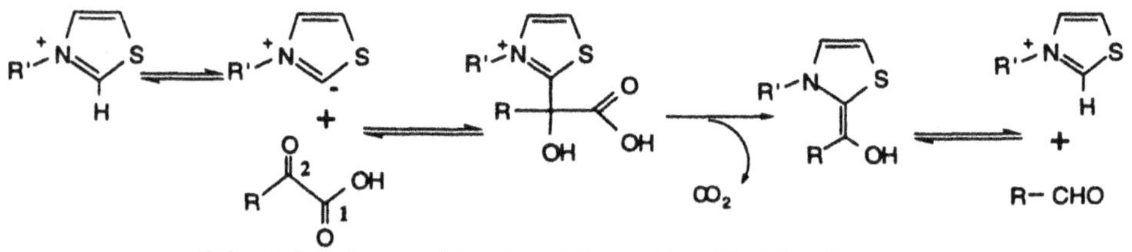

Scheme 5. α-Keto acid decarboxylation catalyzed by thiazolium salts

An important feature to emerge from the crystallographic analysis of the native thiamin-dependent enzymes concerns the local environment of the thiamin coenzyme. For example in PDC, the coenzyme is described as "buried in a cavity approximately 10Å wide and 8Å deep", and the residues that line the cavity are exclusively hydrophobic.[28] This unique hydrophobic environment suggests that the actual decarboxylation process may not require the placement of specific functional groups to aid in catalysis, but rather, a hydrophobic environment to accelerate the destabilization of the highly charged intermediate. The most significant role of the polypeptide architecture in this reaction may therefore be in the provision of a hydrophobic environment for the key C-C bond breaking step. Preliminary studies with the thiazolium amino acid-chimeras have demonstrated that this environment may be reproduced within synthetic, multimeric helical bundle constructs.[29]

Helical bundles have been previously utilized successfully in the design of functional polypeptide motifs.[30] A 15 amino acid helical sequence (TazHelix) was synthesized based on DeGrado's $\alpha_1\beta$ multimeric helical bundle.[31, 32]

$\alpha_1\beta$	AcGELEELLKKLKELLKG-CONH$_2$
TazHelix	AcSALEEXLKLLAELLK-CONH$_2$
	(X=Taz)

The thiazolium amino acid in Tazhelix was placed at position 6, which positions it in proximity to the hydrophobic interface of the amphiphilic helix bundle (Fig. 7). Helical net and wheel representations show that the Taz is surrounded by Leu_7, Leu_{10} and Ala_{11}. The alanine at position 11 replaced the original leucine in $\alpha_1\beta$ to accommodate the bulky Taz residue. The N-terminal *i* to *i+4* capping sequence SALEE was included in the design to provide enhanced helix stability[33] without interfering with the locus of the Taz residue.

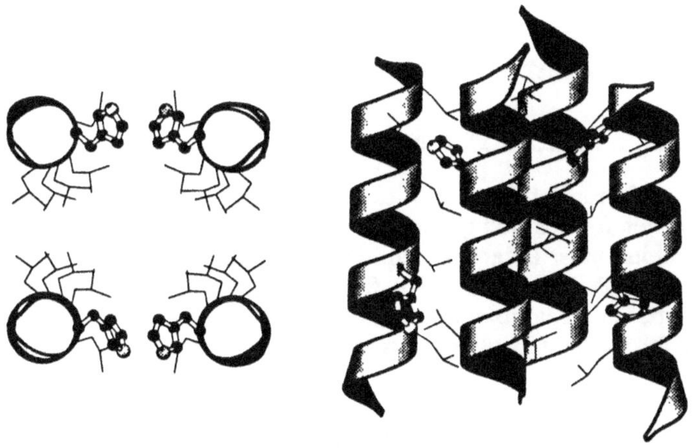

Figure 7. Proposed model of the Tazhelix motif incorporating the Thz residue.

Circular dichroism studies demonstrated that the corresponding methyl (MetTaz-Helix) and benzyl (BzTaz-Helix) thiazolium peptides exhibited different helical contents[34] compared to the parent Thz peptide (Figure 7). The unalkylated Thz helix exhibits 55% helicity. Methylation of this system decreased the helical content to 27%. This erosion in structure is probably due to adverse electrostatic effects of the thiazolium ring, which affect the stabilizing propensity of inter-helix interactions. Interestingly, the benzylated analog exhibits high helicity (63%) which indicates favorable hydrophobic interactions at the helix interface.

EFFECT OF PEPTIDE ARCHITECTURE ON THE EXCHANGE KINETICS OF THE THIAZOLIUM RING SYSTEM

It is well established that the reactivity of the thiazolium heterocycle in thiamine-dependent processes is due to facile ionization of the C-2 methine. This ionization is promoted by a hydrophobic microenvironment which stabilizes the zwitterionic intermediate in preference to the charged precursor.[35, 36] Therefore in a preliminary analysis, hydrogen/deuterium (H/D) exchange rates were measured to evaluate the effect of peptide architecture on the acidity of the C-2 methine. In addition, modulation of hydrophobicity by the N-3 substituent was assessed by a comparison of the methyl and benzyl thiazolium derivatives. The exchange kinetics of the thiazolium salts **9** and **10** served as controls. The benzylated derivatives (entries 2, 4 and 6; Table 6) consistently showed faster rates of exchange than the methylated analogs. The accelerated rates in this case may reflect the increased hydrophobic environment of the thiazolium moiety. As predicted by the CD analyses, optimum exchange rates were observed with BzTaz-Helix, underscoring the importance of a defined protein architecture in enhancing proton exchange kinetics.

Table 6. H/D exchange kinetics at 300 K (pH 5.1, 50mM NaOAc/D_2O, 150mM NaCl).

Entry	Substrate	$t_{1/2}$ (min)	k_{obs} (min^{-1})
1	**9**	> 50	0.014
2	**10**	11.3	0.061
3	Ac-MetTaz-Pro-DSer-Phe-NH_2	5.9	0.12
4	Ac-BzTaz-Pro-DSer-Phe-NH_2	3.3	0.21
5	MetTaz-Helix	3.2	0.22
6	BzTaz-Helix	~0.5	1.38

The rapid rate of C2 proton exchange observed for the BzTaz-Helix system indicates that even for this first generation design, the primary sequence dictates that the thiazolium residue must lie at the helix interface and be poised to associate with the surrounding hydrophobic surface. The first priority in future investigations will be the analysis of pyruvate decarboxylation kinetics and a comparison of these results with data obtained for model thiazolium salts and thiamin diphosphate.

Nicotinamide-Based Coenzyme-Amino Acid Chimera

The redox active nicotinamide heterocycle forms the functional core of the pyridine nucleotides.[37] This functionality provides a mechanism for hydride transfer to electrophilic centers that are typically activated by Lewis acid or general acid catalysis. The nicotinamide amino acid chimera **11** was synthesized to investigate similar redox processes within synthetic polypeptides.

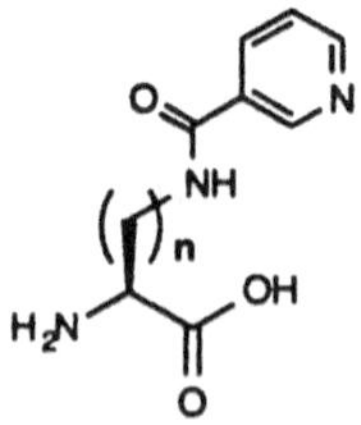

11 n=4 Lys(nic)

This residue incorporates a 3-carboxamido pyridine (nicotinamide) group covalently attached *via* a side chain amide linkage to the ε-amino functionality of L-lysine.

Methodology has been developed to selectively quaternize the pyridyl nitrogen prior to peptide cleavage from the resin. This process yields a 3-carboxypyridinium group in the same functional form as that found in nicotinamide adenine dinucleotide (NAD^+). This residue has been incorporated into peptides, and a preliminary evaluation of its redox properties has been carried out.

SYNTHESIS, PEPTIDE INCORPORATION, AND ALKYLATION OF THE NICOTINAMIDE-AMINO ACID CHIMERA

The logistics of solid phase synthesis hinder the direct incorporation of a pyridinium functionality into peptides by the standard Fmoc regime. In addition to the concerns of solubilizing a charged amino acid, the quaternized pyridinium group is susceptible to degradation in the basic environment of Fmoc synthesis. To circumvent these problems, the NAD chimera was incorporated into peptides as the corresponding nicotinamide. Access to the nicotinamide containing amino acid was obtained from commercially available N^{α}-Fmoc-N^{ε}-Boc-L-lysine-pentafluorophenyl ester (**12**), *via* acidolysis of the Boc group followed by coupling with nicotinoyl chloride hydrochloride (Scheme 6). This scheme allows for the production of gram quantities of enantiomerically pure Fmoc amino acid for use in peptide synthesis. The product, N^{α}-Fmoc-N^{ε}-3'-pyridinecarboxamido-L-lysine pentafluorophenyl ester (**13**), was used to establish the viability of quaternization and cleavage protocols.

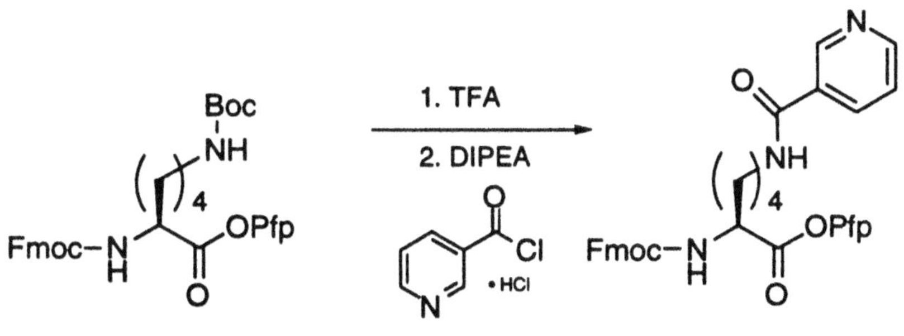

12 **13**

Scheme 6. Synthesis of ε-nicotinoyl-L-lysine

The Lys(nic) Fmoc amino acid has been incorporated into the hexapeptide AcGlyValProDSerLys(nic)GlyNH_2 with high coupling efficiencies (>97%). Alkylation to the Lys(nic-Me) residue was performed prior to peptide deprotection/cleavage with either iodomethane (2-3 days) or, under anhydrous conditions using methyl triflate (1-2 hours). The latter transformation is highly efficient and selective as the other functionalized amino acids remain protected. The structural integrity of the alkylated peptide has been confirmed by high resolution mass spectrometry and NMR.

GENERATION OF THE REDOX ACTIVE 1,4-DIHYDROPYRIDINE SPECIES

It has long been known that 1,4-dihydropyridine (DHP), the reductive moiety in NADH, may be produced from the corresponding pyridinium species with sodium dithionite.[38] Assays for monitoring the corresponding reductive process in the Nic peptide have been developed. Observation of DHP formation in the latter is possible by ultraviolet spectroscopy. Figure 8 shows a typical dithionite reduction with spectra acquired over a period of one hour.

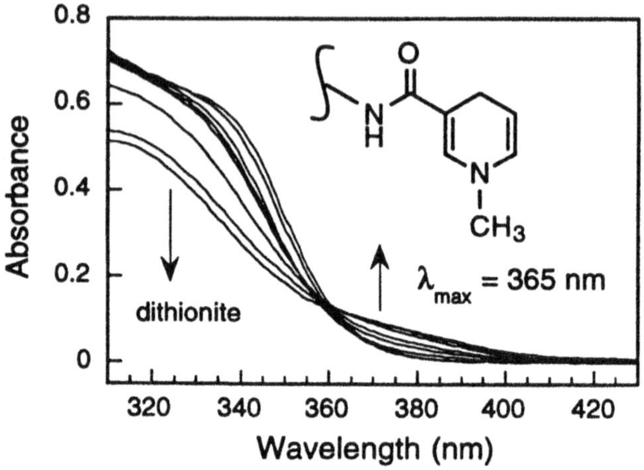

Figure 8. Dithionite reduction of Nic to produce DHP.

The feasibility of producing nicotinamide amino acid chimeras has been demonstrated through the stages of synthesis, peptide incorporation, and generation of the redox active species. Redox reactivity in these systems will be amenable to investigation by many complementary techniques including NMR, UV spectroscopy and gas chromatography.

Conclusion

The inherent reactivity of coenzymes has attracted a great deal of interest in the past three decades. While much emphasis has been placed on the integration of these intriguing chemical constituents into *non-peptidyl* hosts,[38-43] the coenzyme-amino acid chimeras will enhance our understanding of coenzyme function by allowing access to a wide variety of peptide and protein hosts. This versatility in design will allow a detailed analysis aimed at unraveling the complex interplay of structure and function within catalytic polypeptides.

Acknowledgments

This research was supported by the Office of Naval Research and the National Institutes of Health. The award of Glaxo Summer Fellowships to R.S.R and G.K.W. and support from the A. P. Sloan Foundation and the Camille and Henry Dreyfus Foundation are also gratefully acknowledged.

References

1. Kokubo, T., Sassa, S. and Kaiser, E.T. (1987) Flavohemoglobin: A Semisynthetic Hydroxylase Acting in the Absence of Reductase, *J. Am. Chem. Soc.*, **109**, 606-607.
2. Kaiser, E.T. (1988) Catalytic Activity of Enzymes Altered at Their Active Sites, *Angew. Chem. Int. Ed., Engl.*, **27**, 913-922.
3. Aitken, D.J., Alijah, R., Onyiriuka, S.O., Suckling, C.J., Wood, H., C. S. and Zhu, L. (1993) Approaches to Chemically Modified Enzymes as Synthetic Catalysts, *J. Chem. Soc. Perkin Trans. 1*, 597-608.
4. Suckling, C.J. and Zhu, L.M. (1993) Carbon-Carbon Bond Formation Mediated by Papain Chemically Modified by Thiazolium Salts, *Bioorg. and Med. Chem. Lett.*, **3**, 531-534.
5. Imperiali, B. and Sinha Roy, R. (1994) Coenzyme Amino Acid Chimeras: New Residues for the Assembly of Functional Proteins, *J. Am. Chem. Soc.*, **116**, 12083-12084.
6. Imperiali, B. and Sinha Roy, R. (1995) Stereoselective Synthesis and Peptide Incorporation of a Pyridoxal Coenzyme-Amino Acid Chimera, *J. Org. Chem.*, **60**, 1891-1894.
7. O'Donnell, M.J., Bennett, W.D. and Wu, S. (1989) The Stereoselective Synthesis of α-Amino Acids by Phase-Transfer Catalysis, *J. Am. Chem. Soc.*, **111**, 2353-2355.
8. Fatiadi, A.J. (1976) Active Manganese Dioxide Oxidation in Organic Chemistry - Part I, *Synthesis*, 65-104.
9. Vazquez, M.A., Donoso, J. and Munoz, F. (1990) Kinetic and Thermodynamic Parameters for Schiff-Base Formation between 5'-Deoxypyridoxal and Hexylamine, *Helv. Chim. Acta*, **73**, 1991-1998.
10. Imperiali, B., Fisher, S.L., Moats, R.A. and Prins, T.J. (1992) A Conformational Study of Peptides with the General Structure Ac-L-Xaa-Pro-D-Xaa-L-Xaa-L-Xaa-NH_2: Spectroscopic Evidence for a Peptide with Significant β-Turn Character in Water and in Dimethyl Sulfoxide, *J. Am. Chem. Soc.*, **114**, 3128-3188.
11. Bender, M.L., Bergeron, R.J. and Komiyama, M. (1984) *The Bioorganic Chemistry of Enzymatic Catalysis*, Wiley-Interscience, New York.
12. Martell, A.E. (1989) Vitamin B_6 Catalyzed Reactions of α-Amino and α-Keto Acids: Model Systems, *Acc. Chem. Res.*, **22**, 115-124.
13. Iwata, C. (1968) 5-Deoxypyridoxal, *Biochemical Preparations*, **12**, 117-121.
14. Chaiken, I.M. (1981) Semisynthetic Peptides and Proteins, *CRC Crit. Rev. Biochem.*, **11**, 255-301.
15. Richards, F.M. (1958) On the Enzymic Activity of Subtilisin-Modified Ribonuclease, *Proc. Natl. Acad. Sci. U.S.A.*, **44**, 162-166.
16. Hofmann, K., Frances, F.M., Limetti, M., Montibeller, J. and Zanetti, G. (1966) Studies on Polypeptides. XXXIV. Enzymic Properties of Partially Synthetic De(16-20)- and De(15-20)-ribonucleases S', *J. Am. Chem. Soc.*, **88**, 3633-3639.
17. Van Batenburg, O.D., Voskuyl-Holtkamp, I., Schattenkerk, C., Hoes, K., Kerling, K.E.T. and Havinga, E. (1977) The Role of the Imidazolyl Nitrogen Atoms of Histidine-12 in Ribonuclease S, *Biochem. J.*, **163**, 385-387.
18. Irie, M., Ohgi, K., Yoshinaga, M., Yanagida, T., Okada, Y. and Teno, N. (1986) Roles of $Lysine_1$ and $Lysine_7$ Residues of Bovine Pancreatic Ribonuclease in the Enzymatic Activity, *J. Biochem.*, **100**, 1057-1063.
19. Roberts, G.C.K., Dennis, E.A., Meadows, D.H., Cohen, J.S. and Jardetzky, O. (1969) The Mechanism of Action of Ribonuclease, *Proc. Natl. Acad. Sci. U.S.A.*, **62**, 1151-1158.
20. Woodfin, B.M. and Massey, V. (1968) Spectrophotometric Determination of the Dissociation Constant of Ribonuclease S', *J. Biol. Chem.*, **243**, 889-892.
21. Finn, F.M. (1972) Spectrophotometric Measurement of Binding of S-Peptide Analogs to S-Protein, *Biochemistry*, **11**, 1474-1478.
22. Bruice, T.C. and Benkovic, S.J. (1966) *Bioorganic Mechanisms*, W.A. Benjamin Inc., New York.
23. Dunathan, H.C. (1971) Stereochemical Aspects of Pyridoxal Phosphate Catalysis, *Adv. Enzymol.*, **35**, 79-134.
24. Haake, P. (1987) *Thiamin-Dependent Enzymes*, Royal Society of Chemistry, London.
25. Lindqvist, Y. and Schneider, G. (1993) Thiamin Diphosphate Dependent Enzymes: Transketolase, Pyruvate Oxidase and Pyruvate Decarboxylase, *Curr. Opin. Struc. Bio.*, **3**, 896-901.
26. Hsiao, C.N., Leanna, M.R., Bhagavatula, L., De Lara, E., Zydowsky, T.M., Horrom, B.W. and Morton, H.E. (1990) Synthesis of *N*-(tert-butoxycarbonyl)-3-(4-thiazolyl)-L-Alanine, *Synth. Commun.*, **20**, 3507-3517.
27. Katritzky, A.R. and Rees, C.W. (1984) *Thiazoles and their Benzo Derivatives*, Pergammon Press, Oxford.
28. Dyda, F., Furey, W., Swaminathan, S., Sax, M., Farrenkopf, B. and Jordan, F. (1993) Catalytic Centers in the Thiamin Diphosphate Dependent Enzyme Pyruvate Decarboxylase at 2.4Å Resolution, *Biochemistry*, **32**, 6165-6170.
29. Imperiali, B., Sinha Roy, R. and Wang, L. (1995) Peptidyl Models for Coenzyme Catalysis, *Peptides Chemistry, Structure, and Biology, [Proc. 14th Amer. Pep. Symp.]*, ESCOM, Leiden, in press.
30. Johnsson, K., Allemann, R.K., Widmer, H. and Benner, S.A. (1993) Synthesis, Structure and Activity of Artificial, Rationally Designed Catalytic Polypeptides, *Nature*, **365**, 530-532.
31. Ho, S.P. and DeGrado, W.F. (1987) Design of a 4-Helix Bundle Protein: Synthesis of Peptides Which Self-Associate into a Helical Protein, *J. Am. Chem. Soc.*, **109**, 6751-6758.

32. Osterhout Jr., J.J., Handel, T., Na, G., Toumadje, A., Long, R.C., Connolly, P.J., Hoch, J.C., Johnson Jr., W.C., Live, D. and DeGrado, W.F. (1992) Characterization of the Structural Properties of α_1B, a Peptide Designed to Form a Four-Helix Bundle, *J. Am. Chem. Soc.*, **114**, 331-337.
33. Harper, E.T. and Rose, G.D. (1993) Helix Stop Signals in Proteins and Peptides: The Capping Box, *Biochemistry*, **32**, 7605-7609.
34. Scholtz, J.M., Qian, H., York, E.J., Stewart, J.M. and Baldwin, R.L. (1991) Parameters of Helix-Coil Transition Theory for Alanine-Based Peptides of Varying Chain Lengths in Water, *Biopolymers*, **31**, 1463-1470.
35. Breslow, R. (1958) On the Mechanism of Thiamine Action. IV. Evidence from Studies on Model Systems, *J. Am. Chem. Soc.*, **80**, 3719-3726.
36. Crosby, J., Stone, R. and Lienhard, G.E. (1970) Mechanisms of Thiamine-Catalyzed Reactions. Decarboxylation of 2-(1-Carboxy-1-hydroxyethyl)-3,4-Dimethylthiazolium Chloride, *J. Am. Chem. Soc.*, **92**, 2891-2900.
37. Adams, M.J. (1987) *Oxido-Reductases - Pyridine Nucleotide-dependent Enzymes*, Royal Society of Chemistry, Belfast.
38. Kaplan, N.O., Boyer, P.D., Lardy, H. and Myrback, K. (1960) *The Pyridine Coenzymes*, Academic Press, New York.
39. Breslow, R. (1982) Artificial Enzymes, *Science*, **218**, 532-537.
40. Breslow, R., Canary, J.W., Varney, M., Waddell, S.T. and Yang, D. (1990) Artificial Transaminases Linking Pyridoxamine to Binding Cavities: Controlling the Geometry, *J. Am. Chem. Soc.*, **112**, 5212-5219.
41. Diederich, F. and Lutter, H.-D. (1989) Catalytic Cyclophanes. 4. Supramolecular Catalysis of Benzoin Condensations by a Thiazolium Cyclophane, *J. Am. Chem. Soc.*, **1989**, 8438-8446.
42. Murakami, Y., Kikuchi, J.I., Hisaeda, Y. and Nakamura, K. (1990) Functionalized Bilayer Membranes as Artificial Tryptophan Synthase. Characterization of Catalytic Efficiency, Substrate Specificity, and Reaction Selectivity, *Bull. Chem. Soc. Jpn.*, **63**, 2339-2345.
43. Suckling, C.J. (1990) *Enzyme Chemistry. Impact and Applications*, Chapman and Hall, London.

NONCONVENTIONAL APPROACHES TO ENZYME-CATALYZED PEPTIDE SYNTHESIS

H.-D. JAKUBKE, U. EICHHORN, S. GERISCH, M. HÄNSLER, K. SALCHERT, G. ULLMANN, D. ULLMANN
Department of Biochemistry, Leipzig University, Faculty of Biosciences, Pharmacy and Psychology, Talstr. 33, D-04103 Leipzig, Germany

1. Introduction

Peptides play a fundamental role in the function of living systems. Due to their diverse biological activities there is a great interest in pharmaceutically active peptides with many applications and great potential in medicine. Since the demand for peptides is enormous, classical chemical synthesis in solution, solid-phase synthesis and recombinant techniques belong to the most important methods of peptide synthesis. Unfortunately, neither of these methods is completely free of drawbacks. Especially, chemical synthesis of peptides often suffers from problems such as time-consuming side-chain protection/deprotection requirements, racemization and other side-reactions. Furthermore, in recombinant DNA methods, the incorporation of unnatural amino acids is limited in preparative scale. For these reasons, the use of suitable peptide ligases should provide an attractive alternative by combining the flexibility of chemical strategies with the advantages of the regio- and stereospecific enzymatic reactions.

2. What is the Enzyme of Choice for Peptide Bond Formation?

There is no doubt that the enzyme of choice for peptide bond formation should be the ribosomal peptidyltransferase which exhibits its full catalytic action independent of side chain functions of the amino acids being ligated. Since this ideal enzyme requires for its catalytic activity coordination with numerous other ribosomal factors, it is not suitable for simple practical use as a C-N ligase in peptide synthesis.

Last but not least, only those enzymes remain which usually act as hydrolases catalyzing the cleavage of peptide bonds. In principle, they must also catalyze the reversal of hydrolysis if not in actual practice. The main problem using a protease as a ligase is finding optimal conditions for shifting the aminolysis/hydrolysis equilibrium towards aminolysis.

As seen in recent reviews [1-3] two strategies for protease-catalyzed peptide synthesis are distinquished. The equilibrium-controlled approach represents the direct reversal of

C. S. Wilcox and A. D. Hamilton (eds.), Molecular Design and Bioorganic Catalysis, 53–70.

the proteolysis and ends with a true equilibrium, whereas in the kinetic approach the protease acts as a transferase, using a mildly activated acyl donor to rapidly acylate a serine or cysteine protease. The acyl enzyme undergoes a rate-limiting competitive deacylation by water, yielding the undesired acid, whereas the added nucleophilic amino component causes a transient accumulation of the desired peptide product (Fig. 1). The peptide product accumulation is only temporary, since the enzyme will begin to slow-

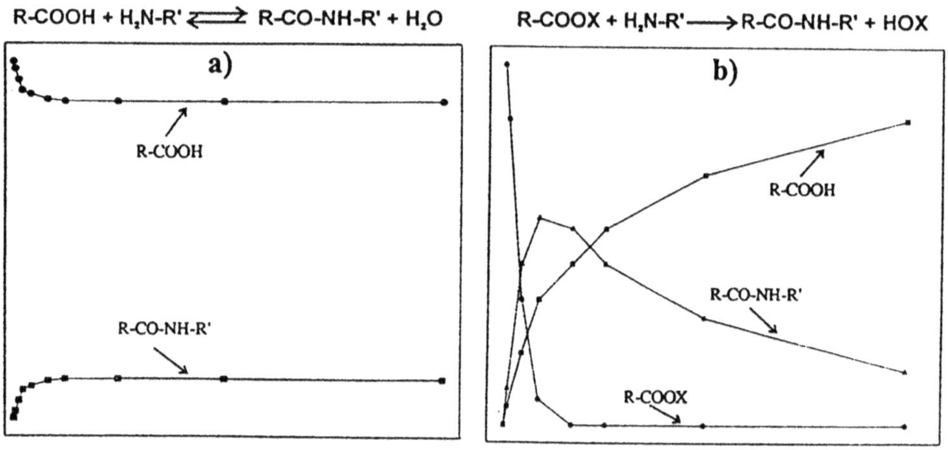

Figure 1. Comparison for the enzymatic synthesis of RCONHR' by the equilibrium (a) and the kinetically controlled (b) route

ly hydrolyze the newly formed peptide bond as the acyl donor ester is consumed. For preparative purposes the enzymatic reaction must be stopped preferably at the point of maximum peptide yield. Contrary to the equilibrium-controlled enzymatic synthesis the kinetic approach needs much less enzyme, the reaction time is significantly shorter and the product yield depends on the properties of the enzyme as well as the substrate specificity. Both approaches can be employed in peptide synthesis on a preparative scale. Since the kinetic approach can be more efficiently influenced than the equilibrium approach, the influence of the protease specificity on the efficiency of peptide bond formation, including routes toward a broad exploitation of serine and cysteine proteases for the C-N bond ligation, guarantee more flexibility in synthesis planning.

3. Prediction of the Peptide Bond Formation by S' Subsite Mapping

The ratio of formation of aminolysis and hydrolysis products is of decisive importance for successful preparative peptide synthesis as shown in Figure 1. Since this ratio is essentially determined by the S' subsite specificity of the protease, a parameter describing the acyl transfer efficiency should be very useful for the prediction of the synthesis course. In contrast to proteolysis the C-N ligation is a two-substrate reaction. Therefore, the efficiency of the nucleophilic attack of the amino component on the acyl enzyme depends essentially on an optimal binding within the active site. Substrate binding is generally not limited to a single amino acid residue flanking the scissile bond. Following the short-hand notation [4] to represent the substrate (P_n...P_3-P_2-P_1 ↓ P_1'-P_2'-P_3'...P_n'), the specificity of most proteases is not only determined by the P_1 residue (primary specificity). However, many proteases have long binding grooves (S_n...S_3-S_2-S_1 ↓ S_1'-S_2'-S_3'...S_n') on either side of the scissile bond (↓). Our group has recently begun to address the important issue of the P'-S' interactions of serine and cysteine proteases by measuring their rates of deacylation by amino acid and peptide nucleophiles using the kinetic scheme given in Figure 2. We have also recently published a review [5]

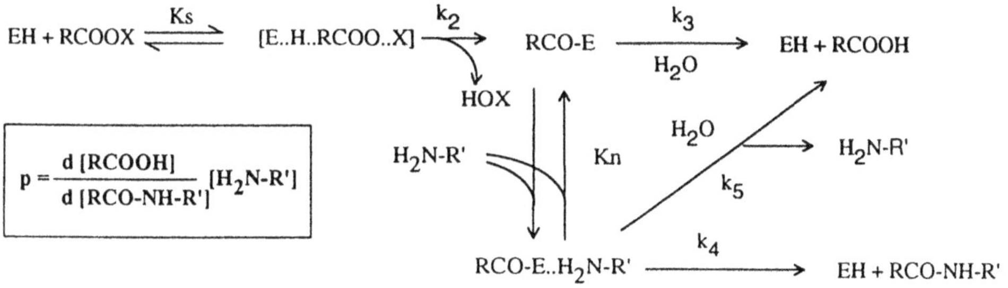

Figure 2. Reaction pathway of protease-catalyzed acyl transfer.
EH, free enzyme; RCOOX, acyl donor; HOX, leaving group; RCO-E, acyl enzyme; RCOOH, hydrolysis product; H_2N-R', nucleophile; RCO-NH-R', aminolysis product

devoted exclusively to the kinetic approach including all details on the introduced p value as an efficiency parameter of kinetically controlled peptide synthesis. For planning and optimization of enzymatic peptide synthesis it is useful to know basic kinetic parameters of the reaction course including the data of S' subsite mapping. This was done starting with a large number of amino acid derivatives and oligopeptides [5,6] up to a pentapeptide library which was used to study the S' subsite of clostripain, a cysteine protease, which is one of only a few enzymes with an acceptance for P'_1 Pro [7,8]. Schellenberger *et al.* [9,10] continued the strategy of S' subsite mapping by the use of mixtures of peptide nucleophiles and data processing by multiple substrate kinetics [11].

4. Manipulations to Suppress Competitive Reactions in Enzymatic Synthesis

The most important factors which limit the widespread routine use of proteases in peptide synthesis are undesired proteolysis of the peptide product as well as starting segments in semisynthesis and undesired hydrolysis of the acyl enzyme in the kinetic approach. Enzymatic synthesis can be influenced at the level of the enzyme by synthesis with chemically modified or mutant enzymes, catalytic antibodies or, on the other hand, by medium engineering and/or improvement in the kinetically controlled methodology [12].

4.1. LEAVING GROUP MANIPULATIONS OF THE ACYL DONOR IN THE KINETIC APPROACH

Undesired side reactions in the kinetic approach are based merely on the fact that the specificity of the enzyme for the acyl donor ester is not sufficiently above its specificity for the peptide product formed during synthesis. Since the sequence of the reactants cannot be changed, the only practical alternative to suppress competitive reactions is to use a highly specific leaving group of the acyl donor ester. For this reason, if the specificity of the enzyme for the acyl donor can be raised several orders of magnitude higher than its specificity for any of the internal peptide cleavage sites by using a suitable ester leaving group (X), peptide bond formation can proceed to high yield with high efficiency, even if the starting segments contain multiple competing cleavages sites. We have recently conducted an experiment designed to test the influence of the ester leaving group of the acyl donor on the outcome of a kinetically controlled synthesis of a model peptide containing a second proteolytically sensitive bond in addition to the bond to be formed. In the chymotrypsin-catalyzed synthesis of Mal-Leu-Phe-pNA starting from Mal-Leu-OX (X = Me, Bzl, 4-nitrobenzyl) and H-Phe-pNA it was found that the relative rates of coupling vs. the competing cleavage of the highly sensitive Phe-pNa anilide bond were dramatically influenced by the nature of the leaving group [13]. Mal-Leu-ONp, for instance, was a far superior ester substrate compared to Mal-Leu-OMe (ca. three hundredfold higher k_{cat}/K_M) and thus gave a much higher peptide yield than Mal-Leu-OMe (80 % vs. 3 %), concurrent with the competing cleavage at the Phe-pNA bond. Beside chymotrypsin, this effect could also be observed using the cysteine proteases papain and ficin as shown in Table 1. Furthermore, the results underline that a decrease of specificity in P_1 position can be compensated by a suitable ester leaving group.

TABLE 1. Influence of the leaving group R on the cysteine protease-catalyzed synthesis of Pht-Xaa-Leu-NH_2 starting from Pht-Xaa-OR and H-Leu-NH_2

Acyl donor	Enzyme	Peptide	Yield (%)
Pht-Ala-OMe	Ficin	Pht-Ala-Leu-NH_2	20
Pht-Ala-OEt(Cl)			59
Pht-Ala-OBzl			87
Pht-Ala-OMe	Papain	Pht-Ala-Leu-NH_2	57
Pht-Ala-OEt(Cl)			74
Pht-Ala-OBzl			97
Pht-Leu-OMe		Pht-Leu-Leu-NH_2	6
Pht-Leu-OBzl			64
Pht-Val-OMe		Pht-Val-Leu-NH_2	0*
Pht-Val-OBzl			42

Reaction conditions: pH 7.8; [acyl donor] = 2 mM; [amino component] = 25 mM (effective concentration); papain: 0.1-1 mg/ml;* no ester conversion

By using nonspecific substrates concerning the P_1 amino acid bearing a highly specific leaving group after the formation of the peptide bond, the resulting low specificity of the protease for the acyl part ends in a practically irreversible synthesis course. This approach could also be verified in trypsin-catalyzed peptide synthesis using nonspecific substrates [14] combined with the carboxamido methyl(Cam)esters (Fig. 3) which were introduced by our group as suitable acyl donors some years ago [15]. A modified type of this ester was used by Jackson *et al.* [16] for a new synthesis of Ribonuclease A, in which an enzyme mutant was used for the first time to couple fragments obtained by solid-phase synthesis to form a protein. Leaving group manipulation was also found in practice for protease-catalyzed condensations of other long peptide segments [18]. A further approach to this principle of maximizing the binding specificity of the ester leaving group has been used by us in recent studies of inverse substrates of trypsin as acyl donors [18].

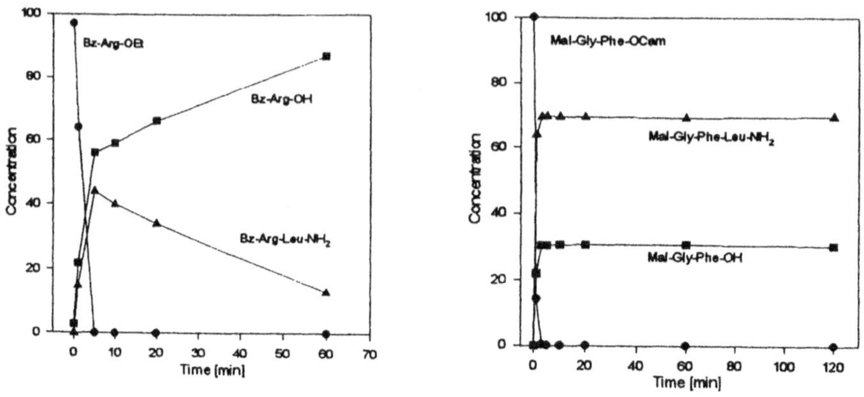

Figure 3. Comparison of trypsin-catalyzed reactions with specific (Bz-Arg-OEt) and nonspecific (Mal-Gly-Phe-OCam) acyl donors.

4.2. ENZYMATIC SYNTHESIS IN FROZEN AQUEOUS SYSTEMS

Competitive reactions in enzymatic peptide synthesis can also be supressed by manipulation of the reaction conditions. Based on literature data that freezing decreases the rate of proteolysis [19] and enhances hydroxylaminolysis [20] we have started enzymatic peptide synthesis in frozen aqueous systems [21] in order to improve product yield of inefficient educts.

For this purpose, the enzyme is added to the reactants precooled to 0°C in a polypropylene tube and immediately inserted into liquid nitrogen. After 20 s the tube is transfered into a cryostat. The course of sample temperature during precooling, shock freezing and the further reaction is shown in Figure 4.

In both serine and cysteine protease-catalyzed synthetic reactions, amino components with low nucleophilic efficiency in solution at room temperature gave high peptide yields in frozen aqueous systems. Examples are shown in Table 2. Secondary hydrolysis of the synthesized peptide bond was not observed in ice. In frozen aqueous systems, the endoproteinase α-chymotrypsin was able to act as a reverse carboxypeptidase catalyzing coupling of free amino acids as amino components [22]. Using amino acid alkyl esters as amino components under these conditions the resulting peptide esters can be applied as acyl donors in further kinetically controlled enzymatic syntheses.

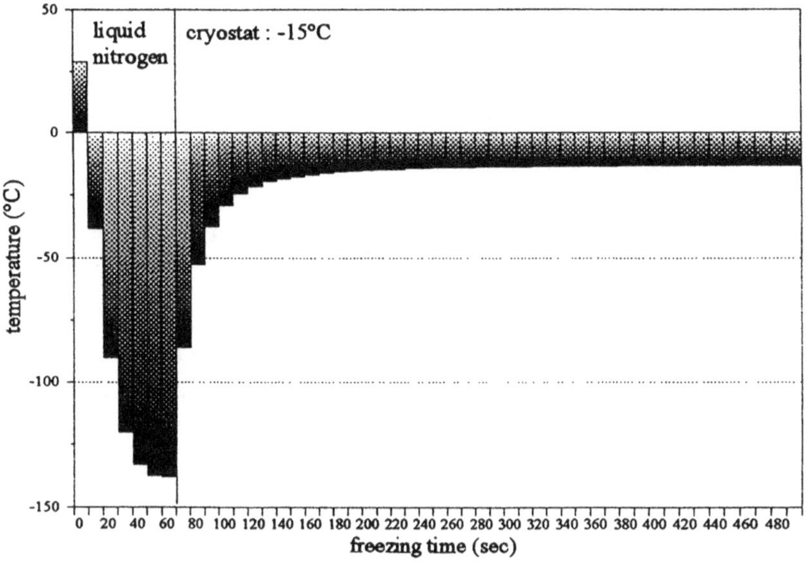

Figure 4. Temperature relations for a model peptide synthesis during shock freezing and transfer into a cryostat.

TABLE 2. Selected protease-catalyzed model peptide synthesis in frozen aqueous solutions and at room temperature

Enzyme	Acyl donor	Amino component	Peptide yield (%)		Ref.
			-15°C	25°C	
Chymotrypsin [a]	Mal-Phe-OMe	H-Ala-OH	70	0	[22]
Trypsin [a]	Pht-Ala-OBzl	H-Leu-NH_2	91	42	
Papain [b]	Bz-Arg-OEt	H-Gly-NH_2	93	20	[23]
Ficin [b]	Bz-Arg-OEt	H-Ala-Leu-OH	73	3	[23]
Clostripain [c]	Bz-Arg-OEt	H-Ala-Leu-OH	74	56	[23]
Subtilisin	Pht-Ala-OMe	H-Gly-NH_2	92	25	
Alcalase	Pht-Ala-OMe	H-Gly-NH_2	95	20	
V8-Protease [a]	Z-Glu-OMe	H-Ala-Ala-OH	76	5	[21]

Reaction conditions: pH 9.0; [acyl donor] = 2 mM; [amino component] =10 mM (effective concentration); [a]: [amino component] = 25 mM (effective concentration) ; [b]: pH 7.8; [c]: pH 7.0 . Enzyme concentrations were chosen to achieve complete ester conversion within at most 24 h.

In order to establish that the observed reactions in ice are not nonspecific reactions at the protein surface, several proteins, e.g. BSA, lysozyme, insulin and casein, have been tested for their catalytic ability in frozen aqueous systems [24]. We could show that peptide bond formation was catalyzed only by the active protease since the other proteins gave no peptide yield at all.

Frozen state enzymology opens completely new possibilities in enzymatic peptide synthesis. The simplest strategy of peptide bond formation which cannot be performed by chemical methods is using N-terminal free amino acid or peptide esters as acyl donors. By this means we could establish that the protection of the α-amino group of

TABLE 3. α-Chymotrypsin-catalyzed tetrapeptide synthesis using N^{α}-unprotected dipeptide esters as acyl donors and H-Ala-Ile-OH as nucleophile at room temperature and in frozen state at -15°C

Acyl donor H-Xaa-Phe-OMe	Peptide product	Peptide yield (%)*)	
Xaa		25°C	-15°C
Gly	H-Gly-Phe-Ala-Ile-OH	23	85
Asp	H-Asp-Phe-Ala-Ile-OH	23	91
Leu	H-Leu-Phe-Ala-Ile-OH	5	88

Reaction conditions: [H-Xaa-Phe-OMe] = 10mM; [H-Ala-Ile-OH] = 60mM (effective concentration); $[CT]_{25°C}$= 0.2-0.4μM; $[CT]_{-15°C}$= 1-8μM; pH = 8.1 (HEPES buffer, 0.05M); *) yields after complete ester consumption

the carboxyl component is not necessary in α-chymotrypsin-catalyzed peptide synthesis reactions in frozen aqueous solutions [3]. This strategy could be extended to N^{α}-unprotected dipeptide esters H-Xaa-Phe-OMe as acyl donors (Tab.3). The data demonstrate the higher efficiency of the reactions at -15°C compared to room temperature. H-Asp-Phe-Ala-Ile-OH was synthesized under semipreparative conditions with 0.075mmol H-Asp-Phe-OMe and 0.3mmol H-Ala-Ile-OH. The tetrapeptide was separated by preparative HPLC in 86% yield based on the donor ester and characterized. Furthermore, the S' subsite specificity of α-chymotrypsin was investigated using a series of di- and tripeptides as nucleophiles (Fig.5). As predicted by subsite mapping [10] the enzyme shows a preference for positively charged amino components in P_1'- and P_3'-position but not in P_2' for reactions at room temperature. The data in Figure 5 underline a modified nucleophilic specificity of α-chymotrypsin for peptide bond formation at -15°C. In frozen aqueous systems high peptide yields can be achieved even with amino components which are inefficient reactants at room temperature.

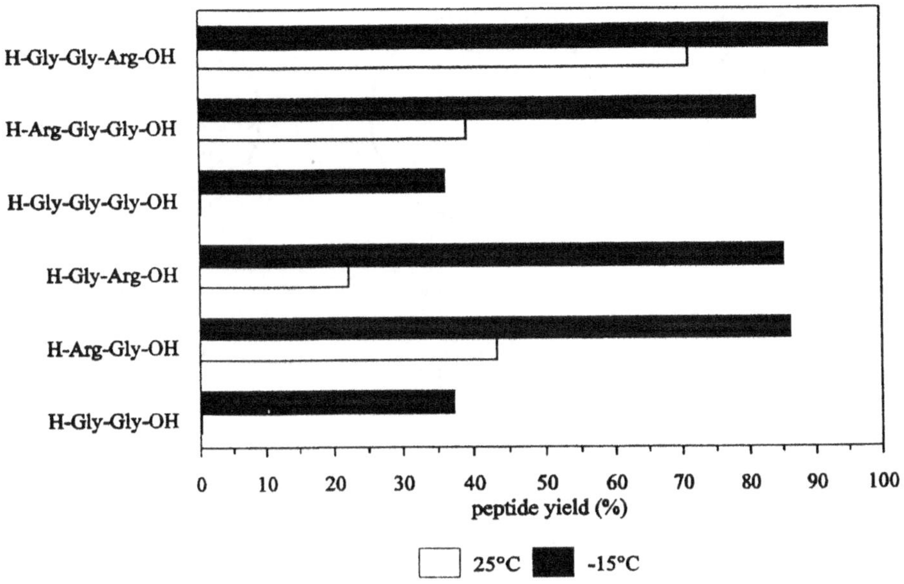

Figure 5. Comparison of the S' specificity of α-chymotrypsin using H-Leu-Phe-OMe as acyl donor and a series of di- and tripeptides as amino components both at 25°C and in frozen solution at -15°C.
[H-Leu-Phe-OMe] = 10mM; [Nucleophile] = 120mM; $[CT]_{25°C}$= 0.4μM; $[CT]_{-15°C}$= 4μM; pH = 8.1 without buffer; results after complete ester consumption.

The high peptide yields obtained in frozen systems can be explained by the assumption that freezing the reaction mixture results in an effective increase in concentration of the reactants in the remaining unfrozen liquid phase, which is in equilibrium with the ice crystals according to the "freeze-concentration-model" [25]. In order to estimate the freeze concentrating factor, the ^{1}H-NMR relaxation time technique has been used to determine the amount of unfrozen water in frozen reaction samples at -15°C. A comparison with the results of relaxation time measurements for the same system at room temperature gave an apparent concentration factor of 50 concerning the unfrozen water concentration. To simulate these concentration relations, acyl transfer experiments were carried out at room temperature with high concentration of the educts. In Figure 6 the conditions and peptide yields are given for a model reaction. From the results, it can be concluded that increasing the educt concentration at room temperature cannot simulate the reaction conditions in frozen systems, i.e. the yield-increasing effect cannot be caused by the freeze concentration effect alone.

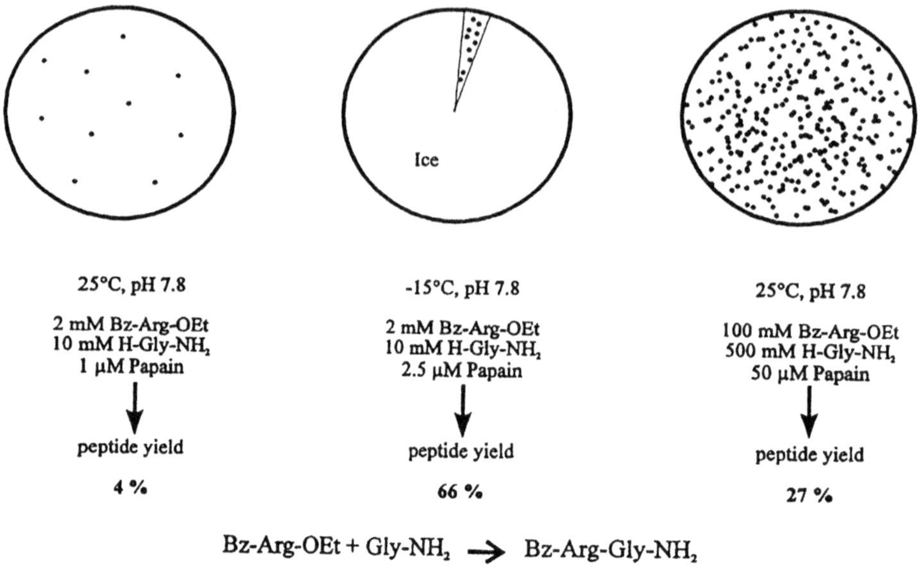

Figure 6. Papain-catalyzed synthesis starting from different substrate concentrations at room temperature and in frozen aqueous state.

4.3. ENZYMATIC SYNTHESIS IN HIGH-DENSITY MEDIA

When proteases are used in direct reversal of their normal hydrolytic function, the equilibrium position is very important in limiting the attainable yield in equilibrium-controlled enzymatic peptide synthesis. Analysis of equilibrium position revealed a favourable shift towards peptide product when starting materials are largely undissolved in the reaction medium and the product precipitates [26]. Recently, heterogeneous systems were successfully applied to enzymatic peptide synthesis [27-30]. Enzymatic catalysis takes place mostly, if not exclusively, in the liquid phase [29]. We have used largely undissolved substrates as the solid phase pool from which consumed substrate can be replaced. Providing the peptide product precipitates at all, the conversion in equilibrium controlled peptide synthesis reactions should continue until at least one of the substrate pools disappears completely. Using such a pool can shift the overall equilibrium far towards peptide product [26]. Therefore, we have performed thermolysin-catalyzed peptide syntheses in aqueous media with solid phase substrate pools and high yields were obtained without any addition of organic solvent [31].

In this work amino acid esters were used to synthesize dipeptide esters, which are suitable substrates for serine or cysteine protease-catalyzed fragment couplings. Among the ester nucleophiles studied benzyl esters were found to give the highest synthetic rates. Initial synthetic rates of thermolysin-catalyzed synthesis of Z-Phe-Leu-Y from Z-Phe-OH and H-Leu-Y were 2.12 ± 0.33, 0.356 ± 0.036, 0.111 ± 0.012 μmol/min*mg for Y = OBzl, OMe and ONb, respectively, compared with 18,6 ± 0,97 for Y = NH_2.

The approach using solid phase substrate pools has now been extended to kinetically controlled enzymatic peptide syntheses. As a model system, the α-chymotrypsin-catalyzed synthesis of Ac-Tyr-X (X = Arg-NH_2, Gly-Gly-OH) from Ac-Tyr-OEt and H-X was selected. The two nucleophiles, H-Arg-NH_2 and H-Gly-Gly-OH, were chosen for the comparison of an efficient and a poor nucleophile, respectively. First, the α-chymotrypsin-catalyzed hydrolysis of Ac-Tyr-OEt in solution with solid ester substrate present was investigated. Surprisingly, the specific reaction rate decreased as the amount of solid ester present increased. Addition of a detergent to ensure complete wetting of ester particles increased reaction velocity to the value found in a saturated solution.

Second, for both H-Arg-NH_2 and H-Gly-Gly-OH the nucleophile efficiency was determined via acyl transfer. In Table 4 partition (p) values are shown by means of weighted linear regression analysis [5].

TABLE 4. Effective p values of α-chymotrypsin-catalyzed acyl transfer of Ac-Tyr-OEt to H-Arg-NH_2 and H-Gly-Gly-OH

Nucleophile	p_o (mM)	p_N
H-Arg-NH_2	0.512 ± 0.039	0.0296 ± 0.0017
H-Gly-Gly-OH	96.7 ± 1.4	0.0045 ± 0.0163

reaction conditions: [Ac-Tyr-OEt] = 2.9 mM, [α-chymotrypsin] = 4 nM, [KCl] = 2 M, ϑ = 25°C, pH 7.9 [Hepes/Na^+] = 20 mM, [H-Arg-NH_2] = 2.9...290 mM, [H-Gly-Gly-OH] = 14.5... 725 mM

p_N values of around 0 for both nucleophiles reveals that the hydrolysis of the acyl enzyme-nucleophile complex does not play a significant role. Thus, if the enzyme can be saturated with nucleophile, high yields can be obtained.

Finally, we combined high nucleophile concentrations with solid phase ester pools. This approach enabled us to supply the ester substrate and nucleophile in equimolar amounts in a high-density medium [31]. Yields of further reactions are compiled in Table 5. Kasche and Galunsky [32] obtained similar results for reactions with precipitated subtrate present, which support our findings.

TABLE 5. Yields of α-chymotrypsin-catalyzed peptide syntheses in high-density media

Substrate	Nucleophile	S/V (mol/l)	Yield (%)	Time (h)
Ac-Tyr-OEt[a]	$H\text{-}Arg\text{-}NH_2$	0.5	90	1
		1	90	1
	H-Gly-Gly-OH	0.5	58	1
		1	63	2
	H-Ala-Ile-OH	1	86	3
Z-His-Phe-OBzl[b]	$H\text{-}Arg\text{-}NH_2$	0.5	77	2
	$H\text{-}Leu\text{-}NH_2$	0.5	73	2
	$H\text{-}Arg\text{-}Trp\text{-}NH_2$	0.55	95	2.5

S/V substrate-volume(buffer)-ratio

[a] [NH_4CH_3COO-buffer, pH 9.15]=0.5 M, 100 μmol of both substrate and nucleophile, 0,22 nmol α-chymotrypsin, ϑ = 25°C, 0.5% (v/v) Triton X-100

[b] [HEPES/Na^+-buffer, pH 8]= 0.5 M, 50 μmol of both substrate and nucleophile, 0,25 μmol α-chymotrypsin, ϑ = 25°C

Both approaches, equilibrium and kinetically controlled reactions, were combined in the synthesis of a tetrapeptide derivative. An example of this reaction sequence, the thermolysin-catalyzed synthesis of Z-His-Phe-OBzl (84% yield) was directly followed by a kinetically-controlled peptide synthesis with α-chymotrypsin in the suspension-to-suspension approach. The tetrapeptide Z-His-Phe-Arg-Trp-NH_2 (ACTH 6-9) was formed in 95% yield.

The proposed reaction system with solid phase substrate pools offers some advantages over other systems: i) organic solvent can be partly or completely avoided; ii) substrates are used in equimolar amounts; iii) high yields are achievable with a minimal need for protection and deprotection steps; iv) approach is compatible with well-established chemical methods.

4.4. ZYMOGEN-CATALYZED PEPTIDE SYNTHESIS

As mentioned above, in kinetically controlled peptide synthesis catalyzed by serine and cysteine proteases, undesired hydrolysis of the acyl enzyme intermediate and the danger of secondary hydrolysis of the formed peptide product are the most important drawbacks because of the amidase activity of the enzymes. The partitioning may be effectively shifted to favor aminolysis by medium engineering [3] or protein engineering techniques [16,33]. However, an effective enhancement of the ligase activity together with an reduced amidase activity can only be achieved with chemically or genetically modified enzymes or by protease-catalyzed peptide synthesis in frozen aqueous systems.

Recently, we established that the zymogens trypsinogen and chymotrypsinogen are capable of catalyzing peptide synthesis under kinetically controlled reaction conditions [34]. This behavior was surprising since zymogens are known as enzymatically inactive precursors of the proteolytic enzymes, but its principal ability to form an acyl zymogen was already demonstrated by the hydrolysis of active-site specific esters with a significantly decreased acylation rate [35].

In order to demonstrate the differences between bovine trypsinogen and trypsin concerning their esterase and amidase activities, respectively, we have synthesized the nonspecific trypsin substrate carboxamido methyl ester, Bz-Arg-Phe-OCam, and compared the results with Bz-Arg-Phe-OMe. As shown in Figure 7 trypsinogen is not capable of cleaving the trypsin-specific Arg-Phe bond. In the case of trypsin, the conventional methyl ester leads to a rapid formation of Bz-Arg-OH which was not observed with trypsinogen. The acylation of trypsinogen proceeds faster with the corresponding carboxamido methyl ester and underlines the importance of leaving group

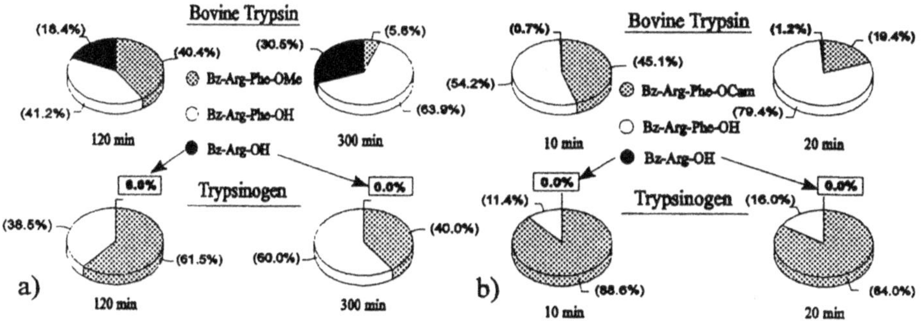

Figure 7. Composition of the reaction mixture in the reaction of a) Bz-Arg-Phe-OMe and b) Bz-Arg-Phe-OCam with bovine trypsin and trypsinogen.
Conditions: [S]=10mM; T=298K; pH 8.3 (0.05M borate, 0.12M NaCl, 0.02M $CaCl_2$); [trypsin]=2μM; [trypsinogen]=53 μM (a)/10 μM (b).

manipulation for zymogen catalysis, too. As a result of these studies, this substrate could be used for a sensitive discrimination between trypsinogen and the active enzyme.

In order to characterize the specificity of the zymogen S' subsites a series of acyl transfer experiments were performed with amino acid amides and peptide nucleophiles and the results were compared with those of trypsin (Fig. 8). From these studies we are able to discuss the data in terms of binding specificity. The partition value p_0 is an

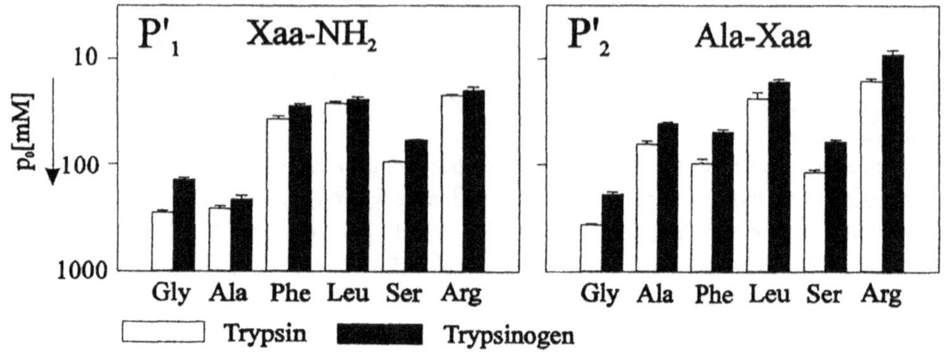

Figure 8. Comparison of the S' subsite specificity of bovine trypsin and trypsinogen for a series of amino acid amide and dipeptide nucleophiles.
Conditions: substrate: Bz-Arg-OEt, [S]=5mM; [HN]=3.12...50mM; pH 9.0 (0.05M veronal, 0.04M Ca^{2+}, 0.03M Na^{+}, I=0.2M); T=303K; [trypsin]=0.2μM; [trypsinogen] =160-312μM.

expression for the affinity of the acyl enzyme to the nucleophile. It is a suitable term for the characterization of the S' specificity of serine and cysteine proteases at any ratio of substrate and nucleophile. Small values of p_0 indicate efficient nucleophiles and efficient P'-S' interactions. The opposite is valid for large values of p_0 which indicate less efficient nucleophiles. Figure 8 shows significant differences for the deacylation of Bz-Arg-trypsin and Bz-Arg-trypsinogen, respectively, by various amino acid amides and Ala-Xaa dipeptides. It can be concluded that in all cases trypsinogen shows a more efficient binding of the nucleophiles, especially P'_2 Arg, which shows a specific and strong binding to the S'_2 subsite of trypsinogen. This behavior is reinforced by an elongation of the peptide chain up to the pentapeptides Arg-$(Ala)_3$-Gly (trypsin p_0=14.0±0.6 /trypsinogen p_0=7.73±0.13) and Ala-Arg-$(Ala)_2$-Gly (trypsin p_0= 11.0±0.6 /trypsinogen p_0=2.90±0.05). The elongation of the peptide chain improves the binding and simultaneously, the efficiency of the binding between the nucleophile and trypsinogen is significantly improved. In trypsin, the crystal structure suggests that P'_2 side chains bind to a region on the enzyme surface that is lined by His-40 and Tyr-151 and it is known that P'_2 and P'_3 side chains point in opposite directions, so that residues with large and specific side chains can be accommodated there [10]. In addition, a comparison of the superimposed structures of complexes between trypsin and BPTI [36] shows a contact region concerning the P'_2-S'_2 interactions. Our results underline that these interactions are much stronger with trypsinogen and demonstrate the extraordinary preference at this position. Furthermore, we could establish that acyl transfer can be successfully applied to the comparative analysis of the S' subsite specificity of proteases and the appropriate proenzymes. The data obtained indicate a significant shift of the partitioning to favor aminolysis using trypsinogen.

Otherwise, the results could be interpreted in terms of real zymogen catalysis, because only deacylation of the already formed acyl enzyme is important to consider. The only difference in the protocol of the acyl transfer studies is the protein concentration of the enzyme and the zymogen, respectively. The use of zymogens in peptide synthesis requires higher protein concentrations compared with the active enzymes due to the lower acylations rates. To avoid misinterpretations we performed a set of acyl transfer reactions with trypsin, trypsinogen and trypsin plus an appropriate amount of human serum albumine (HSA) to get the same protein concentration as with the zymogen (Fig. 9). The reaction rate with trypsin was adjusted to equal those of trypsinogen. It is ob-

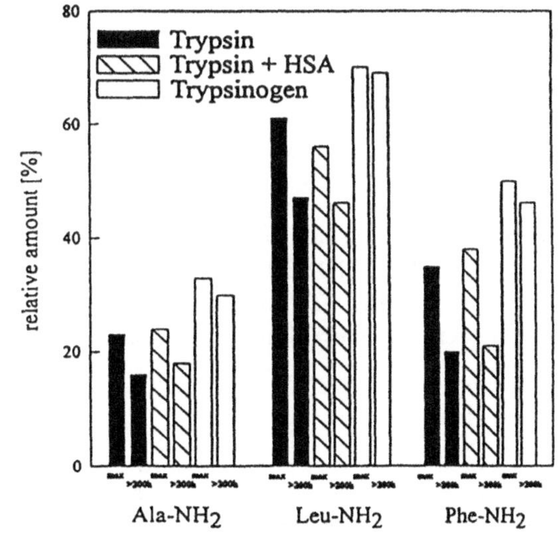

Figure 9. Stability of Bz-Arg-Xaa-NH_2 peptides in trypsin/trypsinogen-catalyzed acyl transfer of Bz-Arg-OEt to various nucleophiles.
Conditions: [S]=2mM; [HN]=50mM; T=298K; pH 9.0; [trypsin]=5nM; [trypsinogen] =8μM; [HSA]=8μM. All experiments were performed in duplicate.

vious that during the long reaction time reaction the trypsin- and trypsin+HSA-catalyzed syntheses resulted in significant secondary hydrolysis. This was not the case for the zymogen-catalyzed acyl transfer where secondary hydrolysis is kept on a significantly lower level after 200h. The small degree of proteolysis after the long reaction may be due to the formation of traces of active trypsin traces by activation of the zymogen.

5. Conclusions

Due to the stereo- and regiospecificity of the proteases, enzymatic peptide synthesis guarantees the formation of stereochemically pure peptide products with the starting reactants requiring only minimal semipermanent protection of side-chain functions for trifunctional building blocks. Furthermore, the determination of S' subsite specificity by acyl transfer studies provides useful parameters for prediction and optimization of the synthetic course. It could be demonstrated that important approaches towards suppressing competitive reactions in the reversal of proteolysis are: leaving group manipulations of the acyl donor ester in kinetically controlled synthesis; peptide synthesis in frozen-aqueous systems and zymogen-catalyzed peptide synthesis. In addition, in both equilibrium-controlled synthesis and in the kinetic approach, synthetic reactions in high-density media provide high peptide yields using equal concentrations of the educts and avoiding high concentrations of organic co-solvents. The use of proteases and even zymogens to perform selective transformations in peptide synthesis is advantageous since chemical ligation methods are prone to racemization and suffer from time-consuming side-chain protection/deprotection necessities.

Acknowledgements

Research was supported by Deutsche Forschungsgemeinschaft (DFG), Bundesministerium für Forschung und Technologie (BMFT), Fonds der Chemischen Industrie, DECHEMA, DEGUSSA AG, HÖCHST AG and E. MERCK. We wish to thank Stan Theophilou (London) for reading and correcting the manuscript.

6. References

1. Jakubke, H.-D. (1987) Enzymatic peptide synthesis, in S. Udendriend and J. Meienhofer (eds.), *The Peptides: Analysis,Synthesis, Biology*, Academic Press, New York, pp. 103-165.
2. Bongers, J., and Heimer, E.P. (1994) Recent applications of enzymatic peptide synthesis, *Peptides* **15**,183-193.
3. Jakubke, H.-D. (1994) Protease-catalyzed peptide synthesis: basic principles, new synthesis strategies and medium engineering, *J. Chin. Chem. Soc.* **41**,355-370.
4. Schechter, I., and Berger, A. (1967) On the size of the active site in proteases. I. Papain, *Biochem. Biophys. Res. Commun.* **27**,157-162.
5. Schellenberger, V., and Jakubke, H.-D. (1991) Protease-catalyzed kinetically controlled peptide synthesis, *Angew. Chem., Int. Ed. Engl.* **30**,1437-1449.
6. Schuster, M., Kasche, V., and Jakubke,H.-D. (1992) Contributions to the S'-subsite specificity of papain, *Biochim.Biophys.Acta* **1121**,207-212.
7. Ullmann, D., and Jakubke, H.-D. (1994) Solid phase synthesis of a peptide library for S' subsite mapping of serine and cysteine proteases, in R. Epton (Ed.), *Innovations and Perspectives in Solid Phase Synthesis, Biological and Biomedical Applications, Collected Papers 3rd International Symposium*,Worldwide Ltd., Birmingham, pp. 693-696.
8. Ullmann, D., and Jakubke, H.-D. (1994). The specificity of clostripain from *clostridium histolyticum.*

Mapping the S' subsites via acyl transfer to amino acid amides and peptides, *Eur. J. Biochem.*, **223**,865-872.

9. Schellenberger, V., Turck, C.W., Hedstrom, L., and Rutter, W.J. (1993) Mapping the S'subsites of serine proteases using acyl transfer to mixtures of peptide nucleophiles, *Biochemistry* **32**,4349-4353.
10. Schellenberger, V., Turck, C.W., and Rutter, W.J. (1994) Role of the S' subsites in serine protease catalysis. Active-site mapping of rat chymotrypsin, rat trypsin, α-lytic protease, and cercarial protease from Schistosoma mansoni, *Biochemistry* **33**,4251-4257.
11. Schellenberger, V., Siegel, R.A., and Rutter, W.J. (1993) Analysis of enzyme specificity by multiple substrate kinetics, *Biochemistry* **32**,4344-4348.
12. Jakubke, H.-D. (1995) Peptide ligases: tools for peptide synthesis, *Angew. Chem., Int. Ed. Engl.* **34**,175-177.
13. Schellenberger, V., Görner, A., Könnecke, A., and Jakubke, H.-D. (1991) Protease-catalyzed peptide synthesis: prevention of side-reactions in kinetically controlled reactions, *Peptide Research* **4**,265-269.
14. Mitin, Y.V., Zapevalova, N.P., and Gorbunova, E.Y. (1991) New approaches to peptide synthesis with the help of trypsin, *Biomed. Biochim. Acta* **50**,74-79.
15. Kuhl, P., Zacharias, U., Burckhardt, H., and Jakubke, H.-D. (1986) On the use of carboxamidomethyl esters in peptide synthesis, *Monatsh. Chem.* **117**,1195-1204.
16. Jackson, D.Y., Burnier, J., Quan, C., Stanley, M., Tom, J., and Wells, J.A. (1994) A designed peptide ligase for total synthesis of ribonuclease A with unnatural catalytic residues, *Science* **266**, 243-247.
17. Bongers, J., Lambros, T., Liu, W., Ahmad, M., Campell, R.M., Felix, A.M., and Heimer, E.P. (1992) Enzymatic semisynthesis of a superpotent analog of human growth hormone-releasing factor, *J. Med. Chem.* **35**,3934-3941.
18. Schellenberger, V., Jakubke, H.-D., Zapevalova, N.P., and Mitin, Y.V. (1991) Protease-catalyzed peptide synthesis using inverse substrates: the influence of reaction conditions on the trypsin acyl transfer efficiency, *Biotechnol. Bioeng.* **38**,104-108.
19. Lineweaver, H. (1939) The energy of activation of enzyme reactions and their velocity below 0°C, *J. Am. Chem. Soc.* **61** ,403-407.
20. Grant, N.H., and Album, H.E. (1966) Acceleration of enzyme reactions in ice, *Nature* **212**, 194.
21. Schuster, M., Aaviksaar, A., and Jakubke, H.-D. (1990) Enzyme-catalyzed peptide synthesis in ice, *Tetrahedron* **46**, 8093-8101.
22. Ullmann, G., and Jakubke, H.-D. (1993) Frozen aqueous systems: New efficient reaction media for enzyme-catalyzed peptide bond formation, in C. Schneider and H. Eberle (eds.), *Peptides 1992*, ESCOM, Leiden, pp. 36-37.
23. Hänsler, M., Ullmann, G., and Jakubke, H.-D. (1995) The application of ficin, papain and clostripain in kinetically controlled peptide synthesis in frozen aqueous solutions, *J.Peptide Science*, in press.
24. Gerisch, S., Ullmann, G., Stubenrauch, K., and Jakubke, H.-D. (1994) Enzymatic peptide synthesis in frozen aqueous systems: Influence of modified reaction conditions on the peptide yield, Biol. Chem Hoppe-Seyler **375**,825-828.
25. Pincock, R.E., and Kiovsky, T.E. (1966) Kinetics of reactions in frozen solutions, *J. Chem. Educ.* **43**, 358-360.
26. Halling P. J., Eichhorn U., Kuhl P., and Jakubke H.-D. (1995) Thermodynamics of solid-to-solid-conversion and application to enzymic peptide synthesis, *Enzyme Microb. Technol.*, in press.
27. Kuhl P., Eichhorn U., Jakubke H.-D., Drauz K., and Bommarius A.S. (1992) Enzymatische Reaktionen und Verwendung einer Vorrichtung zu ihrer Durchführung. DE 42 14 157 C1.
28. Kuhl P., Eichhorn U., and Jakubke H.-D. (1995) Enzymic peptide synthesis in micro-aqueous, solvent free systems, *Biotechn. Bioeng.* **45**,276-278.
29. López-Fandiño R., Gill I., and Vulfson E. N. (1994) Enzymatic catalysis in heterogeneous mixtures of substrates: The role of the liquid phase and the effects of "adjuvants", *Biotechn. Bioeng.* **43**,1016-1023.
30. Čeřovský, V. (1992) Protease-catalyzed peptide synthesis in solvent free system, *Biotechnol. Tech.* **6**, 155-160.
31. Eichhorn, U., Bommarius, A.S., Drauz, K., and Jakubke, H.-D. (1995) Environmentally acceptable synthesis of peptides in high yields by suspension-to-suspension conversion via protease catalysis, in H.L.S. Maia (Ed), *Peptides 1994*, ESCOM, Leiden, pp. 226-227.

32. Kasche, V., and Galunsky, B. (1995) Enzyme catalyzed biotransformations in aqueous two-phase systems with precipitated substrate and/or product, *Biotechn. Bioeng.* **45**,261-267.
33. Wong, C.-H. (1993) Development of enzymes for chemoenzymatic synthesis, *Chimia* **47**,127-132.
34. Ullmann, D., Salchert, K., Bordusa, F., Schaaf, R., and Jakubke, H.-D. (1994) Are zymogens suitable tools for peptide bond formation ?, in E. Wünsch (Ed), *Proc 5th Akabori-Conference*, Max-Planck-Gesellschaft, pp. 70-75.
35. Gertler, A., Walsh, K.A., and Neurath, H. (1974) Catalysis by chymotrypsinogen: increased reactivity due to oxidation of methionine 192, *FEBS Lett.* **38**,157-160.
36. Rühlemann, A., Kukla, D., Schwager, P., Bartels, K., and Huber, R.(1973) Structure of the complex formed by bovine trypsin and bovine pancreatic trypsin inhibitor, *J. Mol. Biol.* **77**,417-436.

ALKALINE-EARTH METAL ION CATALYSIS OF ESTER CLEAVAGE

ROBERTA CACCIAPAGLIA AND LUIGI MANDOLINI
Centro di Studio sui Meccanismi di Reazione
and Dipartimento di Chimica, Università La Sapienza,
00185 Roma, Italy.

ABSTRACT. Alkaline-earth metal ions are efficient promoters of acyl transfer reactions from esters to anionic nucleophiles as a result of transition state stabilization *via* cation-anion electrostatic interactions. The beneficial influence of additional binding energy rendered available by a metal ion binding site proximal to the reaction centre is illustrated by a number of model systems. On this basis, partial transacylase mimics have been designed in which a complexed alkaline-earth metal electrophile assists both the acylation and deacylation steps. These prototype catalysts, although still unable of substrate binding and recognition, cleave simple aryl esters *via* a ping-pong mechanism in which the catalyst takes the acyl group from the substrate and then transfers it to an external nucleophile.

1. Introduction

The basic idea underlying the design of our supramolecular complexes with transacylase activity based on alkaline-earth metal cations is schematically depicted in Scheme 1. As pointed out by Swain [1], if nucleophilic attack of a reactant N at a substrate S is catalyzed by an electrophile E (case I), enhanced reactivity is expected if two of the components, either the nucleophile

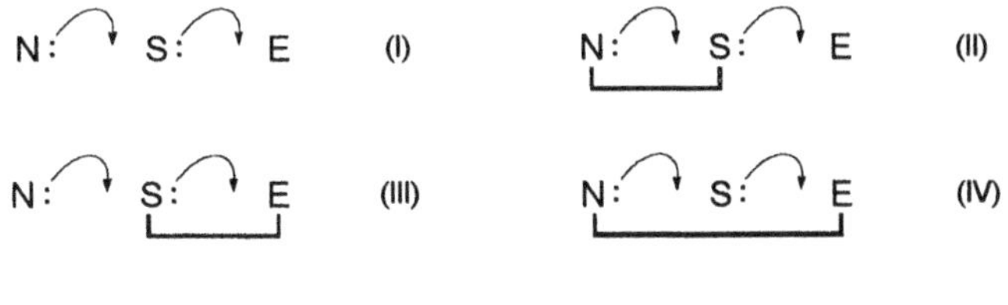

Scheme 1

C. S. Wilcox and A. D. Hamilton (eds.), Molecular Design and Bioorganic Catalysis, 71–86.

and the substrate (case II), the electrophile and the substrate (case III), or the nucleophile and the electrophile (case IV), are joined together in the same molecule or in the same complex by covalent bonding or weak noncovalent interactions, respectively.

The findings that alkaline-earth metal ions perform as electrophilic catalysts in reactions of alkoxides with simple esters (type I reactions), and that significantly larger rate enhancements are found in model systems in which the ester substrate is covalently linked to a metal binding site (type III reactions) [2], led us to design nucleophilic catalysts in which the nucleophilic character of an OH/SH group covalently bound to the host skeleton is combined with the electrophilic character of a complexed alkaline-earth metal ion. These complexes perform as type IV turnover catalysts of acyl transfer processes.

2. Metal Ion Promotion of Ester Cleavage

2.1. NUCLEOPHILICITY OF ALKALINE-EARTH METAL ALKOXIDES

Rates of reaction of $MeONMe_4$ with phenyl acetate in MeOH are enhanced by added alkaline-earth (Sr, Ba) bromides in a way which is consistent with a reaction scheme involving association of methoxide ion with the alkali metal counterion (K_M) and independent contributions to the overall rate from free (k_o) and cation-paired (k_M) methoxide [3,4].

$$MeO^- + M^{2+} \xrightleftharpoons{K_M} (MeOM)^+ \qquad (1a)$$

$$MeOH + CH_3CO_2C_6H_5 \xrightleftharpoons{MeO^-,\ k_o} CH_3CO_2Me + C_6H_5OH \qquad (1b)$$

$$MeOH + CH_3CO_2C_6H_5 \xrightleftharpoons{(MeOM)^+,\ k_M} CH_3CO_2Me + C_6H_5OH \qquad (1c)$$

Analysis of rate data by means of a standard binding isotherm [4] gave K_{Sr}=60 M^{-1}; K_{Ba}=44 M^{-1}; k_{Sr}/k_o=4.7; k_{Ba}/k_o=3.7. Given the widespread notion that ion pairs are less reactive than free anions, the discovery that a metal-bound methoxide species cleaves phenyl acetate more rapidly than free methoxide ion was somewhat surprising. There seems to be little doubt that simultaneous carbonyl activation by the metal electrophile and M-OMe participation are taking place *via* a transition state having the form of the chelate four-membered

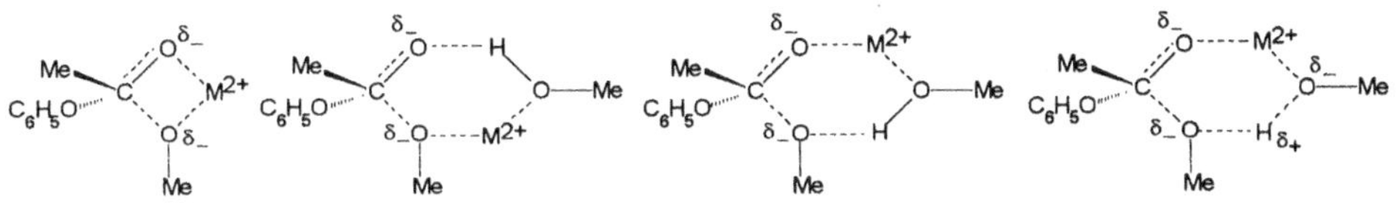

Figure 1. Schematic pictures of the transition state of the metal ion promoted methanolysis of phenyl acetate.

structure I (Fig. 1), but each of the kinetically equivalent solvent-shared forms II-IV is a plausible alternative.

Whatever the detailed structure of the transition state, however, the important conclusion is reached that significant rate-enhancements can be obtained with such a simple catalyst as an alkaline-earth metal ion, which is equivalent to saying that cation-anion electrostatic attraction is an effective force for catalysis.

In the $EtONMe_4/EtOH$ base-solvent system [5] binding of ethoxide ion to strontium and barium ions is so strong ($K_M > 10^4\ M^{-1}$) that upon mixing equimolar amounts of $EtONMe_4$ and MBr_2 in the concentration range of 1-10 mM a metal bound ethoxide species is quantitatively formed (eq 2). It is

$$EtONMe_4 + MBr_2 \longrightarrow EtOMBr + Me_4NBr \qquad (2)$$

unknown to what an extent the metal-ethoxide complex contains a bromide counterion, but it is clear that it behaves kinetically as a single species. It is remarkable that cleavage of phenyl acetate is 62 and 45 times faster with EtOSrBr and EtOBaBr, respectively, than with $EtONMe_4$ [6]. The corresponding figures for deacetylation of *p*-nitrophenyl acetate are 8.0 and 7.0, respectively [5], showing that the need for electrophilic assistance by the metal ion dramatically decreases under the influence of a strong electron withdrawing group.

A selection of the results obtained in an extensive investigation [7] of the effect of cation-complexing agents on the Ba(II)-promoted basic ethanolysis of phenyl acetate is shown in Table 1.

It is remarkable that the reactivity of the crown-complexed ion pair (ternary complex) is significantly higher than that of the uncomplexed ion pair $(EtOBa)^+$. It is also worth noting that the statistically corrected reactivity of $(EtO)_2Ba$ is very similar to that of $(EtOBa)^+$, showing that the behaviour of the complexed ethoxide species in EtOMX is not dramatically affected by the nature of X.

TABLE 1. Ethanolysis of phenyl acetate at 25.0 °C. Reactivity of various metal-bound ethoxide species.[a]

	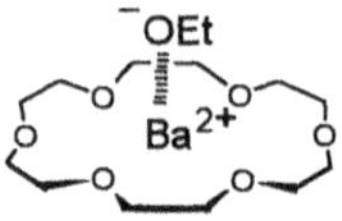	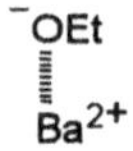		EtO^-
k ($M^{-1}s^{-1}$)	122	48	38[b]	1.44
k_{rel}	85	33	26	1

[a] The source of barium ion was $Ba(SCN)_2$.
[b] Statistically corrected.

2.2. CROWN ETHER SUBSTRATES

In a series of studies [2] we have obtained overwhelming evidence that much larger metal ion effects on rates of ester cleavage are obtained in model substrates in which the ester function is connected to a polyether chain capable of holding the metal ion in close proximity to the ester function. This strategy is illustrated in Fig. 2, where a rate-determining nucleophilic attack at a substrate-metal ion complex is schematically depicted.

A sample of our data, reported in Table 2, shows that the size of the acceleration is not only dependent on the substrate-metal ion combination, but it is also dramatically influenced by the solvent. The largest acceleration is exhibited by 2-AcO-21C6, which reacts with EtOBaBr in EtOH 500,000 times faster than with $EtONMe_4$. Kinetic evidence was obtained that reactions in the

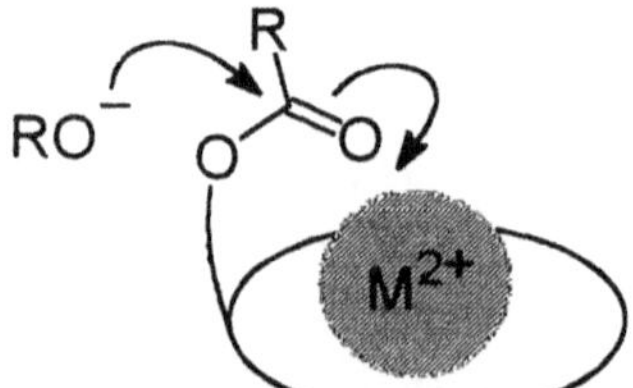

Figure 2. The complexed metal ion as a built-in electrophilic catalyst.

TABLE 2. Metal ion promoted basic alcoholysis of aryl acetates at 25 °C. Rate enhancements (k_M/k_o).

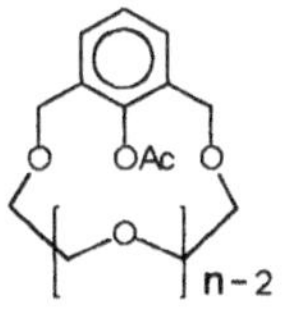

n = 4 2-AcO-15C4
n = 5 2-AcO-18C5
n = 6 2-AcO-21C6

	k_{Ba}/k_o	k_{Sr}/k_o
2-AcO-15C4 (MeOH)	93	40
2-AcO-15C4 (EtOH)	7200	2300
2-AcO-18C5 (MeOH)	760	230
2-AcO-18C5 (EtOH)	46000	48000
2-AcO-21C6 (MeOH)	250	190
2-AcO-21C6 (EtOH)	500000	41000

$EtONMe_4/EtOH$ base-solvent system proceed through a ternary complex (Fig. 3), which has the same composition as the rate-limiting transition state and decomposes into products in a monomolecular step. Whenever this ternary complex does not accumulate to a significant extent, the pathway proceeding through the ternary complex is kinetically indistinguishable from, and consequently operationally identical to the bimolecular mechanism depicted in Fig. 2. It has been suggested that the ternary complex is a plausible intermediate also in those cases in which it does not accumulate to such an extent as to affect the kinetics [8].

In any event, independent of the detailed mechanism by which the transition state is attained, what is really significant in the present context is that transition state stabilization by metal ions arises from cooperation of coordination to the oxygen donors of the bridge and electrostatic binding to the negative charge being

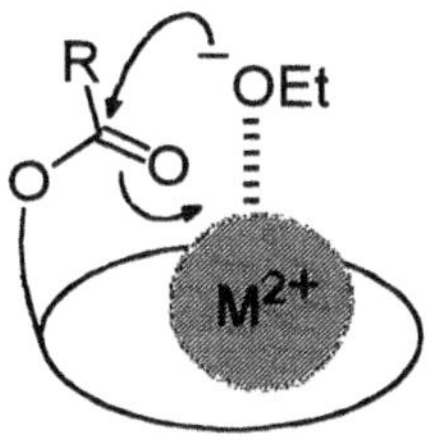

Figure 3. The ternary complex as reactive intermediate in the cleavage of crown ether acetates in $EtONMe_4/EtOH$.

transferred from the incoming nucleophile to the carbonyl oxygen. It appears therefore that the polyether chain increases the available binding energy and, consequently, the portion of this binding energy which can be utilized in catalysis.

3. From Metal Ion Promotion to Metal Ion Catalysis

Partial transacylase mimics based on host structures which like many hydrolytic enzymes bear an OH/SH group to act as nucleophile, have been described by several authors [9-12]. The designed host structures contain substrates binding sites proximal to the nucleophilic group, so that rates of acyl transfer from suitably functionalized ester substrates are greatly enhanced. However, no provision is made to activate the cleavage of the acylated host, with the result that turnover catalysis is generally not observed.

On the basis of our findings that (i) alkaline-earth metal ions are efficient promoters of acyl transfer reactions from esters to anionic nucleophiles and (ii) rate enhancements are particularly large when the ester substrate is covalently linked to a metal ion binding site, we reasoned that a complex between an alkaline-earth metal ion and a crown ether bearing an OH/SH side-arm should act as a transacylation catalyst, provided that the crown ether is an efficient binder of the metal ion. This concept is schematically illustrated in Scheme 2.

By virtue of the acidity enhancing influence of the complexed metal ion, extensive dissociation of the proton-ionizable function XH should take place under moderately basic conditions. The metal ion electrophile should then assist both the acylation of the catalyst and the subsequent transfer of the acyl group to an external nucleophile, thus restoring the active form of the catalyst.

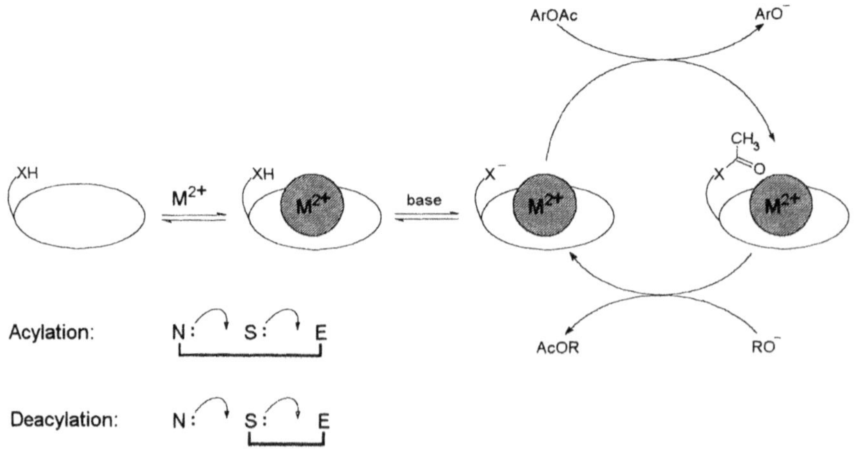

Scheme 2

3.1 CALIXARENES

A first example of nucleophilic catalyst with transacylase activity is offered by an equimolar mixture of $BaBr_2$ and *p-tert*-butylcalix[4]arene-crown-5 (**1**), which catalyzes the methanolysis of aryl acetates in MeCN-MeOH (1:1, v/v) in the presence of a 3:1 diisopropylethylamine-perchlorate salt buffer at 25 °C [13]. Inspection of the UV spectra shows that **1** remains in its unionized form in the given buffer solution, but on addition of 1 mol equiv of $BaBr_2$ it is quantitatively transformed into a mixture of the barium complexes of its singly and doubly ionized forms. Evidence was obtained that the active form of the catalyst is the barium complex of the doubly ionized form (**2**). As shown in Scheme 3, the

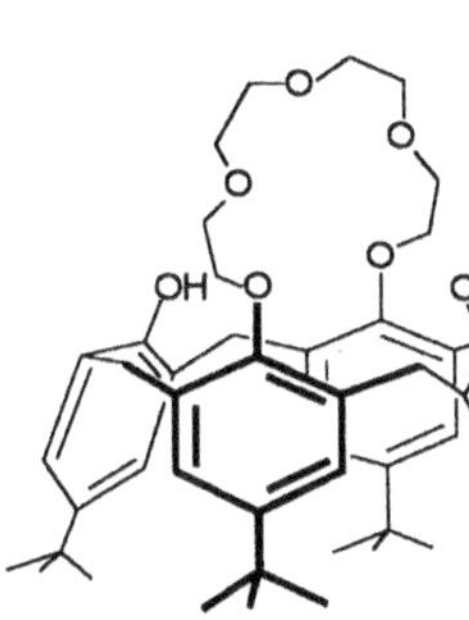

1

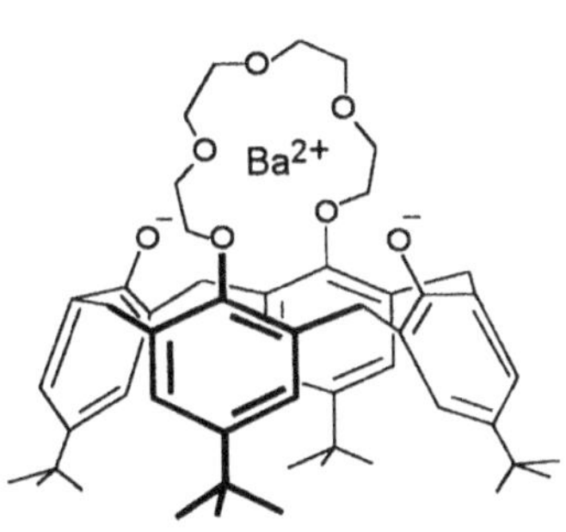

2

catalyst takes the acetyl group from *p*-nitrophenyl acetate and transfers it to methoxide ion, thus restoring its active form and turning over. Biphasic kinetics

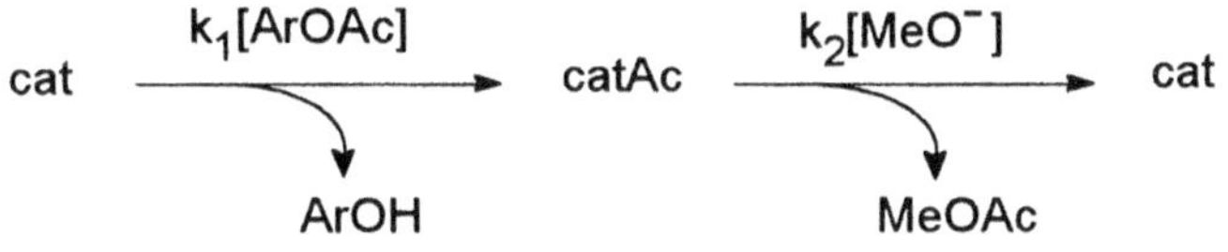

Scheme 3

(burst kinetics) typical of the proteolytic enzymes (e.g. chymotripsin) were observed (Fig. 4). Accumulation of the acetylated catalyst was monitored by HPLC analysis, and found to be fully consistent with the kinetics. Upon replacement of the polyether bridge by two methoxy groups any catalytic activity disappears.

The rate of acetylation of the catalyst by a series of aryl acetates $CH_3CO_2C_6H_4X$, where X is *p*-NO_2, *m*-NO_2, *p*-Cl, and H [14] is affected by the nature of X in such a way that there is a changeover from rate-determining

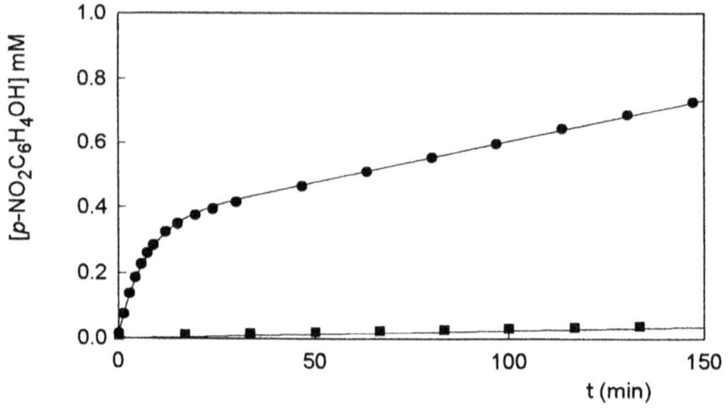

Figure 4. Methanolysis of p-$NO_2C_6H_4OAc$ (3 mM) in MeCN-MeOH (9:1). Liberation of p-$NO_2C_6H_4OH$ in the presence of diisopropylethylamine-perchlorate salt buffer alone (■) and plus 0.4 mM of the catalyst **1**·Ba^{2+} (●).

deacetylation to rate-determining acylation on going from the most reactive p-NO_2 derivative to the least reactive phenyl acetate (Table 3). Strong indication is obtained that the slow step in the acylation of the catalyst is the breakdown of the tetrahedral intermediate. The catalytic efficiency, as conveniently measured by the ratio of the catalyzed process (under steady state conditions) to the rate of the background reaction (Fig. 5) reaches a maximum

TABLE 3. Rate acceleration relative to background methanolysis.[a]

Substrate	pre-steady state (initial rate)	steady state	slow step
p-$NO_2C_6H_4OAc$	272	13.5	deacylation
			⇓
m-$NO_2C_6H_4OAc$	103	21.6	
			⇓
p-ClC_6H_4OAc	51.6	40.9	
			⇓
C_6H_5OAc	17.8	16.9	acylation

[a] [Substrate] = 3 mM, [cat] = 0.4 mM.

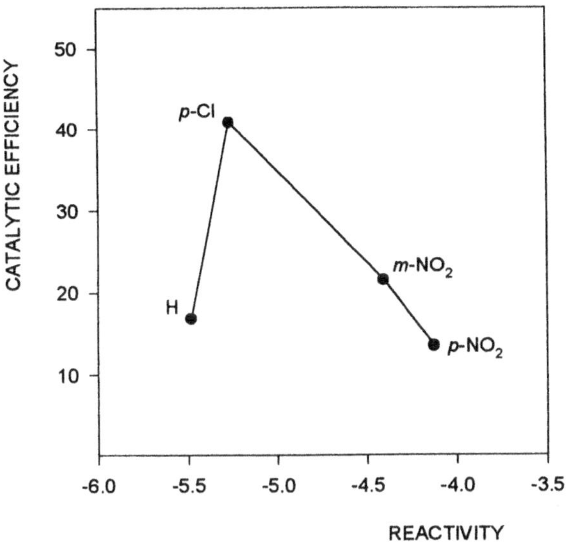

Figure 5. Catalytic efficiency *vs.* ester reactivity. The ratio between the rate of the catalyzed process at steady state and the rate of the background methanolysis is plotted *vs.* log k for background methanolysis.

value in the reaction of the *p*-chlorophenyl ester. It is apparent from Fig. 5 that the scope of **2** as a nucleophilic catalyst is practically restricted to the cleavage of esters more reactive than phenyl acetate and less reactive than *p*-nitrophenyl acetate.

A minor complication which was experienced in the above kinetic study due to backward acylation of the liberated phenol by the acetylated catalyst, suggested an experiment which illustrates well the role of **2** as a transesterification catalyst. As shown in Fig. 6, a mixture of *p*-nitrophenyl acetate and *p*-chlorophenol showed no reaction after two hours, but in the presence of the catalyst a fast production of *p*-chlorophenyl acetate (eq 3) took place.

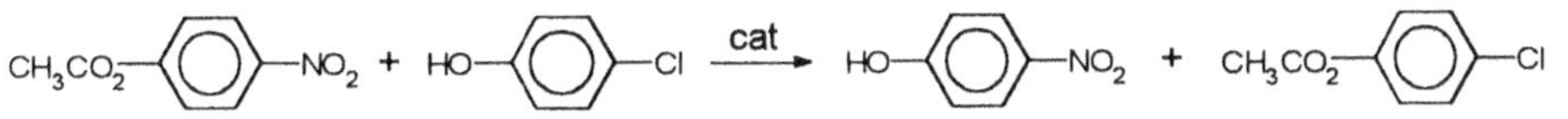

(3)

The Sr^{2+} complex of **1** was also tested for catalytic activity and found to be slightly less efficient than the Ba^{2+} complex, but the Ca^{2+} complex could not be properly tested because of solubility problems. Metal ion complexes of several

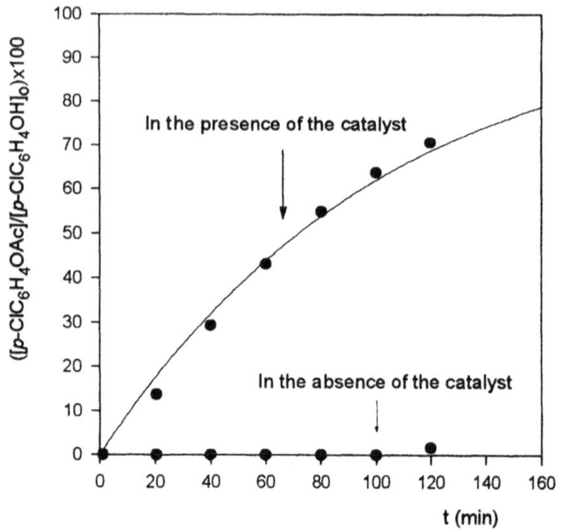

Figure 6. Reaction between *p*-nitrophenyl acetate (3 mM) and *p*-chlorophenol (1 mM) in 9:1 MeCN-MeOH and 3:1 diisopropylethylamine-perchlorate salt buffer at 25 °C, in the absence and presence of the barium complex of **1** (0.4 mM).

polyether-bridged calix[4]- and calix[5]arenes were also screened, but in no case was the catalytic efficiency higher than that obtained with the Ba^{2+} complex of **1**. Eventually the bis-amido derivative **3** was considered [15]. Here the two amide carbonyls serve conveniently as metal ion binding site in the place of the polyether bridge of **1**. Accordingly, the UV spectra revealed that in 9:1 MeCN-MeOH and diisopropylethylamine buffer, and in the presence of alkaline-earth (Ba, Sr, Ca) metal salts, formation of the neutral complexes **4** was significant. At first, the Ba^{2+} complex was tested for catalytic activity in the

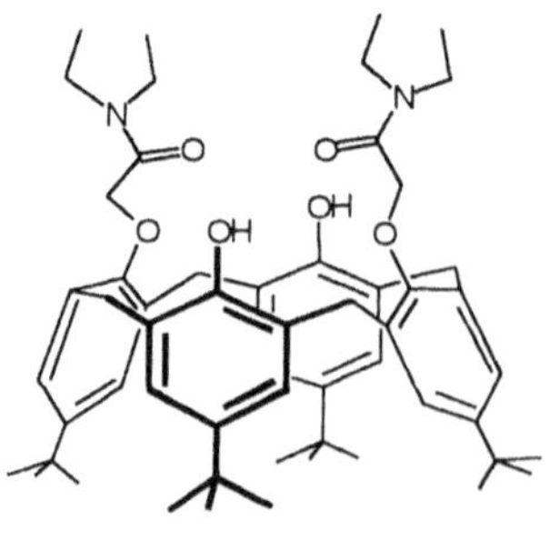

3

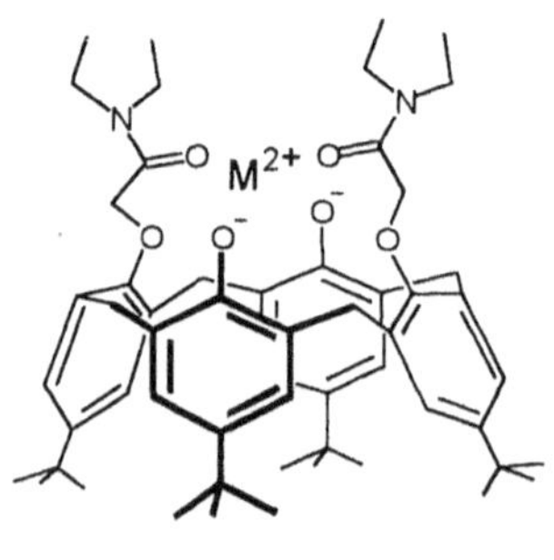

4

TABLE 4. Catalytic efficiency of calixarene-metal complexes in the methanolysis of *p*-nitrophenyl acetate.[a]

	$\mathbf{1}\cdot Ba^{2+}$	$\mathbf{3}\cdot Ba^{2+}$	$\mathbf{3}\cdot Sr^{2+}$	$\mathbf{3}\cdot Ca^{2+}$
k_1' (min^{-1})[b]	0.15	6.2×10^{-3}	0.18	0.20
k_2' (min^{-1})[b]	8.0×10^{-3}	1.3×10^{-3}	5.0×10^{-2}	0.32
turnover frequency (h^{-1})	0.46	6.0×10^{-2}	2.3	7.2

[a] In MeCN-MeOH (9:1), 3:1 diisopropylethylamine-perchlorate salt buffer, 25 °C, concentration of *p*-nitrophenyl acetate 3 mM, concentration of catalyst 0.4 mM.
[b] With reference to Scheme 3, k_1' and k_2' are defined as follows: $k_1' = k_1[ArOAc]$; $k_2' = k_2[MeO^-]$.

cleavage of *p*-nitrophenyl acetate under the same conditions used with **1**, but the results were frustating, in that both acylation and deacylation steps were much slower than with the Ba^{2+} complex of **1** (Table 4). But the Sr^{2+} complex of **3**, and even more so the Ca^{2+} complex, turned out to be much better catalysts, the turnover frequency of the latter being more than 15 times higher than that of the Ba^{2+} complex of **1**. These results show that a search for improved catalytic efficiency of metal ion-calixarene complexes requires much patience and some luck, and critically rests on a proper combination of metal ion and calixarene structure (metal ion binding site).

3.2 THIOL-PENDANT CROWN ETHERS

Additional examples of type IV bifunctional catalysts are provided by the alkaline-earth metal ion complexes of lariat ethers bearing a sulphydryl side-arm. Here the acyl-receiving and acyl releasing unit is a sulphydryl group, instead of the phenolic hydroxyl of a calixarene derivative. Out of the various structures investigated, the lariat ether **5** gave the best results [16]. Molecular models show that a crown-complexed metal ion is in close contact with the sulphydryl group of **5** as well as with the carbonyl oxygen of the acetylated form **6**. Consistently, acetylation of **5** by *p*-nitrophenyl acetate as well as deacetylation of **6** are strongly accelerated by alkaline-earth metal ions (Table 5). Consequently, turnover catalysis is predicted and actually observed. Albeit deacetylation of the

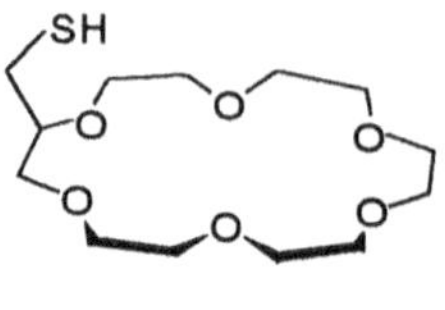

5

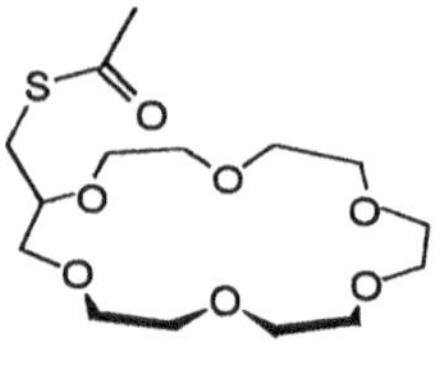

6

catalyst is enhanced by the metal ions to a surprisingly large extent, it is still two orders of magnitude slower than acetylation, and constitutes the slow step of the catalytic cycle. Fig. 7 shows the time-concentration profile for the liberation of *p*-nitrophenol in the **5**·Ca^{2+} catalyzed methanolysis of *p*-nitrophenyl acetate. The full line, which is calculated from the data listed in Table 5, shows that the initial burst of *p*-nitrophenol release coincides with the concentration of the catalyst in what is tantamount to an active-site titration experiment. This is a consequence of the fact that the acetylated catalyst is rapidly formed but solvolyzes slowly. There is a good agreement between experimental data and calculated profile in the pre-steady-state exponential phase, but it is apparent that the slope of the zeroth-order portion of the profile is decidedly higher than the predicted one.

TABLE 5. Metal ion effects on acylation of **5** and deacylation of **6**.[a] Pseudo-first-order rate constants and rate-enhancements (in parentheses).

		Ba^{2+}	Sr^{2+}	Ca^{2+}
Acylation,[b]	**5** → **6**, k_1'	0.116 min^{-1} (134)	0.201 min^{-1} (232)	2.57 min^{-1} (2970)
Deacylation,[c]	**6** → **5**, k_2'	1.86×10^{-3} (>3000)	3.76×10^{-3} (>5000)	5.33×10^{-2} (>70000)

[a] 3:1 Diisopropylamine-perchlorate salt buffer in MeCN-MeOH (9:1, v/v) at 25 °C. $[M^{2+}]$ = 2.7 mM. For the definition of the quantities k_1' and k_2' see footnote *b* to Table 4. In the absence of metal ions, $k_1' = 8.65 \times 10^{-4}$ min^{-1} and $k_2' \leq 7 \times 10^{-7}$ min^{-1}.

[b] [**5**] = 2.7 mM; [*p*-$NO_2C_6H_4OAc$] = 0.05 mM.

[c] [**6**] = 2.7 mM.

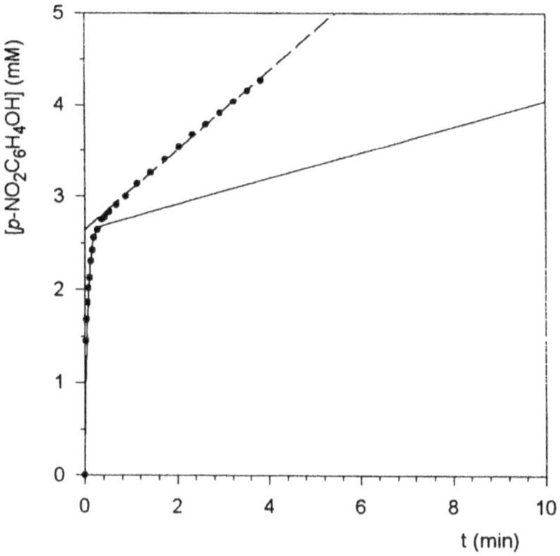

Figure 7. Methanolysis of p-$NO_2C_6H_4OAc$ (17 mM) in MeCN-MeOH (9:1). Liberation of p-$NO_2C_6H_4OH$ in the presence of diisopropylethylamine-perchlorate salt buffer and 2.7 mM **5**·Ca^{2+}. The burst of p-$NO_2C_6H_4OH$ release is 2.7 mM. The full line is calculated on the basis of the rate constants listed in Table 5. The points are experimental.

Mechanism A:

Thiol mediated methanolysis

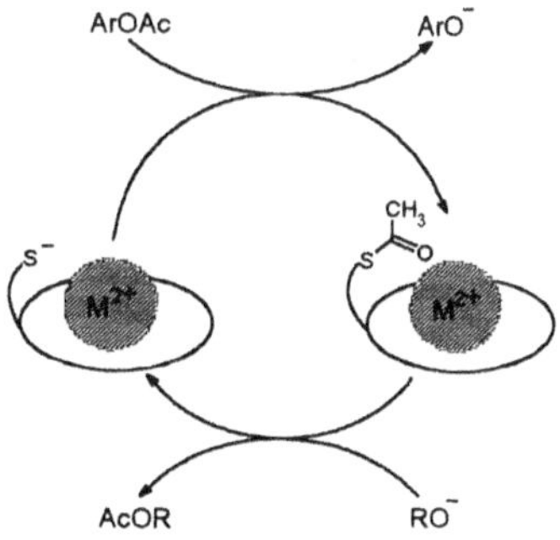

Mechanism B:

Metal ion assisted direct methanolysis

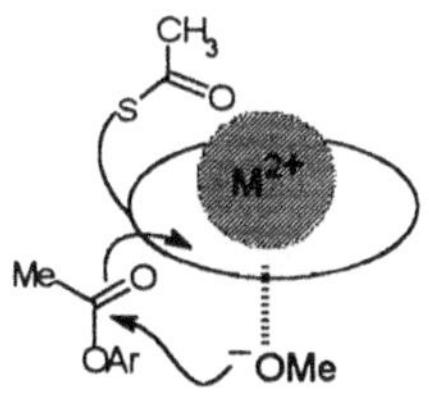

Scheme 4

Similar results are observed with the Ba^{2+}- and Sr^{2+}-complexes. Since the above discrepancies are obviously too large to be ascribable to experimental errors, we suggest that besides a thiol-mediated methanolysis involving an acylation-deacylation cycle (mechanism A in Scheme 4), a direct metal ion assisted methanolysis takes place as schematically depicted in mechanism B. Dissection of the overall turnover frequency into separate contributions from the two competing mechanisms (Table 6) shows that mechanism B is responsible for a major fraction, namely, *ca.* 2/3 of the overall reaction in the case of the Ca^{2+}-assisted reaction, and for even larger fractions with Ba^{2+} and Sr^{2+}.

TABLE 6. Methanolysis of *p*-nitrophenyl acetate catalyzed by alkaline-earth metal complexes of **5**. Turnover frequencies (h^{-1}).

	Ba^{2+}	Sr^{2+}	Ca^{2+}
overall	1.8	3.6	9.8
mechanism A	0.12	0.24	3.2
mechanism B	1.7	3.4	6.6

4. Concluding Remarks

The results presented in this account widen considerably the scope of alkaline-earth metal ions as electrophile catalysts of acetyl transfer reactions. We have provided a number of examples in which turnover catalysis of ester cleavage can be achieved with alkaline-earth metal ion complexes of multidentate ligands bearing a proton ionizable nucleophilic group (OH/SH) in close proximity to the complexed metal ion. Although the scope of our catalysts is limited so far to activated esters, and no secondary binding site is present for substrate recognition, we believe that supramolecular complexes based on alkaline-earth metal ions have potential for further development in the design and construction of synthetic catalysts with transacylase activity.

5. Acknowledgements

We wish to acknowledge the contributions of the colleagues and coworkers whose names appear in the cited references. Financial support from MURST, from EEC Science and HCM programmes, and from COST D7 is gratefully acknowledged.

6. References

1. Swain, C.G. (1950) Concerted Displacement Reactions. V. The Mechanism of Acid-Base Catalysis in Water Solution, *J. Am. Chem. Soc.* **72**, 4578-4853.
2. Cacciapaglia, R. and Mandolini, L. (1993) Catalysis by Metal Ions in Reactions of Crown Ether Substrates, *Chem. Soc. Rev.* **22**, 221-231.
3. Cacciapaglia, R., Lucente, S., Mandolini, L., van Doorn, A.R., Reinhoudt, D.N. and Verboom, W. (1989) Catalysis by Alkali and Alkaline-Earth Metal Ions in Nucleophilic Attack of Methoxide Ion on Crown Ethers Bearing an Intra-Annular Acetoxy Group, *Tetrahedron* **45**, 5293-5304.
4. Ercolani, G. and Mandolini, L. (1990) Alkali and Alkaline-Earth Metal Ion Catalysis in the Reaction of Aryl Acetates with Methoxide Ion. Effect of a Poly(oxyethylene) Side Arm, *J. Am. Chem. Soc.* **112**, 423-427.
5. Kraft, D., Cacciapaglia, R., Böhmer, V., El-Fadl, A.A., Harkema, S., Mandolini, L., Reinhoudt, D.N., Verboom, W. and Vogt, W. (1992) New Crown Ether-like Macrocycles Containing a Nitrophenol Unit. Synthesis and Metal Ion Effects on the Reactivity of Their Acetates in Transacylation Reactions, *J. Org. Chem.* **57**, 826-834.
6. Cacciapaglia, R., Mandolini, L., Reinhoudt, D.N. and Verboom, W. (1992) Alkaline-Earth Metal Ion Catalysis of Alcoholysis of Crown Ether Aryl Acetates. Effect of the Base-Solvent System, *J. Phys. Org. Chem.* **5**, 663-669.
7. Cacciapaglia, R., Mandolini, L. and Van Axel Castelli, V. (1995) The Effect of Cation Complexing Agents in the Ba(II) Promoted Basic Ethanolysis of Phenyl Acetate, in G. Savelli (ed), *Macrocyclic and Supramolecular Chemistry in Italy*, Centro Stampa Università degli Studi, Perugia, pp. 229-234.
8. Cacciapaglia, R., van Doorn, A.R., Mandolini, L., Reinhoudt, D.N. and Verboom, W. (1992) Differential Metal Ion Stabilization of Reactants and Transition States in the Transacylation of Crown Ether Aryl Acetates, *J. Am. Chem. Soc.* **114**, 2611-2617.

9. Lehn, J.M. and Sirlin, C. (1987) Catalyse Supramoléculaire: Coupure des Esters Activés d'Aminoacides Liés a un Récepteur Macrocyclique Portant des Résidus Cystéinyles, *New J. Chem.* **11**, 693-702.

10. Diederich, F. (1991) *Cyclophanes*, The Royal Society of Chemistry, Cambridge, UK.

11. Tecilla, P., Jubian, V. and Hamilton, A.D. (1995) Synthetic Hydrogen Bonding Receptors as Models of Transacylase Enzymes, *Tetrahedron* **51**, 435-448.

12. Cram, D.J. and Cram, J. M. (1995) *Container Molecules and Their Guests*, The Royal Society of Chemistry, Cambridge, UK.

13. Cacciapaglia, R., Casnati, A., Mandolini, L. and Ungaro, R. (1992) The Barium(II) Complex of *p-tert*-Butylcalix[4]arene-crown-5: A Novel Nucleophilic Catalyst with Transacylase Activity, *J. Am. Chem. Soc.* **114**, 10956-10958.

14. Bracchini, C., Cacciapaglia, R., Casnati, A., Mandolini, L. , Spadola, F. and Ungaro, R. (1995) Supramolecular Complexes as Catalysts with Transacylase Activity, in G. Savelli (ed), *Macrocyclic and Supramolecular Chemistry in Italy*, Centro Stampa Università degli Studi, Perugia, pp. 257-264.

15. Cacciapaglia, R., Casnati, A., Mandolini, L., Schuurman, R. and Ungaro, R., manuscript in preparation.

16. Perla, B. (1995) Dissertation, Università la Sapienza.

A STUDY OF THE CATALYTIC PROPERTIES OF SOME HYDROGEN BONDING RECEPTORS

C. Raposo, M. Crego, M. Almaraz, M. L. Mussons, M. Martín,
V. Weinrich, M. C. Caballero, J. R. Morán
Dpto. de Química Orgánica, Universidad de Salamanca, Plaza de los Caídos 1-5, 37008 Salamanca, Spain.

Introduction

Nature shows us that complexed molecules can be much more reactive than free ones. Initially one would expect just the opposite; a molecule taking part in a large complex should react slower; fewer molecular collisions should take place and steric hindrance will probably further reduce reactivity. The right choice of substituents in the active site of an enzyme clearly overrides these facts, affording a host with strongly increased reactivity. Little is known, however, about how an artificial receptor should be built to increase the reactivity of its guests. This is essentially the main goal of this work. We shall explore small artificial receptors, attempting to maximize their catalytic properties towards a reaction by slightly changing their geometries and the nature of the catalytic groups.

C. S. Wilcox and A. D. Hamilton (eds.), Molecular Design and Bioorganic Catalysis, 87–110.

1. Malonic Acid Receptors: Decarboxylases

1.1 RECEPTORS

Malonic acid is an attractive guest for hydrogen bonding receptors owing to the many possible hydrogen bonds it can set. Additionally, a molecular receptor able to catalyze the decarboxylation of a suitably substituted aminomalonic acid[1] could provide a new approach to enantioselective α–aminoacid synthesis, rendering the project even more attractive.

Molecular models show that the readily available spacer 3,3'-diaminobenzophenone fits in well with the structure of malonic acid , making the preparation of this kind of receptors very easy (FIG. 1).

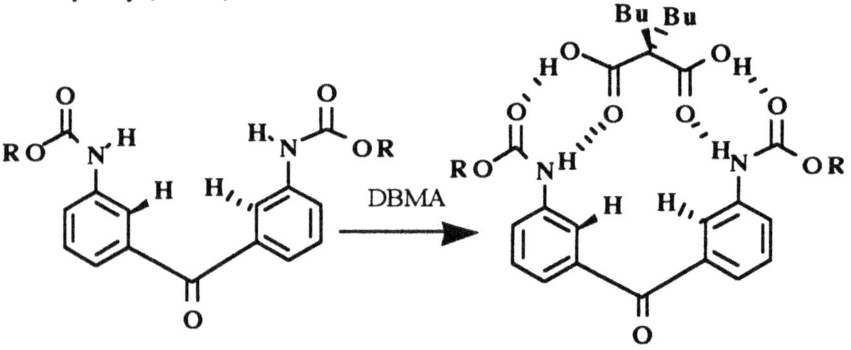

FIGURE 1. Receptor **1** and the expected structure for its complex with dibutylmalonic acid (DBMA). R= $C_{10}H_{21}$

To increase solubility in chloroform, long decyl chains were included in the structure of a first receptor **1,** in which carbamates were to provide the necessary binding arms (FIG. 1). However, we were unable to detect any reasonable complex formation with the chloroform-soluble dibutylmalonic acid (DBMA).

Urea binding arms were also tested. These functional groups yielded chloroform insoluble receptors, even when long saturated alkyl chains were included in their structures. Chloroform solubility was achieved with receptor **2** by including long chains with *cis* double bonds (FIG. 2).

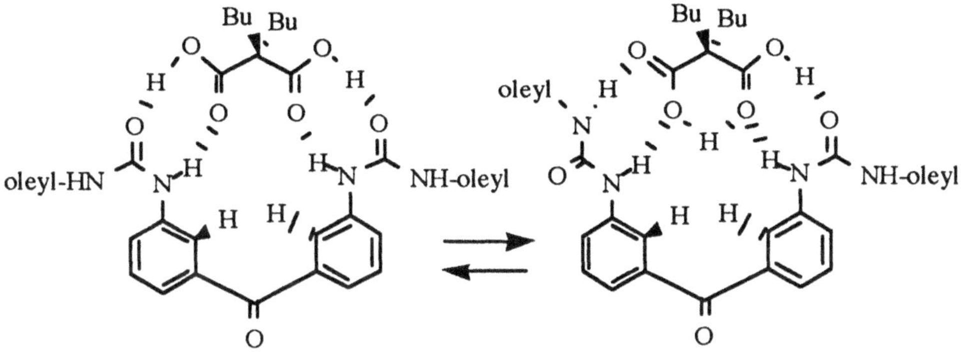

FIGURE 2. Two possible geometries for the complex of receptor **2** and DBMA

This oleyl derivative was able to complex DBMA with Kass= $1.3x10^3M^{-1}$. Two possible reasons explain why ureas are better binding arms than carbamates. First, the more negative carbonyl group of the urea allows better hydrogen bonds with the malonic acid OH. Second, in chloroform malonic acid probably shows an intramolecular hydrogen bond[2] in which only the urea receptor **2** is able to provide a conformation that fits malonic acid well (FIG. 2).

Both decanoic acid and malonic acid monoethyl ester show far smaller association constants in the unit range. Therefore, cooperativity from both binding arms takes place, showing that the benzophenone spacer is suitable for malonic acid receptors.

1.2 CATALYTIC ACTIVITY

In the next step, we attempted to check the catalytic activity of receptor **2**. Dibutylmalonic acid decarboxylates at 180°C, probably through an electrocyclic pathway. This reaction can be easily followed by measuring the amount of CO_2 evolved. Receptor **2** was essentially unable to influence the reaction rate. Furthermore, the intramolecular malonic acid hydrogen bond sets the right conformation to undergo decarboxylation and it is therefore difficult to understand what the benefit of a complex in this electrocyclic pathway would be. This is not the only way in which malonic acids can decarboxylate. It is well known that pyridine and other aromatic amines strongly accelerate ketoacid decarboxylations[3] through a mechanism in which nucleophilic addition of pyridine to one of the carbonyl groups takes place. Dibutylmalonic acid decarboxylates in the presence of pyridine at 125°C; 55°C below the electrocyclic mechanism. The following reaction sequence seems reasonable:

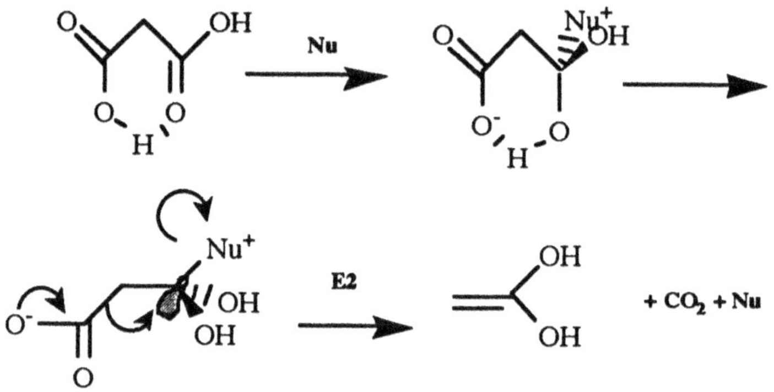

FIGURE 3. A possible mechanism for the malonic acid decarboxylation in the presence of nucleophiles

This reaction mechanism involves two main steps; one is the addition of the nucleophile to a malonic acid carbonyl, probably facilitated by intramolecular general acid catalysis. As a second step, the carboxylate undergoes like an E2 type elimination due to the presence of a good leaving group in the β position.

In order to catalyze this reaction, it is necessary to develop a complex which better associates the transition than its ground state, in agreement with the hypothesis of L. Pauling. For this, it is very important to know which is the rate limiting step of the reaction. Receptors which stabilize a low energy transition state will probably only have a minor impact on the reaction rate. In our opinion nucleophilic addition is the rate limiting step while elimination proceeds fast. This assumption can be made because oxalacetic acid decarboxylates to piruvic acid readily at room temperature[4] in the presence of pyridine, while dibutylmalonic acid requires a temperature of 125°C. It is easy to accept that nucleophilic addition to a malonic acid carbonyl must be difficult because the conjugation energy of the COOH group will be lost, while the elimination step should take place at a similar rate in both substrates.

If indeed the nucleophilic addition step has tetrahedral geometry, a receptor with a complementary tetrahedral geometry would be desirable. Sulfurylamides and phosphoramides (FIG. 4) were included in the benzophenone spacer to yield receptors **3** and **4**.

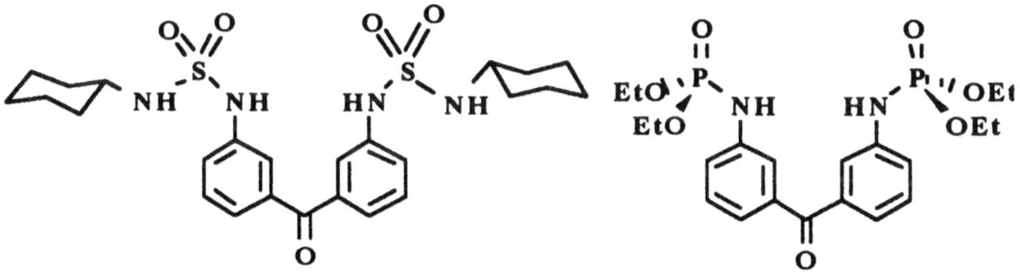

3 **4**

FIGURE 4. Structure of receptors **3** and **4**

While sulfurylamide **3** was unable to associate malonic acids, phosphoramide **4** showed a $3.3x10^{2}\,M^{-1}$ association constant with dibutylmalonic acid. Kinetic experiments also showed that receptor **4** has low but completely reproducible catalytic activity (TABLE 1) when isoquinoline was used as the nucleophile in the DBMA decarboxylation.

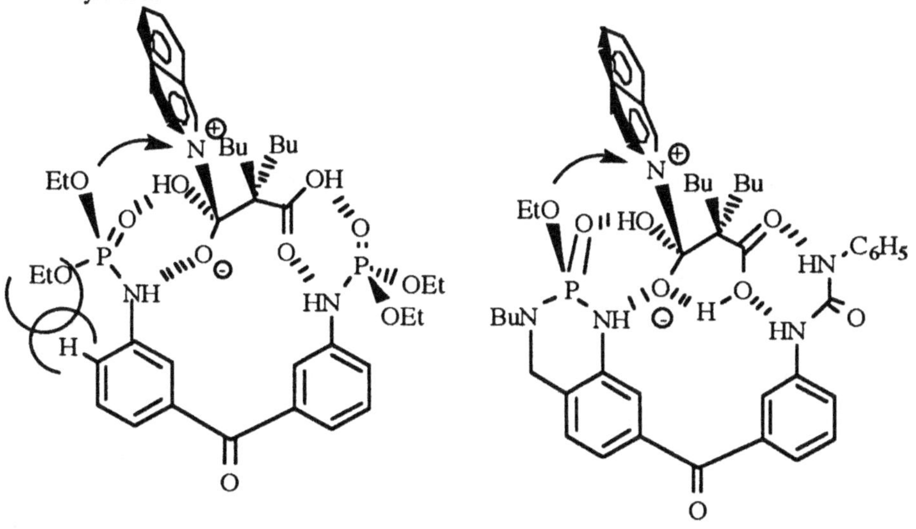

FIGURE 5. Proposed structure for the transition state complexes of receptors **3** and **9** in malonic acid decarboxylation in the presence of isoquinoline

The low catalytic activity of receptor **4** is however not surprising. For complex formation, the malonic acid hydrogen bond must be broken and therefore general acid catalysis is lost in the addition step. Furthermore, CPK models of the transition state complex indicate steric hindrance between the phosphorus ethoxy group and the aromatic ortho hydrogen (FIG. 5).

To overcome these drawbacks receptors from **5** to **9** were prepared. A phosphorinane ring sets the right conformation to complex the tetrahedral transition state without steric interferences (FIG. 5), while a urea binding arm should allow the complexation of malonic acids with the desirable intramolecular hydrogen bond (FIG. 6).

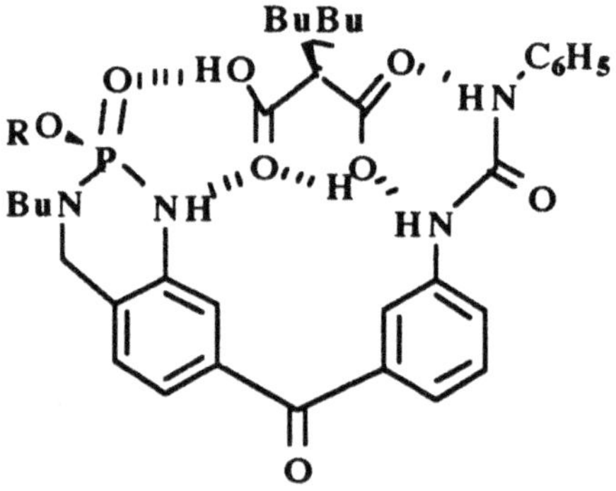

5 to **9**

Compound	R
5	Phenyl
6	Naphthyl
7	4-Methoxyphenyl
8	3-Dimethylaminophenyl
9	Ethyl

FIGURE 6. DBMA complexes with receptors **5** to **9**

The association constants of phosphorinane receptors are only slightly better than the previous ones (TABLE 2) while the catalytic activity is strongly increased. The best catalysis is performed by the ethoxy compound **9**; a 1/16 molar ratio of receptor **9** to DBMA decreases the decarboxylation $t_{1/2}$ in the presence of isoquinoline by a factor of 27 (TABLE 1). Conjugation of the nonbonding electrons of the phosphorus oxygen atom with an aromatic ring diminishes the catalytic activity (compounds **5** to **8**). The more extended aromatic system of the naphthyl ring in compound **6** lowers the catalytic effect slightly more than a simple phenyl ring (receptor **5**), while an electron-releasing oxygen substituent on the aromatic ring increases the reaction rate again in receptor **7**. The dimethylamino derivative **8** probably undergoes protonation, yielding a less active catalyst. All these data are consistent with the proposed transition state in which the interaction of the nonbonding oxygen electrons and the positive isoquinoline nitrogen (FIG. 5) reduces the transition state energy.

TABLE 1. DBMA decarboxylation half lives at 125°C in nitrobenzene with isoquinoline as the nucleophile and several receptors. DBMA/receptor ratio 16/1

Compound	none	**4**	**5**	**6**	**7**	**8**	**9**
Half life (min.)	110	75	8	10	7	18	4

TABLE 2. Association constants of receptors and DBMA in chloroform at 20°C

Receptor	**2**	**4**	**5**	**9**
Kass (M^{-1})	1.3×10^3	3.3×10^2	2.4×10^3	3.8×10^3

A less electronegative atom, such as nitrogen, could interact better with the positively charged pyridinium cation and consequently could increase the reactivity of this malonic acid. Receptor **10** (FIG. 7) does have a phosphotriamide, but it seems to be less efficient ($t_{1/2}$ = 17 min.). This can be due to the presence of the hydrogen atom on the nitrogen which introduces steric interaction in the transition state.

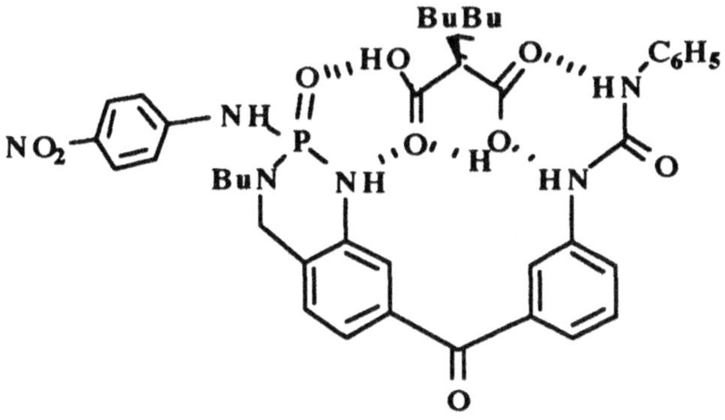

FIGURE 7. DBMA complex with receptor **10**

The association constants of the preceding receptors are not high and hence if under the reaction conditions the receptor is not completely associated, activity will be lower than expected. A new receptor **11** was designed in which an additional hydrogen bond can be set with malonic acid (FIG. 8).

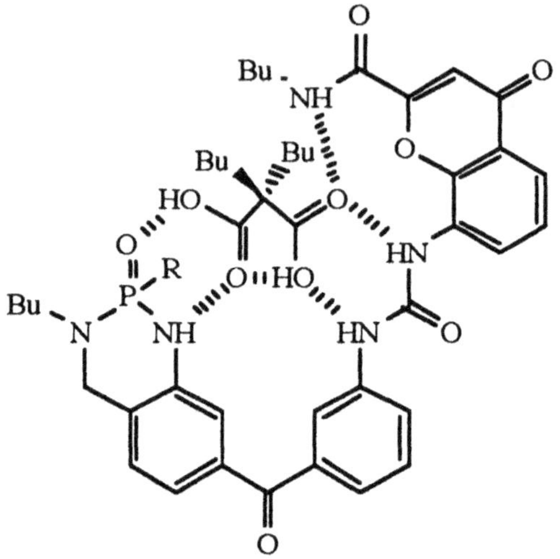

FIGURE 8. Complex for receptor **11** and DBMA

The association constant of this receptor with DBMA is probably higher than the foregoing ones. Broad signals in the ^{1}H-NMR spectrum did not allow accurate measurement. The catalytic activity is however, not higher than the previous ones ($t_{1/2}$= 5 min.), probably because the geometry of the transition state has not been improved with respect to the ground state.

An interesting aspect about this geometry was obtained when different groups were included in the urea aromatic ring. Electron withdrawing substituents on this aromatic ring like chlorine or nitro, should strengthen the urea hydrogen bonds. Since these bonds are set with the negative carboxylate group in the transition state, better activity is to be expected. Thus, receptors **12** and **13** were prepared. The association constants of **12** and **13** with dibutylmalonic acid did not change significantly with this new substitution, the more accurate kinetic studies showed that in these receptors activity was surprisingly reduced (FIG. 9).

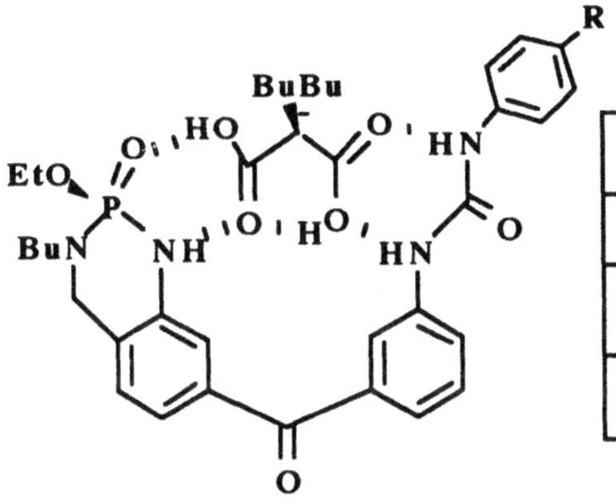

Compound	R	$t_{1/2}$ min.	Kass (M^{-1})
12	Cl	7.5	$5.0x10^3$
13	NO_2	8.0	$4.6x10^3$
14	OCH_3	4.0	$4.5x10^3$

FIGURE 9. Proposed complexes for receptors **12** to **14**. Association constants and reaction half lives under the foregoing conditions

To ascertain that there was indeed a trend between the aromatic ring substituents and catalytic activity, receptor **14**, with an electron releasing group was prepared. A methoxy group was chosen because of the possibility of dimethylamino protonation. Once again, the association constant was similar to the preceding ones (FIG. 9). Activity, however was as for the unsubstituted aromatic ring.

A possible explanation of these results is based on a non-ideal geometry of these receptors for the transition states. Molecular models show that the tetrahedral carbon in the transition state reduces the width of the malonic acid, and that a narrower cleft is necessary. If the cavity were too wide for the transition state, the urea hydrogen bond would probably be longer than in the ground state, thus handicapping the catalytic activity.

To reduce the width of the cleft receptor **15** (FIG. 10) was prepared, replacing the urea function by a sulfuryl amide. The tetrahedral sulfur atom slightly reduces the size of the cleft, now providing a NH to P=O distance of only 6.5 Å instead of the preceding 6.9 Å.

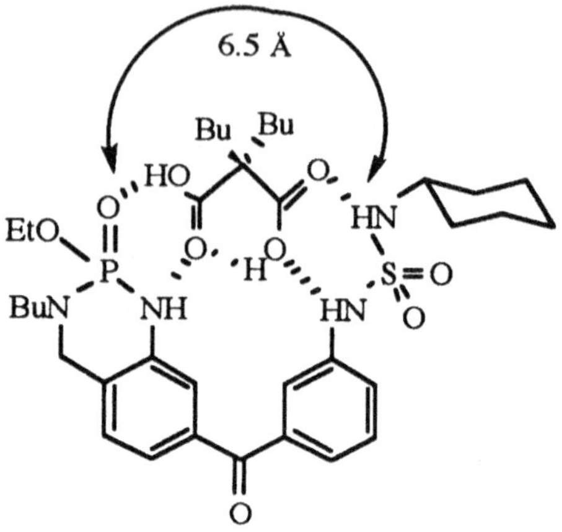

FIGURE 10. Receptor **15** and DBMA complex

The ground state association constant Kass = $1.3x10^3$ M^{-1} is smaller than the previous ones. However, catalytic activity is indeed increased, making **15** the best of these receptors with a reaction half life of 3.5 min. under the same conditions.

2. Receptors for Enolate Like Transition States

Another interesting receptor as regards catalytic activity is the xanthone derivative **16**[5] (FIG. 11). In this receptor, the dinitrotolyl residue affords a kind of second floor that is almost parallel to the aromatic ring of the guest and can therefore undergo stacking and charge-transfer interactions.

The association constants of this receptor with aromatic acids or amides rapidly increase when the aromatic ring shows electron-releasing substituents (TABLE 3). A 1000 fold increase is obtained on passing from isophthalic acid monoethyl ester to dimethylaminobenzoic acid (TABLE 3). Reactions in which a negatively charged transition state is able to reinforce the complex due to charge transfer to the dinitrotolyl residue should be accelerated by this receptor.

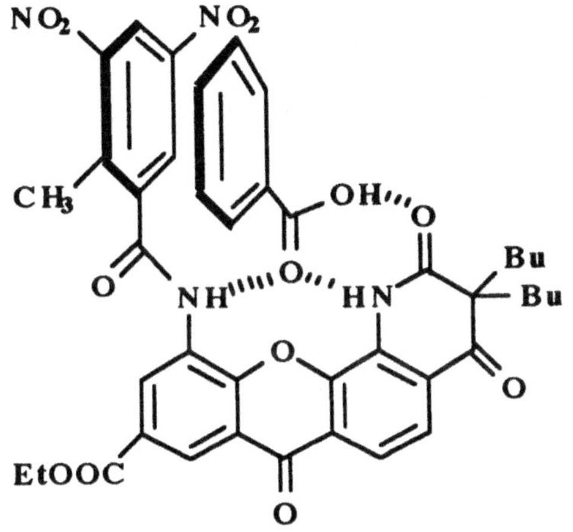

FIGURE 11. Complex of receptor **16** and benzoic acid

TABLE 3. Receptor **16** association constants with several aromatic guests in $CDCl_3$ at 20°C

GUEST	Kass (M^{-1})	GUEST	Kass (M^{-1})
3-Ethoxycarbonylbenzoic ac.	1.47×10^3	Benzamide	2.78×10^4
2-Toluic ac.	3.52×10^4	4-Ethoxybenzamide	4.63×10^4
Benzoic ac.	6.00×10^4	2,4-Diethoxybenzamide	7.75×10^4
4-Ethoxybenzoic ac.	1.53×10^5	3,4,5-Triethoxybenzamide	6.24×10^3
3,4-Methylenedioxybenzoic ac.	2.41×10^5	4-dimethylaminobenzaldehide	2.60×10^2
3,4,5-Trimethoxybenzoic ac.	2.41×10^4		
3-Dimethyaminobenzoic ac.	8.21×10^5		
4-Dimethylaminobenzoic ac.	1.56×10^6		

2.1 VINYLACETAMIDE DEUTERATION

To keep the reaction mixture as simple as possible we opted to initially study the transformation of vinylacetamide to crotonamide. This rearrangement should take place under basic conditions via the negatively charged enolate. CPK models show that the enolate geometry is suitable for undergoing charge-transfer to the dinitrotolyl group[6].

Kinetics were monitored in ^1H NMR with chloroform as solvent and DBU as the base. A slight degree of catalysis was observed, but the results were difficult to evaluate because under these conditions $CDCl_3$ is able to deuterate vinylacetamide.

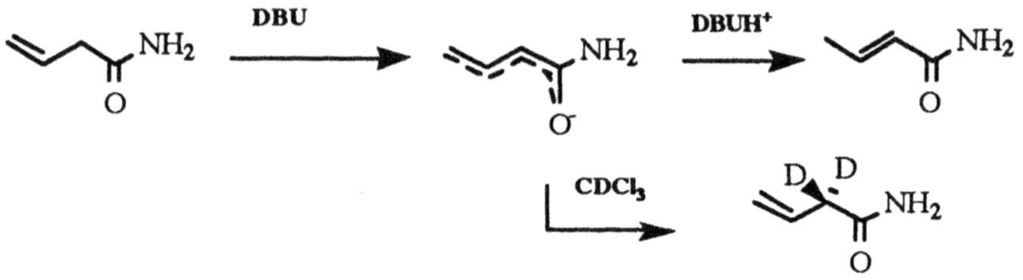

To avoid a situation in which two competing reactions occur simultaneously, 2% acetone$_{d6}$ was added to the reaction mixture. This favours the deuteration pathway, which is also expected to take place through the enolate. Under these conditions 30% of receptor **16** reduced the reaction half life from 535 to 160 minutes, while a similar receptor, lacking the nitro groups was essentially inactive ($t_{1/2}$= 530 min.).

This result is perhaps not very impressive, but it is important to note that the receptor geometry was designed to associate the amide in its ground state, not as the enolate. Therefore, the transition state complex is not much more stable than the one with the ground state. In fact, amide enolates have geometries which are quite different from their amide precursors. An X ray study showed the nitrogen atom of the enolate to be pyramidalized and to accept a strong hydrogen bond from a triMEDA unit[7]. A receptor with a suitable geometry could show this additional hydrogen bond in an enolate-like transition state and therefore should show catalysis. Receptor **17** was prepared to check this possibility.

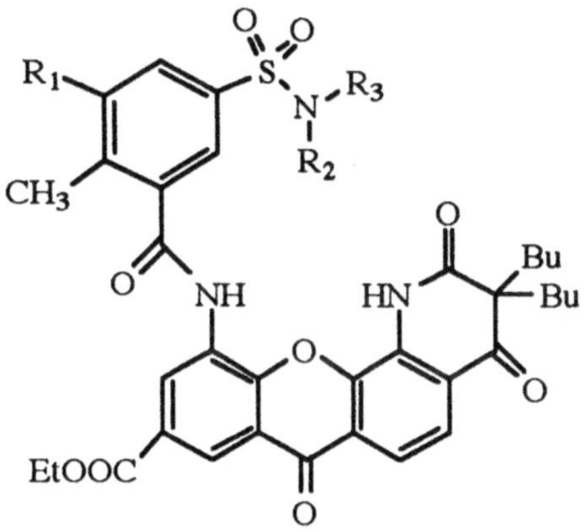

Receptor	R_1	R_2	R_3
17	H	H	Butyl
18	H	Ethyl	Ethyl
19	NO_2	H	Butyl

FIGURE 12. Structures of receptors **17** to **19**

In the complex, the sulfonamide group is close to the amide nitrogen and should be able to set a new hydrogen bond if the amide pyramidalizes. While the association constant of this receptor **17** with vinylacetamide is smaller than in previous cases (TABLE 4), it does show significant catalytic activity, reducing the reaction half life from 535 to 180 minutes. No catalysis was found for receptor **18** (FIG. 12), in which an ethyl group substitutes the sulfonamide proton, or for N-butyltoluensulfonamide. This suggests that catalysis due to complexation would indeed be feasible.

An increase in the catalytic activity can be obtained if the additional hydrogen bond is combined with the charge-transfer effect. Receptor **19** includes a nitro group in its aromatic ring (FIG. 12) and is able to reduce the reaction half life to 115 minutes.

TABLE 4. Association constants in $CDCl_3$ at 20°C and reaction half life of vinylacetamide deuteration in the presence of several receptors. Reactions were carried out in $CDCl_3$/acetone$_{d6}$ 98/2 at 20°C with DBU as the base and 30% of receptors. The reaction half life in the absence of receptor was 535 min.

Receptor	**16**	**17**	**18**	**19**
Kass (M^{-1})	$1.7x10^3$	$3.0x10^2$	$4.0x10^2$	$4.4x10^2$
Half life (min.)	160	480	180	115

2.2 DEUTERATION OF 4-METHOXY-3-PYRROLIN-2-ONE

With a stronger base such as the Schwesinger phosphacene (P1-t-Oct)[8], these receptors provide similar catalytic activities (TABLE 5).

TABLE 5. Reaction half lives of vinylacetamide and 4-methoxypyrrolinone in the presence of receptors and phosphacene P1-t-Oct

Receptor	none	**18**	**19**
Vinylacetamide $t_{1/2}$ (min.)	2050	360	260
Methoxypyrrolinone (min.)	160	110	100

In order to check other substrates, 4-methoxy-3-pyrrolin-2-one was chosen (FIG. 13). The association constants with this new substrate were found to be better than with vinylacetamide and the charge-transfer effect was readily observed (TABLE 6). However, the catalytic activities were poorer. This is easily explained if one takes into account that charge-transfer already exists in the ground state and that pyramidalization of the amide nitrogen atom must necessarily be poor due to the pyrrole enolate structure.

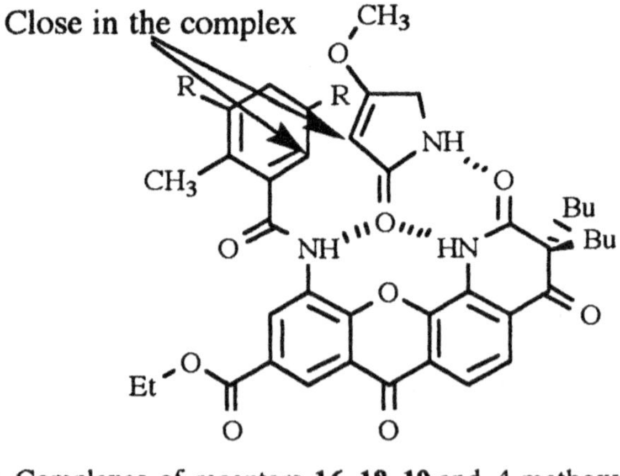

FIGURE 13. Complexes of receptors **16**, **18**, **19** and 4-methoxypyrrolinone

TABLE 6. Association constants and reaction half life in the deuteration of 4-methoxypyrrolinone with several receptors under the conditions of table 4

Receptor	none	**16**	**18**	**19**
Kass (M^{-1})		1.3×10^5	4.0×10^3	1.4×10^4
Half life (min.)	150	50	60	40

2.3 NUCLEOPHILIC ADDITION TO ACRYLAMIDE

Another interesting reaction which should take place through an enolate-like structure is the nucleophilic addition of amines to unsaturated amides. At room temperature pyrrolidine reacts readily with acrylamide in chloroform solution to yield the corresponding addition compound:

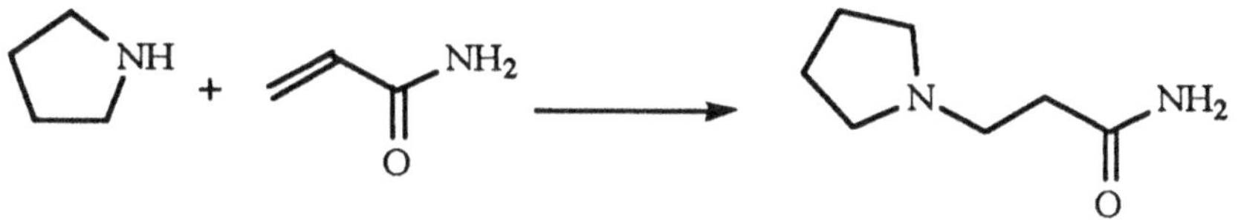

FIGURE 14. Addition of pyrrolidine to acrylamide

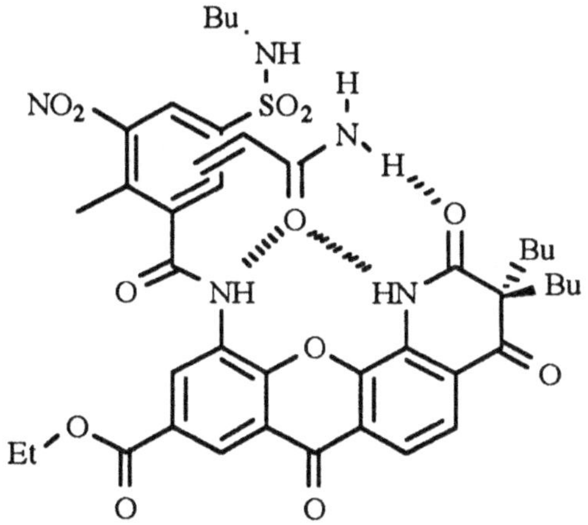

FIGURE 15. Receptor **19** and acrylamide complex

The previous receptors readily form complexes with acrylamide (TABLE 7), and they are also active in this reaction. The catalytic activities (TABLE 7) are similar to those shown in the deuteration reaction. Again, in the absence of nitro or sulfonamide groups receptor **17** shows low catalytic activity, while the best catalyst **19**, combines the sulfonamide and the nitro group in its structure.

TABLE 7. Association constants of receptors with acrylamide and reaction half lives in the presence of pyrrolidine at 20°C. Receptor 0.03 M, acrylamide 0.3 M, pyrrolidine 0.33 M in $CDCl_3$

Receptor	none	**16**	**18**	**19**
Kass (M^{-1})		2.1×10^4	2.8×10^2	3.0×10^3
Reaction half life (min.)	150	35	40	25

Another attempt to set additional hydrogen bonds with the transition state was made with receptors **20** and **21** (FIG. 16).

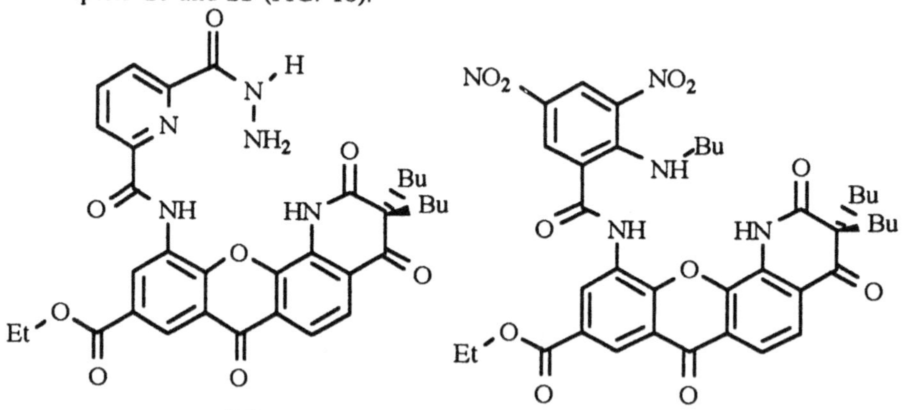

20 **21**

FIGURE 16. Structures of receptors **20** and **21**

Ester, ketone or amide enolates[9] have three non-bonding electron pairs on the carbonyl oxygen, and therefore they should be able to set three strong hydrogen bonds, while only two appear in the ground state. The amine groups in either receptor **20** or **21** are set to try to exploit this kind of hydrogen bond with the carbonyl oxygen. Both receptors associate acrylamide only weakly (TABLE 8) due to the lack of the ortho methyl group on the aromatic ring[5]. They show similar catalytic activities, reducing the reaction half life to one third of its original value, while a similar receptor **22** (FIG. 17) with a dimethylhydrazine residue displays almost no activity (150 to 130 min., TABLE 8). On the other hand activation of the hydrazide hydrogen bond as the corresponding receptor **23** again increases the reaction rate (TABLE 8) despite a very small association constant with acrylamide.Another interesting question is the effect of acidic hydrogens in the receptor. These nucleophilic additions show general acid catalysis due to pyrrolidinium ion. Either dodecanoate, acetate, benzoate or tosylate pyrrolidinium ions produce the same rate increase yielding a reaction half life of 8 minutes when used in the same amount as the receptors (10%); therefore, receptors with acidic hydrogens should also be active in this reaction. The sulfurylamide receptor **24** (FIG. 18) was however too acidic and its proton was lost in the presence of pyrrolidine. A less acidic receptor **25** was obtained by reduction of the xanthone carbonyl group. The imidazolone receptor **26** was also prepared. In both cases catalytic activities were disappointingly low.

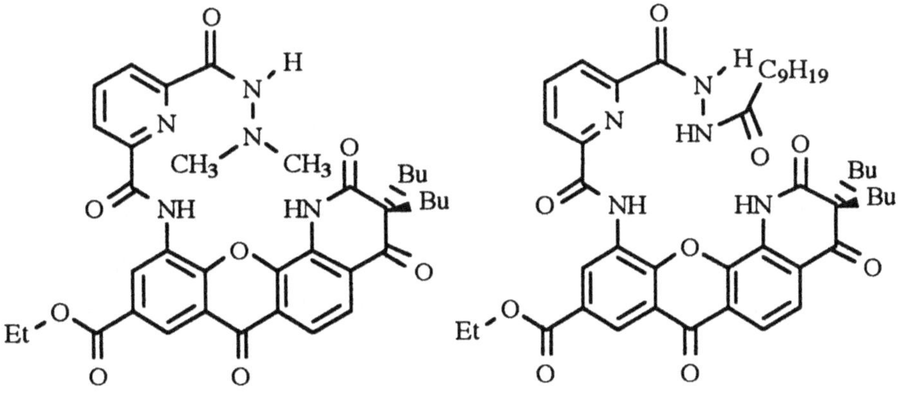

22 **23**

FIGURE 17. Structure of receptors **22** and **23**

TABLE 8. Association constants and catalytic activities of receptors **20-23** under the conditions shown in table 7

Receptor	none	**20**	**21**	**22**	**23**
Kass (M^{-1})		$2.2x10^2$	$2.6x10^2$	$1.5x10^2$	$1.0x10^1$
Half life (min.)	150	50	50	130	25

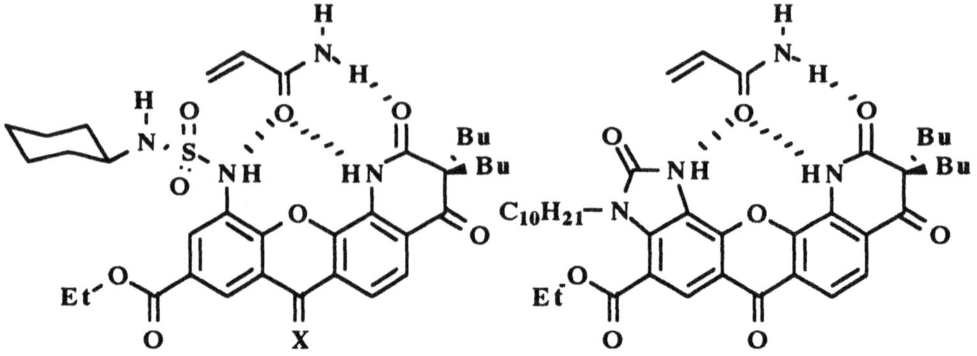

X=O, **24**
X=H_2, **25** $t_{1/2}$= 42 min. **26** $t_{1/2}$= 29 min.

FIGURE 18. Structures and catalytic acitivities of receptors **24** to **26**

2.4 NUCEOPHILIC ADDITION TO 2-(*5H*)-FURANONE

In our opinion the foregoing catalytic systems have a drawback, on going to the transition state two of the hydrogen bonds in which the receptor acts as a donor will be stronger, but the third hydrogen bond in which the receptor acts as acceptor will be weaker, compensating somehow for the gain in stability. Another possibility to explain the small catalytic properties could be an early proton transfer from the nitrogen to the amide carbonyl during the reaction coordinate. It would be appropriate to find a system in which no such proton transfer is possible for geometrical reasons. Due to these facts we decided to study the addition of pyrrolidine to 2-(*5H*)-furanone[10]. In this reaction (FIG. 19) the nitrogen atom is always far from the lactone carbonyl, making proton transfer difficult.

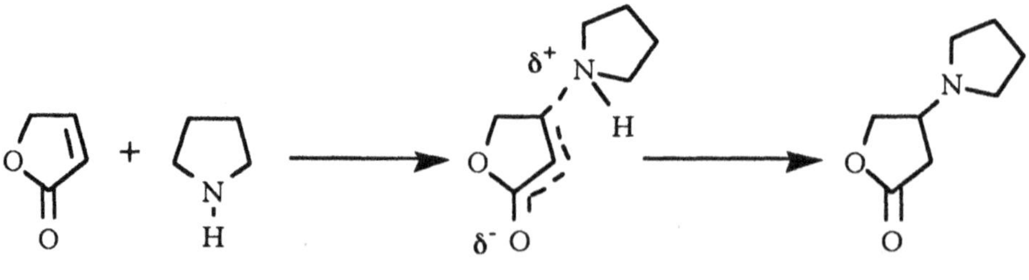

FIGURE 19. Reaction between pyrrolidine and a butenolide

Here, receptors to complex the butenolide should have three hydrogen bond donors. All these hydrogen bonds should be stronger on going to an enolate-like transition state. A new receptor **27** was prepared, this time based on a chromenone skeleton.

Receptor **27** is able to weakly associate butyrolactone in $CDCl_3$ (Kass = 30 M^{-1}). However, this aromatic sulfuryl amide is not compatible with pyrrolidine owing to its high acidity. Accordingly the chromenone skeleton was reduced[11] to the corresponding benzopyrane to yield receptor **28** (FIG. 21). **28** forms only very weak complexes with butenolides in chloroform, and therefore benzene was used as the reaction solvent.

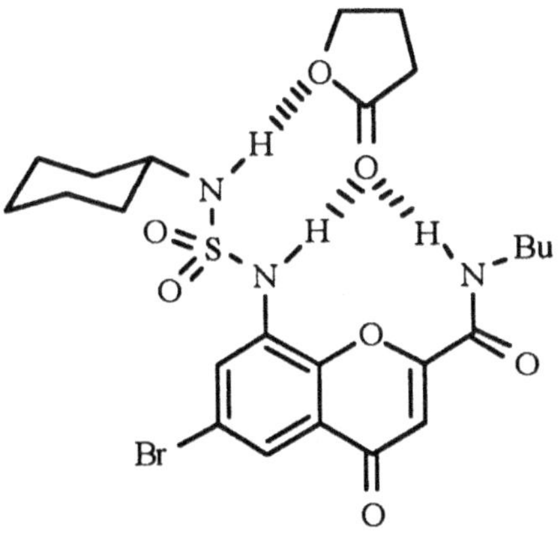

27

FIGURE 20. Butyrolactone and receptor **27** complex

A low degree of catalysis was found when a 10% amount of the receptor was added to the reaction mixture of 2-(5*H*)-furanone and pyrrolidine (TABLE 9). Once again, with the malonamide derivative **29** we tried to set an additional hydrogen bond in the transition state. This receptor shows a very weak association constant with the butenolide even in benzene, but the catalytic activity was better than with the foregoing receptor **28**. Probably, the malonamide forms an intramolecular hydrogen bond which hinders complex formation. In order to weaken this intramolecular hydrogen bond and to obtain more acidic hydrogens, the thioamide **30**[12] was prepared. In this case, both the association constant and the catalytic activity were improved (TABLE 9).

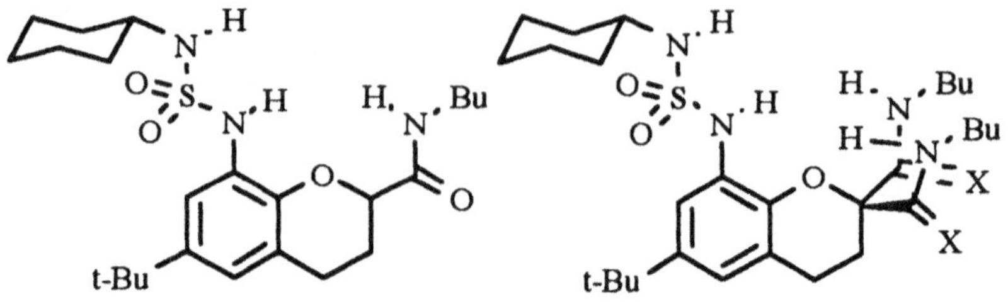

28 X=O **29** **30** X=S

FIGURE 21. Structure of receptors **28, 29** and **30**

TABLE 9. Association constants of receptors **28, 29** and **30** with 2-(*5H*)-furanone in benzene and with tetrabutylammonium benzoate in DMSO. Reaction half lives of the reaction of furanone and pyrrolidine in the presence of receptors

Receptor	none	**28**	**29**	**30**
Furanone (benzene) (M^{-1})		68	4	19
Benzoate (DMSO) (M^{-1})		41	96	210
Half life (min.)	120	51	22	16

The stability of the complexes between the previous receptors and the butenolide enolate cannot be easily measured due to the presence of acidic protons on the receptors. To overcome this inconvenience tetrabutylammonium benzoate was used as an enolate analog and the results showed that indeed the catalytic activities are parallel to the benzoate association constants (TABLE 9).

Some experiments indicated that chromenone receptors were more promising than the benzopyrane ones, and that it would be better to overcome the problem of the acidity of the sulfuryl amide by changing to another functional group. Like sulfurylamides ureas can act as double hydrogen bond donors. CPK models predict that simple ureas will lead to receptors which strongly dimerize and will therefore not associate butenolides:

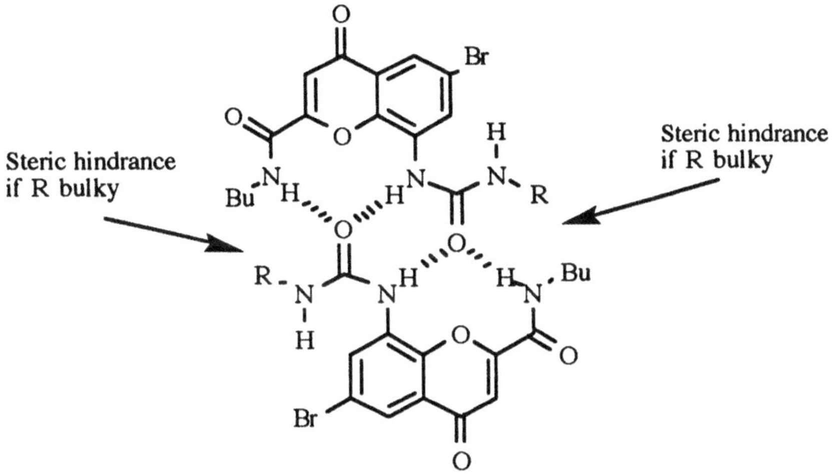

FIGURE 22. Structure of the fourfold hydrogen bonded dimer of a chromenone receptor with a urea side chain

A very bulky substituent on the urea nitrogen can make self-association difficult due to steric hindrance with the butylamide residue.

Two bulky amines were selected to prepare lactone receptors; t-octylamine, to yield receptor **31**, and trityl amine, which leads to receptor **32**.

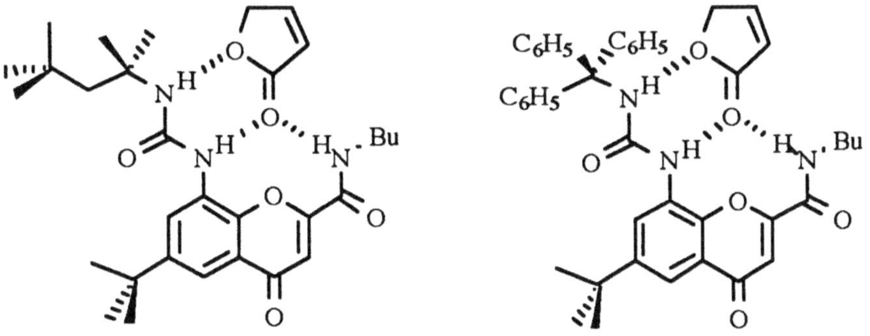

31 **32**

FIGURE 23. Complexes of receptors **31** and **32** with 2-(*5H*)-furanone

The t-octyl derivative **31** was selected because of its higher $CDCl_3$ solublility. **31** shows a Kass = 80 M^{-1} with 2-(*5H*)-furanone and has better catalytic properties than the previous benzopyranes, reducing the half life of the addition reaction from 130 to only 6 minutes. This result is very encouraging because in this system acid catalysis due to pyrrolidinium ions (tosylate) yields a reaction half life of 40 min.; this time catalysis due to the receptor is more efficient.

Stronger hydrogen bonds in the transition state have been postulated to be responsible for most of the catalytic power of enzymes[13]. It is thus reasonable to increase the acidity of the receptor hydrogens to obtain better reaction rates.

Compounds **33** and **34** were prepared to increase the acidity of the amide. The nitroaniline derivative reduces $t_{1/2}$ to 4 min. while the isophthalic analog reaches the reaction half life in only 2 minutes.

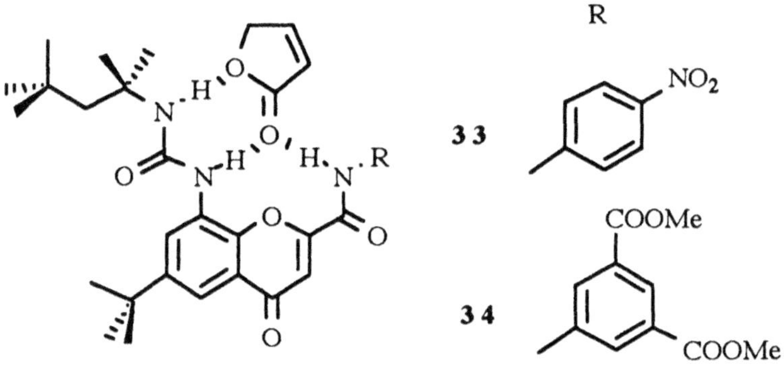

FIGURE 24. Complexes of receptors **33**, **34** with 2-(*5H*)-furanone

Receptors in which complexation is difficult, due to the lack of a hydrogen bond, as in ester **35**, or due to the presence of intramolecular hydrogen bonds, as in stuctures **36** to **38**, show a greatly reduced activity, supporting catalysis through complexation.

In the future, based on this chromenone skeleton, we hope to be able to develop chiral receptors able to induce high enantioselectivities in nucleophilic additions to butenolides.

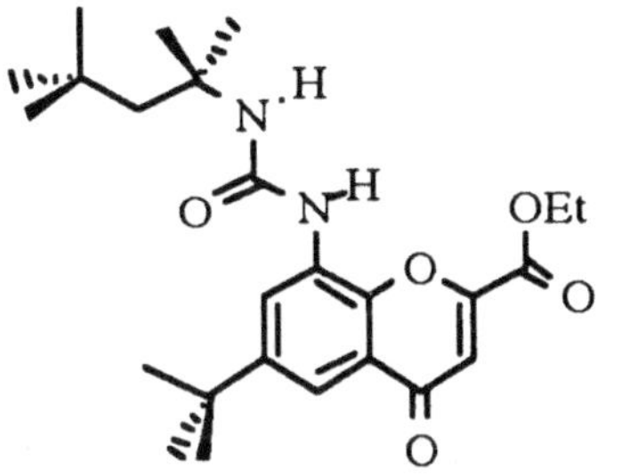

35 Half-life = 70 min.

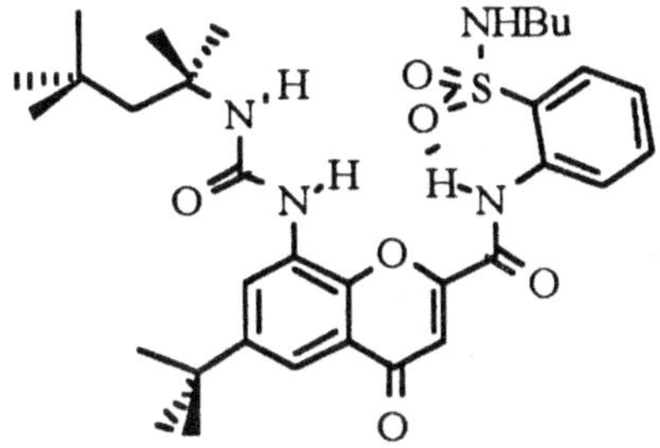

36 Half-life = 60 min.

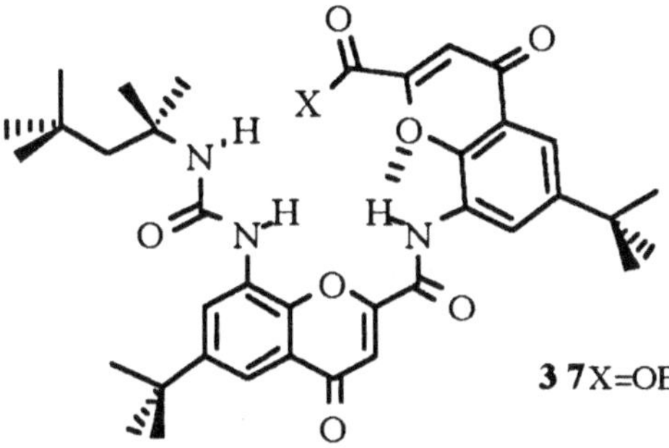

Half-life

37 X=OEt, 23 min. **38** X=NHBu, 19 min.

FIGURE 25. Receptors with a reduced catalytic activity

Acknowledgments: We thank the "Dirección General de Investigación Científica y Técnica" (DGICYT Grant PB 92-0286) for its support of this work. Three of us (M.C., M.L.M. and C.R.) also thank the MEC for a fellowship.

References

1. Snyder, H. R., Shekleton, J. F. and Lewis, C. D. (1945) Synthetic amino acids. Synthesis from acetamidomalonic ester, *J. Am. Chem. Soc.* **67**, 310-312.
2. Eberson, L. (1969) in The chemistry of carboxylic acids and esters; S. Patai; Ed. Wiley-Interscience. London, pp 211-293.
3. Henecka, H. (1952) in Houben-Weyl. Methoden der organischen Chemie, Ed. E. Muller, Vol. VIII, pp 490-494, Georg Thieme Verlag, Stuttgart .
4. Wohl, A. (1907) Über Oxyfumar und Oxymaleinsäure, *Chem. Ber.* **40**, 2282-2293.
5. Crego, M., Raposo, C., Caballero, C., García, E., Saez, J. and Morán, J. R.

(1992) A receptor for aromatic acids and amides, *Tetrahedron Lett.* **33**, 7437-7440.

6. For an interesting discussion on the effect of π complexes on reactivity see: Bender, M. L., Bergeron, R. J. and Komiyama, M. (1984) The bioorganic chemistry of enzymatic catalysis, J. Wiley and Sons, New York.
7. Laube, T., Dunitz, J. D. and Seebach, D. (1985) Über die Wechselwirkung zwischen Lithium-enolaten und sekundaren Aminen in Lösung und im Kristall, *Helv. Chim. Acta* **68**, 1373-1393.
8. Schwesinger, R. and Schlemper, H. (1987) Extremly strongly neutral nitrogen bases peralkylated polyaminophosphacenes, *Angew. Chem., Int. Ed. Engl.* **26**, 1167-1169.
9. Williard, P. G. and Hintze, M. (1987) The first structural characterization of a dimeric lithium ketone enolate-lithium diisopropylamide complex, *J. Am. Chem. Soc.* **109**, 5539-5541.
10. Faber, W. S., Ben de Lange, J. K. and Feringa, B. L. (1994) Catalytic kinetic resolution of 5-alkoxy-2(*5H*)-furanones, *Tetrahedron* **50**, 4775-4794.
11. Hill, J. and Ramage, G. R. (1965) Chromeno-oxazines, *J. Chem. Soc.* **87**, 7348-7351.
12. Fan, E., Van Arman, S. A., Kincaid, S. and Hamilton, A. D. (1993) Molecular Recognition: Hydrogen-bonding receptors that function in highly competitive solvents, *J. Am. Chem. Soc.* **115**, 369-371.
13. a) Davidson, M. M., Gould, I. R. and Hillier, I. H. (1995) Contribution of transition-state binding to the catalytic activity of Bacillus subtilis chorismate mutase, *J. Chem. Soc. Chem. Commun.* 63-64. b) Cleland, W. W. and Kreevoy, M. M. (1994) Low-barrier hydrogen bonds and enzymic catalysis, *Science* **264**, 1887-1890. c) Frey, P. A., Whitt, S. A. and Tobin J. B. (1994) A low-barrier hydrogen bond in the catalytic triad of serine proteases, *Science* **264**, 1927-1930.

DESIGN, SYNTHESIS, AND STUDY OF ARTIFICIAL β-SHEETS

JAMES S. NOWICK, IN QUEN LEE,
GILBERT MACKIN, MASON PAIRISH,
A. J. SHAKA, ERIC M. SMITH, and JOSEPH W. ZILLER
University of California, Irvine
Irvine, CA 92717-2025
USA

With the goal of developing models with which to study β-sheet structure, we are creating a set of building blocks to induce β-sheet formation in peptide derivatives. We have developed oligourea "molecular scaffolds" to hold peptide strands in proximity, and thus facilitate β-sheet formation. We have created a second template, based upon 2-methoxy-5-hydrazinobenzoic acid, to mimic a β-strand conformation and to induce this conformation in an adjacent peptide strand. This paper discusses the design, synthesis, and structural studies of the oligourea and hydrazinobenzoic acid templates, and the application of these templates to the creation of artificial β-sheets.

1. Introduction and Background

β-Strands and β-sheets play a significant role in the structure and biological activity of many peptides and proteins. Almost all proteins contain β-sheets as key structural elements. Proteolytic enzymes, such as HIV-1 protease and renin, bind and cleave peptide substrates in β-strand conformations. HIV-1 protease is, itself, a dimer held together in part by interdigitating β-strands. β-Amyloid plaques, which are found in the brains of individuals with Alzheimer's disease, are aggregates of polypeptide β-sheets.

In a β-strand, a polypeptide chain adopts an extended conformation with a pleated structure (Figure 1). The side-chains of the amino acids lie along the faces of the β-strand, and the edges of the chain offer an alternating array of hydrogen-bond-donor and hydrogen-bond-acceptor functionality. β-Strands can hydrogen bond to each other to form β-sheets (Figure 2). If the peptide strands are oriented in a parallel fashion, a series of twelve-membered hydrogen-bonded rings is formed. If the peptide strands are oriented in an antiparallel fashion, an alternating array of ten- and fourteen-membered hydrogen-bonded rings is formed.

Recently, there has been considerable interest in molecules that mimic the structures of peptides and proteins or act as templates to induce desired conformations

C. S. Wilcox and A. D. Hamilton (eds.), Molecular Design and Bioorganic Catalysis, 111–136.

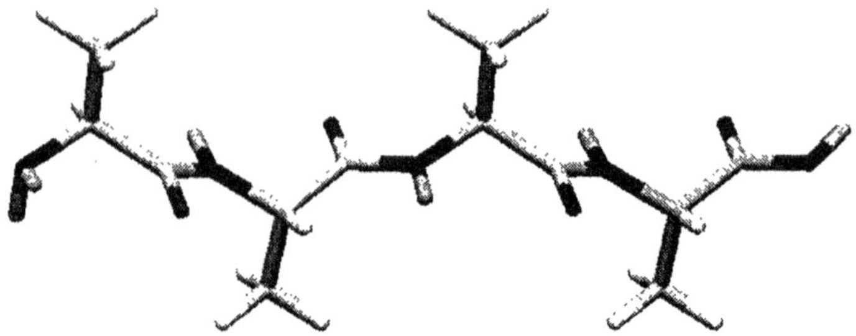

Figure 1. A peptide in a β-strand conformation.

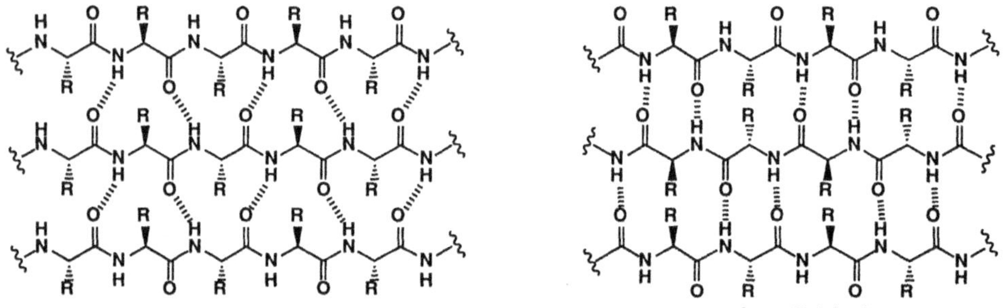

Figure 2. Parallel and antiparallel β-sheets.

in peptide derivatives.[1] Compounds with these features are often termed *peptidomimetics*. Peptidomimetics have become important targets for synthetic, structural, and biological studies, because they offer the potential for the development of pharmacologically active analogs of endogenous peptides and proteins. Peptidomimetic analogs of enzyme substrates or transition states can function as enzyme inhibitors; peptidomimetic analogs of biological ligands can serve as agonists or antagonists. With the growing role of combinatorial chemistry in drug discovery, peptidomimetic compounds are likely to play an increasing role in medicinal chemistry, since many combinatorial strategies rely upon the use of peptides or peptidelike compounds. Fundamental insights into the factors affecting protein structure and stability may also be achieved through the development of peptidomimetic compounds that serve as chemical models with which to study protein structure. An enhanced understanding of the factors affecting the three-dimensional structures of proteins would improve the prediction of the structure and function of a protein from its sequence of amino acids or the sequence of the corresponding gene.

Peptidomimetics that duplicate various aspects of protein secondary structure have been developed as mimics and templates of α-helices, β-turns, β-strands, and β-sheets. Noteworthy examples of β-strand mimics include epindolidione derivatives,[2] 1,2,3-trisubstituted cyclopropanes,[3] 3,5-linked pyrroline-4-ones,[4] and vinylogous polypeptides[5] (Figure 3). Epindolidione derivatives duplicate the array of hydrogen-bond-donor and acceptor functionality of one edge of a tripeptide. 1,2,3-Trisubstituted cyclopropanes provide conformationally-constrained placement of peptide side-chains in similar orientations to those found in β-strand enzyme substrates. 3,5-Linked pyrroline-4-ones mimic the main-chain and side-chain conformations of peptide β-strands with a high degree of precision. In these compounds the NH groups are displaced from their normal position in a peptide backbone and are situated so that they can intramolecularly hydrogen bond to an adjacent carbonyl group. Vinylogous polypeptides have an elongated "main-chain", which adopts a conformation similar to that of a β-strand and provides a pattern of hydrogen-bonding functionality similar to that of a peptide. None of these β-strand mimics can be prepared as linear oligomers of varying lengths that duplicate the pattern of hydrogen-bond-donor and acceptor groups of one edge of a peptide β-strand.

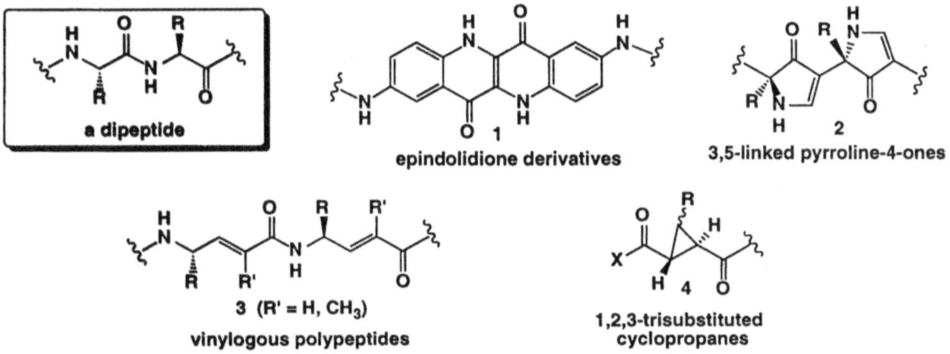

Figure 3. β-Strand mimics: conformationally-constrained molecules similar in structure to β-strands.

Several chemical models for two-stranded β-sheets that offer the promise of enhanced insight into protein folding have been developed. Leading examples are shown in Figure 4. Kemp and coworkers have used the epindolidione β-strand mimic in conjunction with a β-turn unit and a peptide strand to create artificial antiparallel β-sheets (**5**).[2] Parallel β-sheets based upon this template have also been described.[6] Although these antiparallel and parallel β-sheet mimics ostensibly resemble four-stranded β-sheets, they may better be described as C_2 symmetric "dimers" of isolated two-stranded antiparallel β-sheets. Feigel and coworkers have employed several aromatic templates to hold two peptide strands in proximity and stabilize antiparallel and parallel β-sheet structures.[7] Macrocyclic parallel β-sheet mimic **6** provides an example of one such structure.[7e] Kelly and coworkers have extensively studied antiparallel β-sheets in which a dibenzofuran template holds two peptide strands in proximity, while hydrophobic interactions between the template and peptide side-chains help stabilize β-sheet formation.[8] A number of other templates capable of holding two peptide strands in proximity and inducing β-sheet structure have also been reported.[9] Mutter and coworkers introduced a related approach in which a polypeptide template is used to bring structurally biased polypeptide strands into proximity.[10] Detailed structural studies of artificial β-sheets based upon this approach are lacking, however, and it is unclear whether these compounds adopt well-defined hydrogen-bonded β-sheet structures. Complementary to these strategies, which rely upon templates to help induce β-sheet structure, are the use of small naturally occurring β-sheet proteins[11] and de novo designed proteins[12] as systems in which to study β-sheet structure and stability.

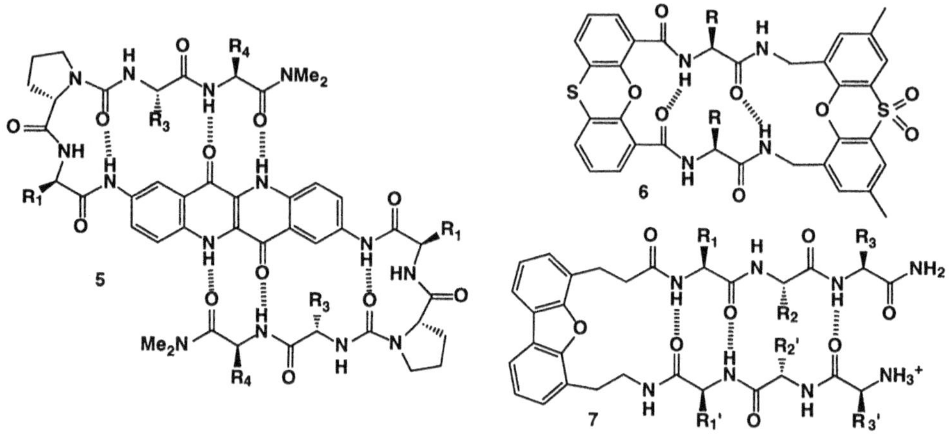

Figure 4. Artificial β-sheets: compounds in which synthetic templates induce β-sheet structure.

A few years ago, we became interested in creating synthetic chemical models for β-sheets. We set out to develop a strategy for the creation of artificial β-sheets larger than those shown in Figure 4, in which the peptide strands were not limited in length or in number. The benefits of this research would include the invention of new peptidomimetic building blocks, which might be of value in drug development, and an enhanced understanding of the factors that affect protein structure and stability.

We conceived a strategy in which two complementary templates, an oligourea "molecular scaffold" and a 5-hydrazinobenzoic acid β-strand mimic, organize attached peptide strands (Figure 5). The molecular scaffold is designed to hold peptide strands in proximity, while the β-strand mimic is designed to stabilize a β-strand conformation in an adjacent peptide strand and prevent fraying of the artificial β-sheet. The β-strand mimic should also minimize the propensity of the artificial β-sheets to aggregate and help maintain the solubility of these structures, because it bears hydrogen-bonding functionality along only one of its edges. Intramolecular hydrogen bonding helps provide the templates with well-defined structures, while minimizing the synthetic efforts required to create large well-structured molecules. The cartoons shown in Figure 5 illustrate the application of these templates to the creation of artificial β-sheets containing multiple peptide strands.

This paper provides an account of our ongoing efforts to implement this strategy: Section 2 details the development of the molecular scaffold. Section 3 describes the creation of a small parallel artificial β-sheet. (We chose to focus initially upon parallel β-sheets because these are less well understood than antiparallel β-sheets.) Section 4 reports the development of the β-strand mimic. Section 5 tells of our preliminary efforts to combine the molecular scaffold, the β-strand mimic, and peptide strands.

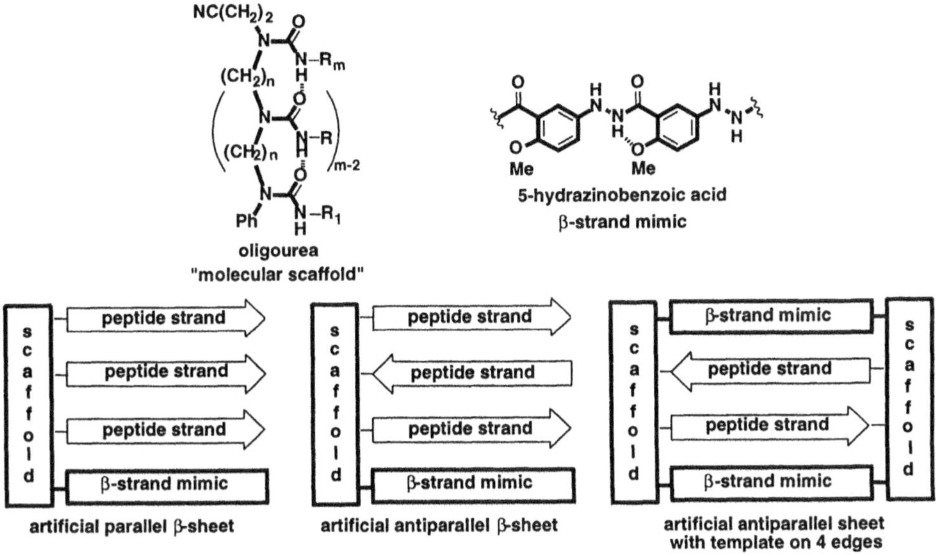

Figure 5. A unified approach to the creation of artificial β-sheets.

2. Structural Studies of Diurea Molecular Scaffolds[13]

In designing the oligourea molecular scaffold we have focused our efforts on urea derivatives of oligomers of 1,3-diaminopropane and 1,2-diaminoethane, with the following rationale: Urea derivatives of 1,3-diaminopropane can form ten-membered hydrogen-bonded rings. Hydrogen-bonded rings of this size are often present in β-turns, and therefore might be particularly well suited to juxtaposing peptide strands to form β-sheets. In contrast, the 1,2-diaminoethane derivatives can form nine-membered hydrogen-bonded rings, which should be less entropically unfavorable and more conformationally homogeneous. To this end, work by Gellman and coworkers indicated nine-membered hydrogen-bonded rings formed by certain di- and triamides to be especially stable.[14] Molecular modeling raised concerns, however, that the nine-membered rings of 1,2-diaminoethane diureas would orient the peptide strands in a divergent fashion, thus preventing the formation of hydrogen-bonded β-sheets. With these considerations in mind, we turned to experiment to determine which oligoamine backbone was best suited for the molecular scaffold.

We began by preparing a number of simple urea derivatives (e.g., **8**-**14**) and studying their structures by infrared (IR) and proton magnetic resonance (^{1}H NMR) spectroscopy and by X-ray crystallography. In the NH stretching region of the IR spectrum in chloroform solution, 1,3-propanediamine diurea **8** exhibits bands at 3483, 3460, and 3325 cm^{-1} (Figure 6). The peak at 3460 cm^{-1} corresponds to the NH stretch of the bottom half of the molecule. This assignment is evident from the position and shape of the peak, which are similar to those of monourea **14** (Figure 6). The peaks at 3483 and 3325 cm^{-1} arise from NH stretches of the top half of the molecule. The peak

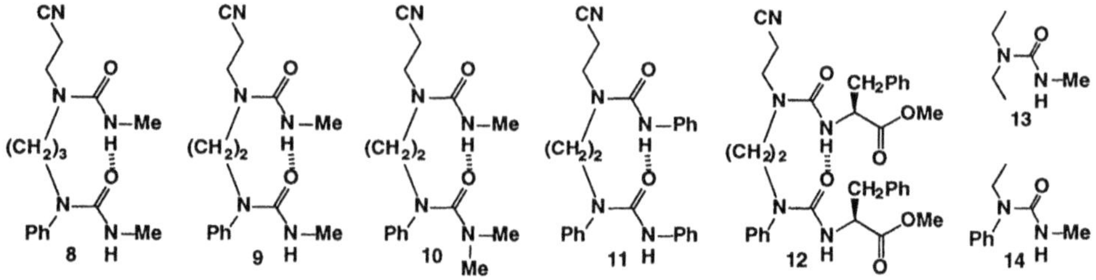

at 3483 cm^{-1} comes from non-hydrogen-bonded conformers and is similar in position to that of monourea **13** (Figure 6); the broad peak at 3325 cm^{-1} comes from hydrogen-bonded conformers. Whereas 1,3-propanediamine diurea **8** is a mixture of hydrogen-bonded and non-hydrogen-bonded conformers, 1,2-diaminoethane diurea **9** is largely or wholly hydrogen bonded. Thus, diurea **9** has peaks at 3456 and 3338 cm^{-1} in the infrared spectrum, corresponding to the free NH stretch of the "lower" urea group and the hydrogen-bonded NH stretch of the "upper" urea group, respectively (Figure 6). The absence of a peak at ca. 3487 cm^{-1} indicates the absence of a significant population of non-hydrogen-bonded conformers. When the NH group of the bottom half of **9** is replaced with a methyl group, only a band at 3340 cm^{-1} is seen (compound **10**, Figure 6). This observation provides further evidence that simple urea derivatives of 1,2-diaminoethane are fully hydrogen bonded in chloroform solution.

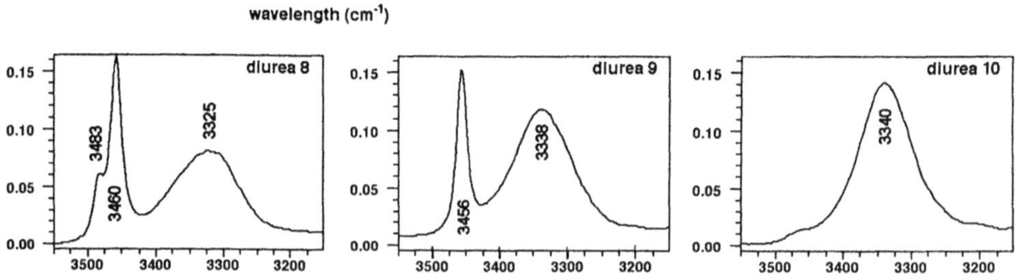

Figure 6. Absorption IR spectra of diurea derivatives of 1,3-diaminopropane and 1,2-diaminoethane (10 mM in $CHCl_3$, 295 K).

X-ray crystallography shows that 1,2-diaminoethane diureas **10, 11**, and **12** form hydrogen-bonded nine-membered ring structures in the solid state (Figure 7). In these structures, the 1,2-diaminoethane backbone adopts an anti conformation with a 159-169° N–C–C–N dihedral angle. The phenyl group that is attached to the

1,2-diaminoethane backbone is s-trans to the carbonyl of the "lower" urea group. (In designing the molecular scaffold, we had incorporated the phenyl group with the goal of orienting the carbonyl groups and controlling interactions between attached peptide strands.[15])

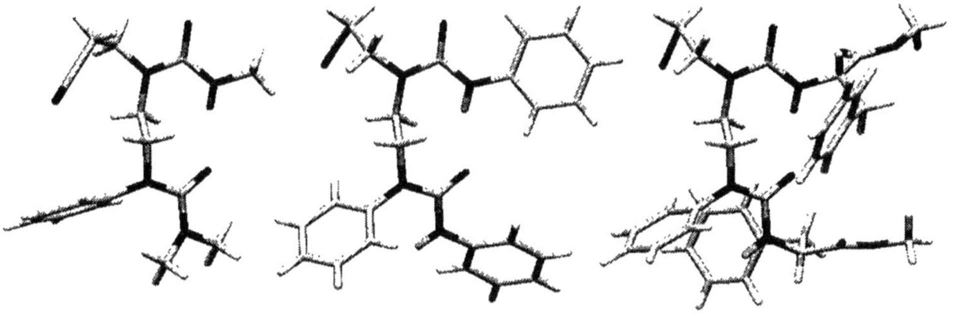

Figure 7. X-ray crystallographic structures of diureas **10**, **11**, and **12**.

^{1}H NMR spectroscopy indicates that the conformation of 1,2-diaminoethane diureas in chloroform solution is similar to the solid-state conformation. Chemical shift studies corroborate the IR spectroscopic studies, indicating that the ureas are hydrogen bonded and the carbonyl groups are directed "upwards" by the phenyl group on the 1,2-diaminoethane backbone. Thus, the "upper" and "lower" NH resonances of diurea **9** appear at 6.81 and 4.25 ppm, respectively, while the NH resonances of reference compounds **13** and **14** appear at 4.32 and 4.06 ppm. The "upper" NH resonance of **9** is shifted downfield by ca. 2.5 ppm with respect to **13**, while the "lower" NH resonance exhibits no significant difference in chemical shift from that of **14**. The vicinal coupling constants, determined by analyzing the non-first-order multiplets associated with the 1,2-diaminoethane backbone, indicate that the diurea derivatives adopt an anti conformation. By comparing the AA'BB' pattern associated with 1,2-diaminoethane backbone of **11** to simulated spectra with various coupling constants, we determined J_{AB} and $J_{A'B'}$ to be 5 Hz, and $J_{AB'}$ and $J_{A'B}$ to be 10 Hz (Figure 8). These coupling constants were compared to those calculated for models in which the 1,2-diaminoethane backbone adopted either a gauche or an anti conformation. The gauche conformation was considered to be a racemic mixture of two rapidly-equilibrating enantiomeric conformers with N–C–C–N torsion angles of ±50°, while the anti conformation was considered to be a racemic mixture of two rapidly equilibrating enantiomeric conformers with N–C–C–N torsion angles of ±160°. (The values of the torsion angles were calculated for hydrogen-bonded gauche and anti conformers using molecular mechanics.) Coupling constants of 5 and 6 Hz were calculated for the gauche conformer, and 4 and 11 Hz coupling constants were calculated for the anti conformer. The experimentally determined coupling constants are in reasonable agreement with those calculated for the anti conformation. These studies suggest that urea derivatives of 1,2-diaminoethane are largely or wholly in anti conformations in chloroform solution and confirm that the solution-state conformations of these compounds are similar to those observed in the crystal structures.

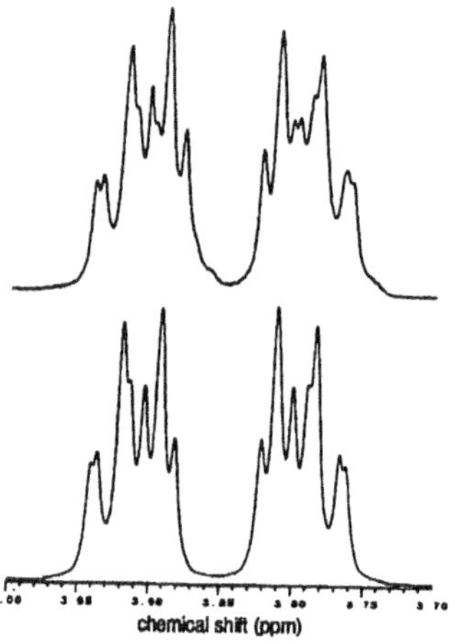

Figure 8. Experimental (upper) and simulated (lower) 300.13 MHz ^{1}H NMR spectra of the diaminoethane group of diurea **11** in $CDCl_3$. The simulated spectrum was calculated using the GEMSIM program of a General Electric GN-500 NMR spectrometer. The following values were used in the simulation: $\nu_A = \nu_{A'} =$ 1174 Hz, $\nu_B = \nu_{B'} =$ 1138 Hz, $J_{AB} = J_{A'B'} = 5$ Hz, $J_{AB'} = J_{A'B} = 10$ Hz, $J_{AA'} = J_{BB'} = -15$ Hz, Lorenzian line width = 2 Hz. The shoulders at δ 3.74 and 3.86 in the experimental spectrum are artifacts associated with the shimming of the spectrometer.

3. Synthetic and Structural Studies of Artificial Parallel β-Sheets[16]

Having established that diurea derivatives of 1,2-diaminoethane adopt hydrogen-bonded conformations that juxtapose pendant groups, we next turned our attention to determining whether these diureas could serve as templates for the creation of parallel β-sheets. To achieve this goal, we required a means by which to attach peptide strands to the 1,2-diaminoethane backbone. Ureas are readily prepared by reaction of amines with isocyanates. To prepare urea-peptide conjugates, in which peptides are coupled to the oligourea molecular scaffold, we required isocyanates of peptides.

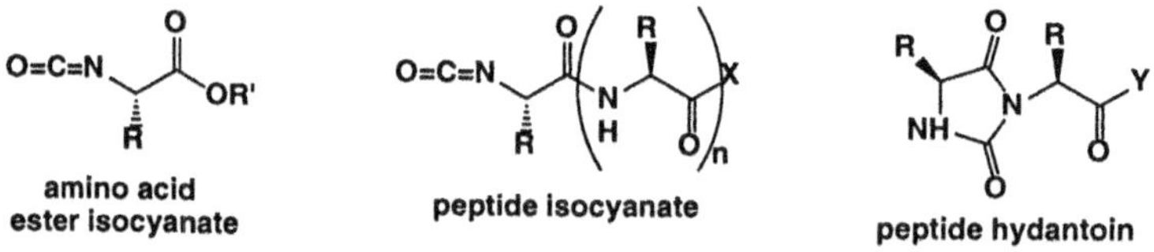

Peptide isocyanates had not previously been reported in the literature, and the stability of this class of compounds was uncertain since cyclization to the isomeric hydantoins is possible. One reported attempt to synthesize glycylglycine ethyl ester isocyanate resulted in the formation of the corresponding peptide hydantoin.[17] We had previously developed an improved method for the preparation of amino acid ester isocyanates, to facilitate the synthesis of urea-peptide conjugates (e.g., **12**).[18] By modifying this method, we were able to prepare peptide isocyanates in high yield and with little or no hydantoin formation. To prepare peptide isocyanates, a solution of phosgene in toluene is added to the corresponding peptide in a mixture of methylene chloride and aqueous sodium bicarbonate solution at 0 °C.[19]

The peptide isocyanates were used to synthesize artificial parallel β-sheet **21**, as shown in Figure 9. Michael addition of *N*-phenylethylenediamine (**15**) to acrylonitrile occurs exclusively at the aliphatic amino group to generate diamine **16**.[20] Diamine **16** reacts with dipeptide isocyanate **17** at ambient temperature with high regioselectivity to afford monourea **18**. Upon heating with dipeptide isocyanate **19**, the monourea is converted to diurea **20**. Aminolysis with methylamine converts the methyl ester groups of **20** to methylamide groups, affording **21**. We have nicknamed this compound a "2x2", because it contains *two* peptide strands that are each *two* amino acids in length.

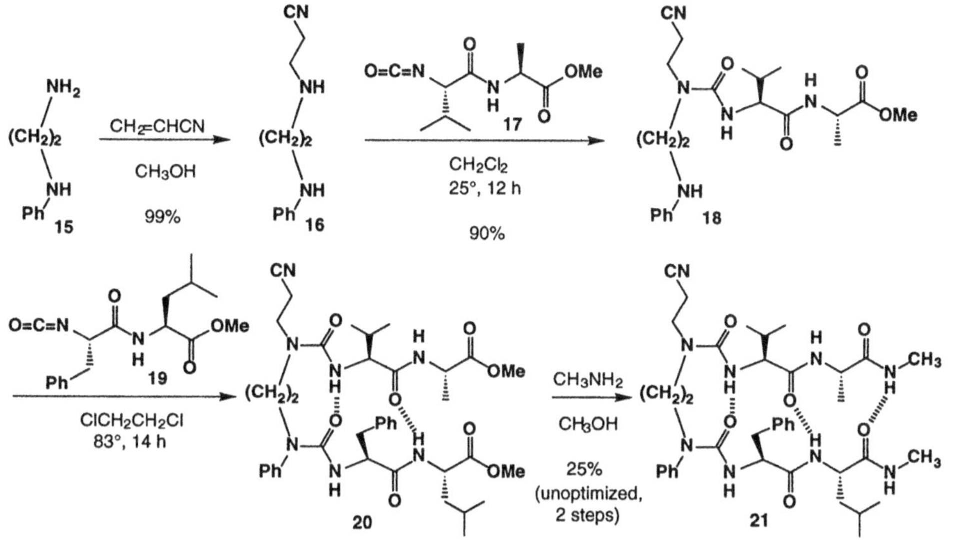

Figure 9. Synthesis of a "2x2" artificial parallel β-sheet.

Proton magnetic resonance nuclear Overhauser effect (NOE) studies reveal proximity between the Val-Ala and Phe-Leu peptide strands of the "2x2" in chloroform solution. The following data are noteworthy: Irradiation of the valine NH enhances phenylalanine α-CH resonance. Irradiation of the alanine α-CH enhances the leucine NH and β-CH_2 resonances. Irradiation of each of the leucine β-CH_2 protons enhances the alanine α-CH resonance. These data are consistent with a parallel β-sheet conformation in which the 1,2-diaminoethane diurea group forms a hydrogen-bonded nine-membered ring, the two urea groups juxtapose the Val-Ala and Phe-Leu peptide strands, and the peptide strands hydrogen bond to each other. A model of this conformation is shown in Figure 10.

Comparison of the ^1H NMR chemical shifts of the amide and urea protons of **21** to those of controls **22-25** provides further evidence for hydrogen-bonded parallel β-sheet structure in chloroform solution (Figure 11). The valine NH, leucine NH, and alanine methylamide NH resonances of **21** appear substantially downfield of the valine NH of **22**, the leucine NH of **24**, and the alanine methylamide NH of **23** (1.93, 1.00, and 0.93 ppm, respectively). In contrast, the phenylalanine NH, the alanine NH, and the leucine methylamide NH resonances of **21** are only slightly downfield of the phenylalanine NH

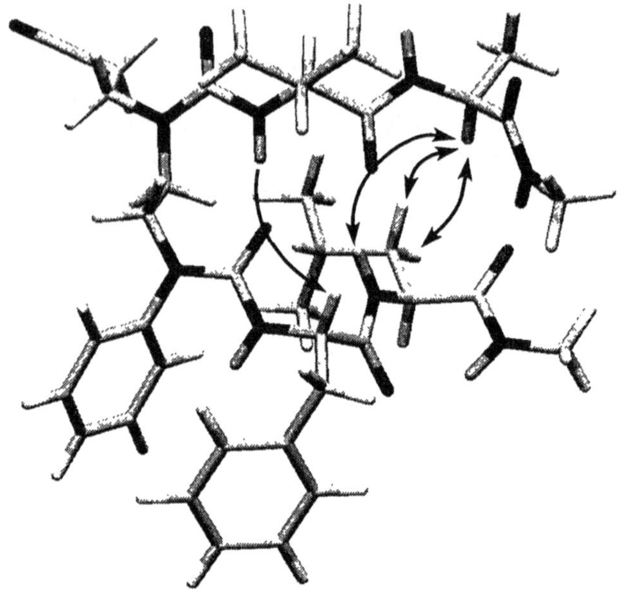

Figure 10. Model of parallel β-sheet conformation of **21**. NOEs observed between the peptide strands are shown with arrows. The model is in a minimum energy conformation (local minimum), as calculated using MacroModel V4.5 with the AMBER* force field.[21]

of **24**, the alanine NH of **22**, and the leucine methylamide NH of **25** (0.18, 0.41, and 0.22 ppm, respectively). This pattern of downfield shifting is consistent with the hydrogen-bonded parallel β-sheet conformation shown in Figure 10. ^{1}H NMR titration studies, in which $CDCl_3$ solutions of controls **22-25** were titrated with $DCON(CD_3)_2$ and 1:1 binding isotherms were fitted to the observed chemical shifts, indicate that non-hydrogen-bonded amide NH groups shift downfield by ca. 2 ppm upon hydrogen bonding to another amide group. From this value and the observed downfield shifts of the leucine NH and alanine methylamide NH of **21**, we estimate the "2x2" to be roughly 50% in a parallel β-sheet conformation.

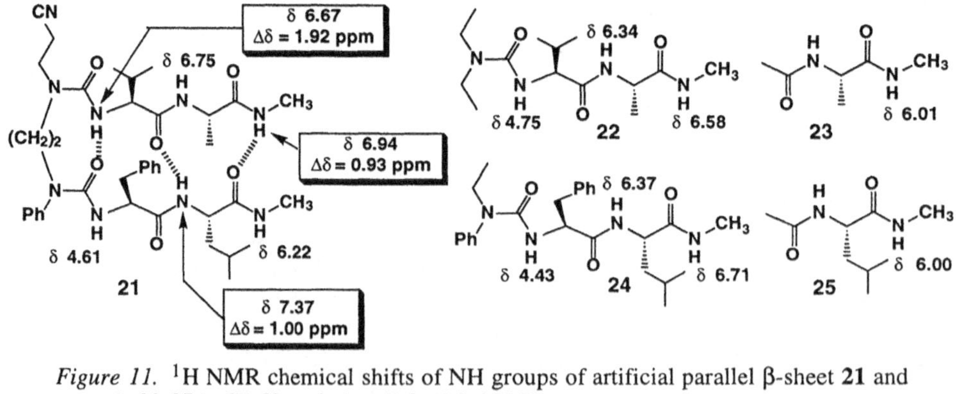

Figure 11. ^{1}H NMR chemical shifts of NH groups of artificial parallel β-sheet **21** and controls **22-25** in $CDCl_3$ solution (1.0 mM, 298 K).

With a small artificial parallel β-sheet in hand, we are now developing procedures for the synthesis of larger artificial β-sheets comprising more peptide strands and peptide strands of greater length. We have prepared a "2x3" oligourea-peptide conjugate by using tripeptide isocyanates in the sequence of reactions shown in Figure 9. We have also synthesized an oligourea-peptide conjugate containing three amino acid strands (a "3x1", Figure 12). In this synthesis, reductive amination of aminourea **28** and Boc-glycinal to form amine **29**,[22] reaction of **29** with valine methyl ester isocyanate to form urea **30**, and removal of the Boc protective group to form aminourea **31** constitute a key series of steps. Repetition of this three-step sequence should allow the preparation of oligourea-peptide conjugates comprising more than three peptide strands. Future structural studies will focus upon the larger artificial β-sheets that we can now prepare.

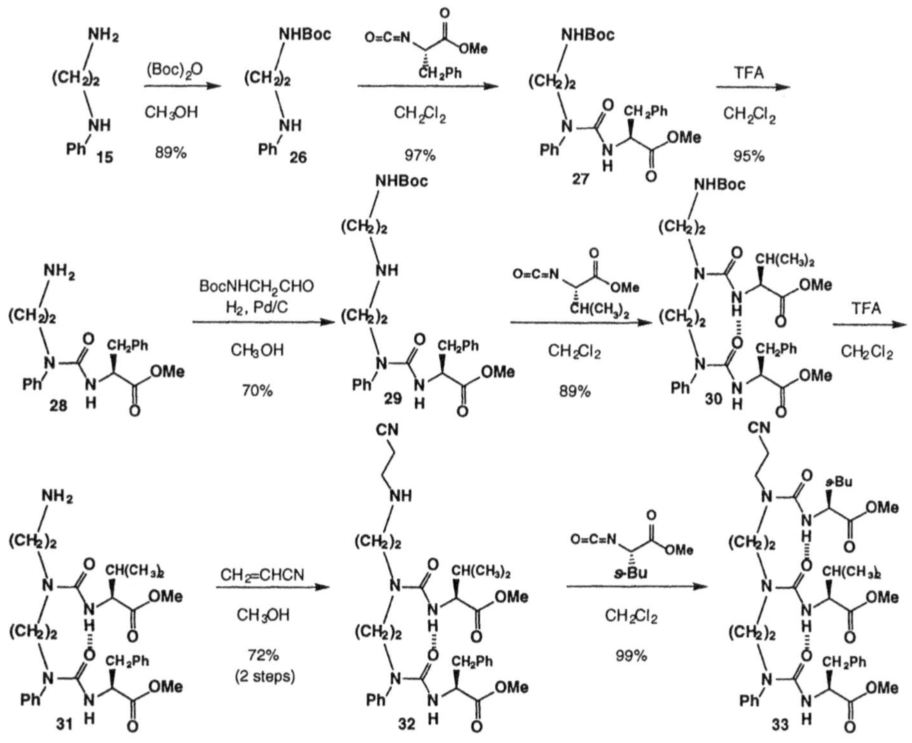

Figure 12. Synthesis of a "3x1" artificial parallel β-sheet.

4. Development of a β-Strand Mimic

To complement the oligourea molecular scaffold in the creation of artificial β-sheets, we have developed a β-strand mimic based upon derivatives of 2-methoxy-5-hydrazino-benzoic acid. This section describes the design, synthesis, and structural evaluation of the β-strand mimic. The subsequent section describes our initial efforts to combine the molecular scaffold and the β-strand mimic in the preparation of a doubly-templated artificial β-sheet.

We envisioned methoxy-substituted derivatives of 5-hydrazinobenzoic acid as β-strand mimics that duplicate the pattern of hydrogen-bond-donor and acceptor groups of one edge of a peptide β-strand and can be prepared as linear oligomers of varying lengths. The 5-hydrazinobenzoic acid unit is composed of a linear array of six atoms that may be thought of as isostructural with a dipeptide. Its derivatives offer the same pattern of hydrogen-bond-donor and acceptor functionality as one edge of a peptide β-strand. Because the N→C directionality of the hydrazinobenzoic acids is opposite that of a peptide, the relationship of 5-hydrazinobenzoic acid to a peptide β-strand may best be seen in the comparison of its oligomers to a polypeptide, as shown in Figure 13.

Figure 13. Relationship of 5-hydrazinobenzoic acids to peptides.

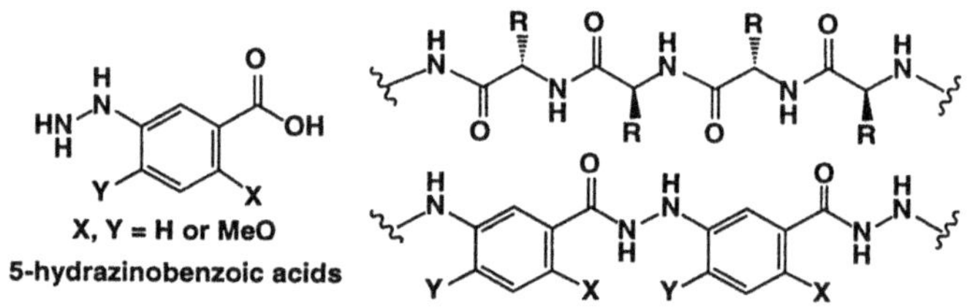

Conformational control is critical in the design of the β-strand mimic. Each amino acid unit in a polypeptide has two flexible dihedral angles, ψ and ϕ. Because of the geometrical constraints of the aromatic ring, 5-hydrazinobenzoic acid derivatives have only three flexible dihedral angles, whereas an analogous dipeptide has four. Methoxy groups at the 2- and 4-positions offer the potential for further conformational control by means of intramolecular hydrogen bonding. Figure 14 illustrates this analysis. As will be seen subsequently, methoxy substitution at the 4-position was found to stabilize a conformation different than that envisioned in Figure 14.

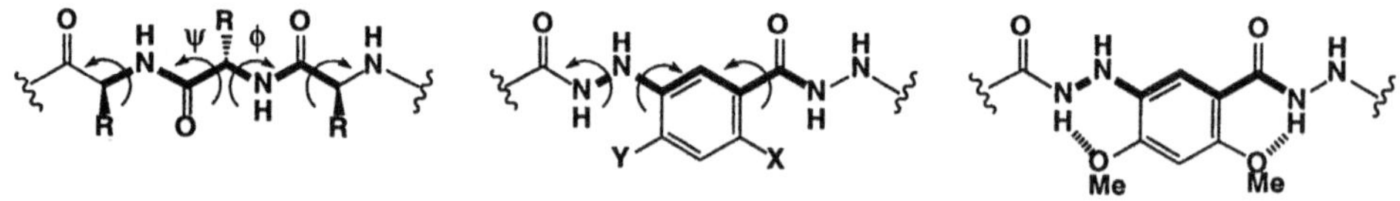

Figure 14. Comparison of the conformational freedom of a dipeptide to that of 5-hydrazinobenzoic acid derivatives.

To gain insight into the structures of methoxy-substituted 5-hydrazinobenzoic acid derivatives, we surveyed the Cambridge Structural Database. Phenylhydrazines and 2-methoxybenzamides were found to be nearly planar. In hydrazide derivatives, little conformational preference about the N-N bond was evident, and C-N-N-C torsion angles ranging from 80° to 180° were observed (Figure 15).[23] The conformation of hydrazide derivatives is apparently sensitive to hydrogen bonding and other crystal-packing forces. Thus, diformyl hydrazide (HCO-NH-NH-CHO) adopts C-N-N-C torsion angles of 180° in the pure state and 105° in a cocrystal with 18-crown-6.[24]

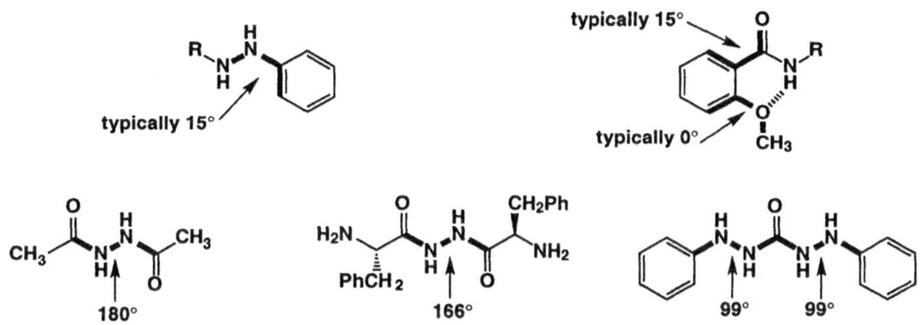

Figure 15. Torsion angles observed in a survey of the Cambridge Structural Database, as illustrated by generalizations and examples.

With these structural considerations in mind, we set out to learn more about the conformations of 5-hydrazinobenzoic acid derivatives through empirical studies. We began our studies by developing syntheses of 2-methoxy-5-hydrazinobenzoic acid, "4-methoxy-5-hydrazino-" benzoic acid, and 2,4-dimethoxy-5-hydrazinobenzoic acid to examine the effects of methoxy substitution upon the structures of 5-hydrazinobenzoic acid derivatives (Figure 16). Bromination of methoxybenzoic acids **34** afforded bromomethoxybenzoic acids **35**. Halogen-metal exchange, followed by reaction of the resulting organolithium compounds with di-*tert*-butylazodicarboxylate, generated *tert*-butyloxycarbonyl derivatives **36**.[25] Removal of the *tert*-butyloxycarbonyl groups afforded the 5-hydrazinobenzoic acid derivatives as their hydrochloride salts (**37**).

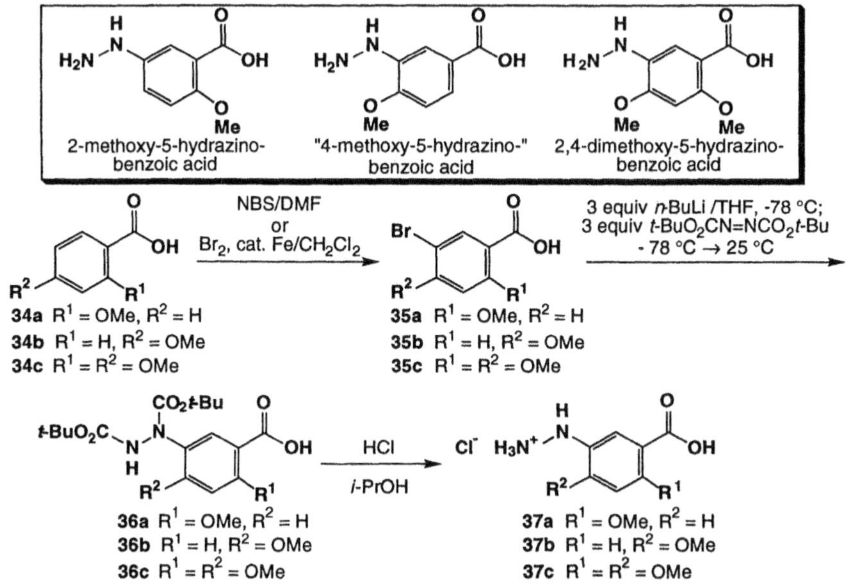

Figure 16. Syntheses of methoxy-substituted 5-hydrazinobenzoic acids.

To determine the effects of methoxy substituents upon the conformations of 5-hydrazinobenzoic acid derivatives, we synthesized and spectroscopically studied hydrazides **38-43**. *tert*-Butyloxycarbonyl compounds **36** were converted to hydrazides **38-43** by successive methyl esterification with diazomethane, removal of the *tert*-butyloxycarbonyl groups with HCl, and acylation with benzoyl chloride or *o*-anisoyl chloride. In the infrared spectrum, the amidic NH_a stretch appears as a band centered at ca. 3437 cm^{-1} in benzoyl derivatives **38-40**, and at ca. 3403 cm^{-1} in *o*-anisoyl derivatives **41-43** (Figure 17). The weakened NH stretch in *o*-anisoyl derivatives **41-43** provides evidence that the adjacent *o*-methoxy group is hydrogen bonding to the amide NH group. In the ^1H NMR spectrum the amidic NH_a group appears at 7.89-7.95 ppm in **38-40** and at 9.43-9.45 ppm in **41-43** (Figure 17). The substantial downfield shift of this resonance in **41-43** provides further evidence for hydrogen bonding of the *o*-methoxy group. These studies indicate that methoxy substitution at the 2-position stabilizes the desired hydrogen-bonded conformation, but that methoxy substitution at the 4-position does not.

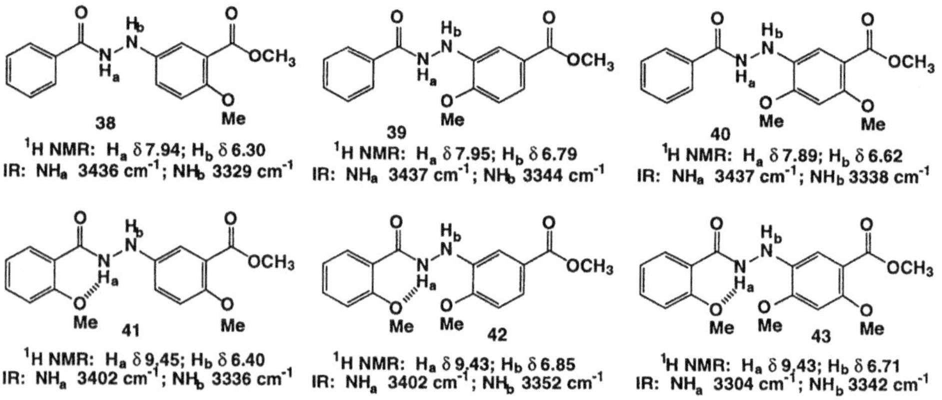

Figure 17. ^1H NMR and IR data for the NH groups of hydrazides **38-43**. Studies were performed on 10 mM solutions in $CDCl_3$ or $CHCl_3$ at 295 K.

To gain further insight into the conformations of the methoxy-substituted 5-hydrazinobenzoic acid derivatives, we turned to X-ray crystallography. We synthesized and crystallized a variety of hydrazide derivatives. Compounds **38, 39**, and **44-48** afforded crystals suitable for X-ray crystallography. The X-ray crystal structures

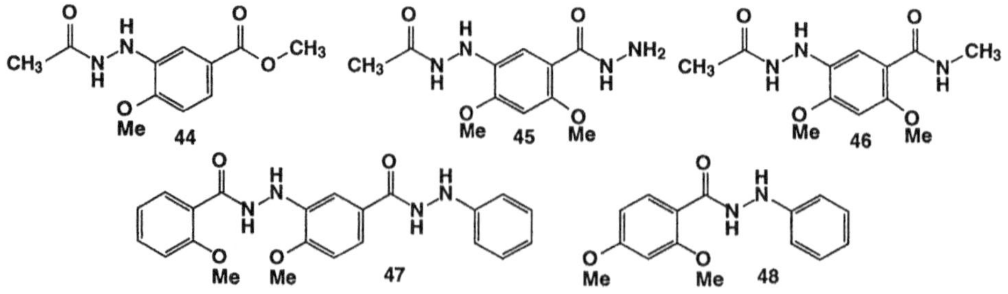

of these compounds are shown in Figure 18. The C-N-N-C torsion angles of the hydrazide groups in these molecules range from 70° to 97°. When a methoxy group is at the 2-position of an aromatic ring, an amide group at the 1-position is hydrogen bonded to this group. When a methoxy group is at the 4-position of an aromatic ring, a hydrazide group at the 5-position is rotated so that the amine NH group is directed toward the methoxy group. These observations are consistent with the solution-phase studies described above and indicate that methoxy substitution at the 2-position stabilized a desired conformation, while methoxy substitution at the 4-position stabilizes an undesired conformation.

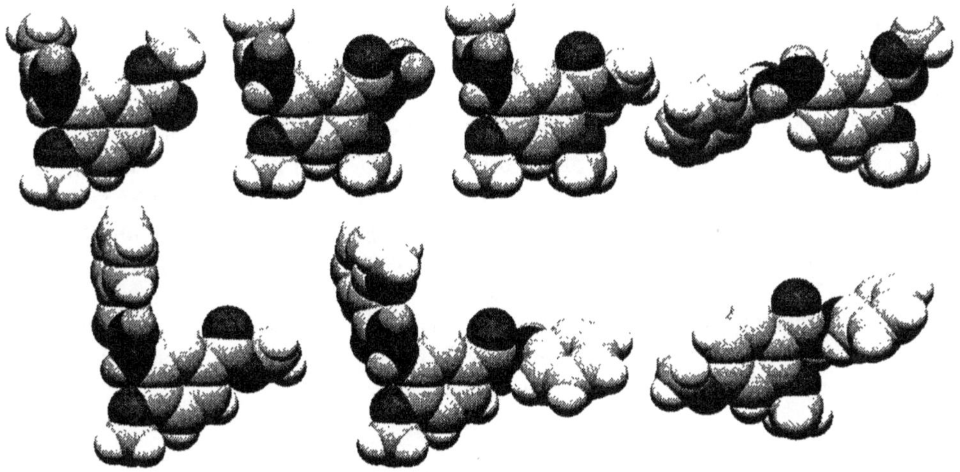

Figure 18. X-ray crystallographic structures of diureas **44**, **45**, **46**, and **38** (top row), and **39**, **47**, and **48** (bottom row).

Examination of the crystal structures and crystal packing interactions reveals that hydrazide derivatives **38**, **47**, and **48** form β-sheetlike hydrogen-bonded dimers containing hydrogen-bonded ten-membered rings (Figure 19). In each of these dimers the amine group that participates in the ten-membered ring lacks *ortho* methoxy substitution (at what we have been calling the 4-position). (Structures with methoxy substitution *ortho* to an amine group form hydrogen-bonded chains in the crystal lattice.) A methoxy substituent is at the *ortho* position of the benzoyl group in the dimer of **48**, indicating this pattern of dimerization tolerates methoxy substitution at the 2-position. From these observations, we conclude that methoxy-substituted 5-hydrazinobenzoic acid derivatives form β-sheetlike hydrogen-bonded dimers if there is a methoxy substituent at the 2-position, but not if there is a methoxy substituent at the 4-position.

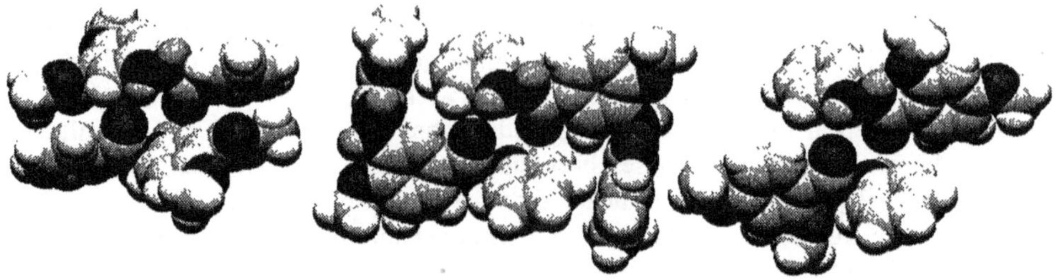

Figure 19. Hydrogen-bonded dimers in the crystal structures of **38**, **47**, and **48**.

The hydrogen-bonded dimeric structures of **38**, **47**, and **48** suggest that oligomeric 2-methoxy-5-hydrazinobenzoic acid hydrazides may also form β-sheetlike hydrogen-bonded dimers. To model the structure of such a dimer, we began with the crystallographically-determined structure of the hydrogen-bonded dimer of **48**. Using the crystallographic coordinates of the dimer, we extended the structure to generate a model for the hydrogen-bonded dimer of an oligomer (Figure 20). The alternating array of ten- and fourteen-membered hydrogen-bonded rings in this structure is similar to that of an antiparallel β-sheet, and the distance between the carbonyl groups of each oligomer (6.7 Å) is comparable to that found in β-sheets.[26] To experimentally determine whether such structures form in the solid state, we are now synthesizing various oligomeric derivatives of 2-methoxy-5-hydrazinobenzoic acid and attempting to grow crystals for X-ray crystallographic studies.

Figure 20. Crystallographically-based model of the hydrogen-bonded dimer of a 2-methoxy-5-hydrazinobenzoic acid oligomer.

The similarity of this model to an antiparallel β-sheet suggests that oligomeric derivatives of 2-methoxy-5-hydrazinobenzoic acid may be suitable as templates to induce β-sheet structure in peptides. Docking a peptide β-strand to the 2-methoxy-5-hydrazinobenzoic acid hydrazide oligomer from Figure 20 reveals excellent structural complementarity. The docked structure is shown in Figure 21.

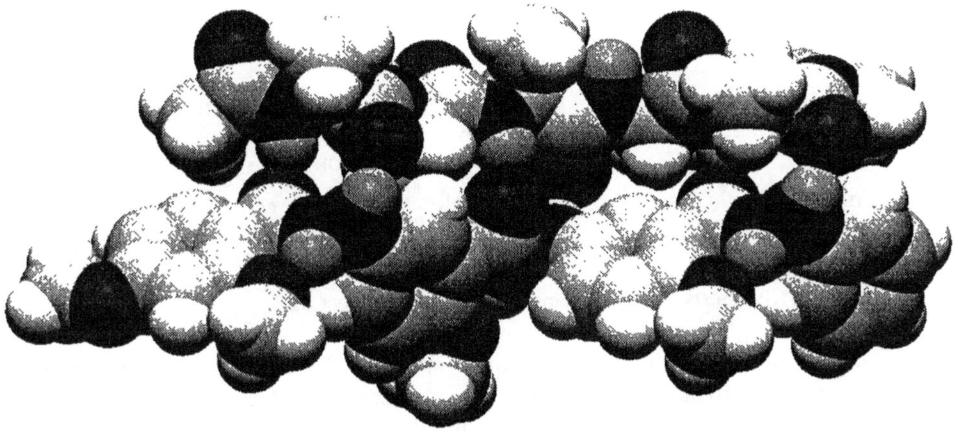

Figure 21. Model for the interaction of a 2-methoxy-5-hydrazinobenzoic acid oligomer (bottom) with a peptide β-strand (top). Energy minimization was performed using MacroModel V4.5 with the AMBER* force field.[21] The geometry of the 2-methoxy-5-hydrazinobenzoic acid oligomer was held fixed, while the geometry of the peptide strand was allowed to vary during minimization.

The studies described in this section establish that oligomeric hydrazide derivatives of 2-methoxy-5-hydrazinobenzoic acid can mimic the array of hydrogen-bond-donor and acceptor functionality of one edge of a polypeptide in a β-strand conformation. Although the hydrazide group is severely twisted about the N-N bond, the hydrazides that we have studied form hydrogen-bonded dimers in the solid state, and we anticipate that β-strand mimics based upon 2-methoxy-5-hydrazinobenzoic acid will be able to form hydrogen-bonded β-sheet structures with peptides.

5. Artificial β-Sheets Containing the Molecular Scaffold and the β-Strand Mimic

We are now beginning to combine the oligourea molecular scaffold and the 2-methoxy-5-hydrazinobenzoic acid β-strand mimic, to realize the goal of preparing artificial β-sheets in which the two templates act in conjunction to stabilize β-sheet structure. Our first endeavor in this area involves the combination of a 1,2-diaminoethane diurea unit, a 5-amino-2-methoxybenzoic acid unit (a version of the 5-hydrazino-2-methoxybenzoic acid β-strand mimic that is truncated to allow suitable connection to the diurea unit), and a dipeptide strand. Thus, doubly-templated artificial β-sheet **51** was synthesized by sequential treatment of diamine **16** with isocyanate **49**, isocyanate **19**, and methylamine, as shown in Figure 22.

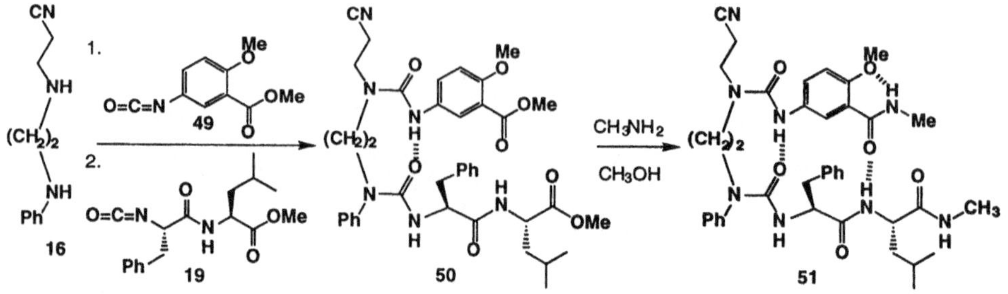

Figure 22. Synthesis of doubly-templated artificial β-sheet **51**.

Comparison of the ^{1}H NMR chemical shifts of the amide and urea protons of doubly-templated artificial β-sheet **51** to those of controls **52** and **24** suggests that this compound adopts a pattern of hydrogen bonding similar to that of an antiparallel β-sheet (Figure 23). Most notably, the "upper" urea NH and leucine NH resonances of **51** appear substantially downfield of the corresponding resonances of **52** and **24** in chloroform solution (3.69 and 1.83 ppm, respectively). Other smaller differences are also observed. For example, the leucine methylamide NH resonance of **51** appears 1.28 ppm upfield of that of **24**. We attribute this difference to the propensity of **24** to adopt a hydrogen-bonded β-turn conformation, in which the methylamide NH is hydrogen bonded to the urea carbonyl group, and to anisotropic effects of the aromatic rings in **51**. One additional noteworthy feature of the ^{1}H NMR spectrum of **51** is that one of the leucine methyl resonances appears unusually upfield at 0.44 ppm. In contrast, the other leucine methyl resonance and those of control **24** appear at 0.8-0.9 ppm.

Figure 23. ^{1}H NMR chemical shifts of the NH groups of doubly-templated artificial β-sheet **51** and controls **52** and **24** in $CDCl_3$ solution (1.0 mM, 294 K).

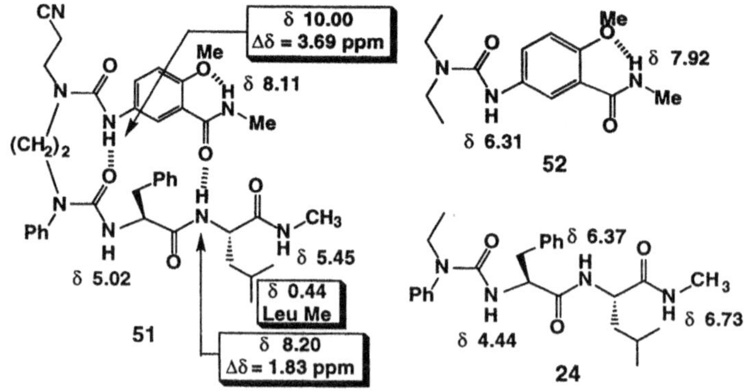

^{1}H NMR DPFGSE NOE (double pulsed field gradient spin echo NOE) experiments shed light on the unusual chemical shift of one of the leucine methyl groups of **51** and provided insight into the conformation of this molecule.[27] Irradiation of the upfield methyl group leads to enhancements of the "upper" urea NH group, and the protons at the 4- and 6-positions of the "upper" aromatic ring. These data indicate that the methyl group sits over the "upper" aromatic ring. (Large enhancements of the various leucine protons and small enhancements of other protons associated with the "upper" aromatic ring and the 1,2-diaminoethane backbone are also seen.) Irradiation of the proton at the 6-position of the "upper" aromatic ring leads to enhancements of the phenylalanine α-proton, the leucine NH group, the leucine γ-proton, and the "upper" urea NH group. Collectively, the chemical shift and NOE data suggest that the 1,2-diaminoethane diurea and 2-methoxy-5-aminobenzamide templates act in conjunction to stabilize a β-strand conformation in the peptide strand. A model for this structure, consistent with these data and with the X-ray crystallographic data of related diureas and 5-hydrazinobenzoic acid derivatives, is shown in Figure 24.

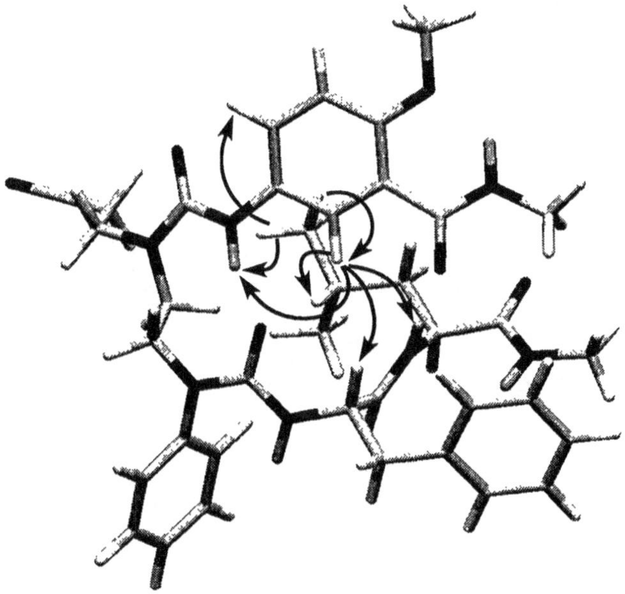

Figure 24. Model of **51** in a β-sheet conformation, based upon observed chemical shift data, NOEs, and the crystallographic structures shown in Figures 7 and 18. Key NOEs are shown with arrows. The model is in a minimum energy conformation (local minimum), as calculated using MacroModel V4.5 with the AMBER* force field.[21]

We are now preparing larger artificial β-sheets, which are composed of longer peptide strands and more peptide strands. Figure 25 shows a model of one such structure, based on the structural information gained from the IR, NMR, X-ray crystallographic, and molecular modeling studies described in this paper. In this model, a triurea molecular scaffold and a "tetrapeptide" β-strand mimic stabilize β-sheet structure in two tetrapeptide strands, which are parallel to each other and "antiparallel" to the β-strand mimic. It remains to be determined experimentally whether oligomeric hydrazinobenzoic acid derivatives will act as templates that induce β-sheet formation in polypeptide strands and whether oligoureas containing three or more urea groups will be suitable templates to hold multiple peptide strands in proximity. Studies aimed at answering these questions are now underway.

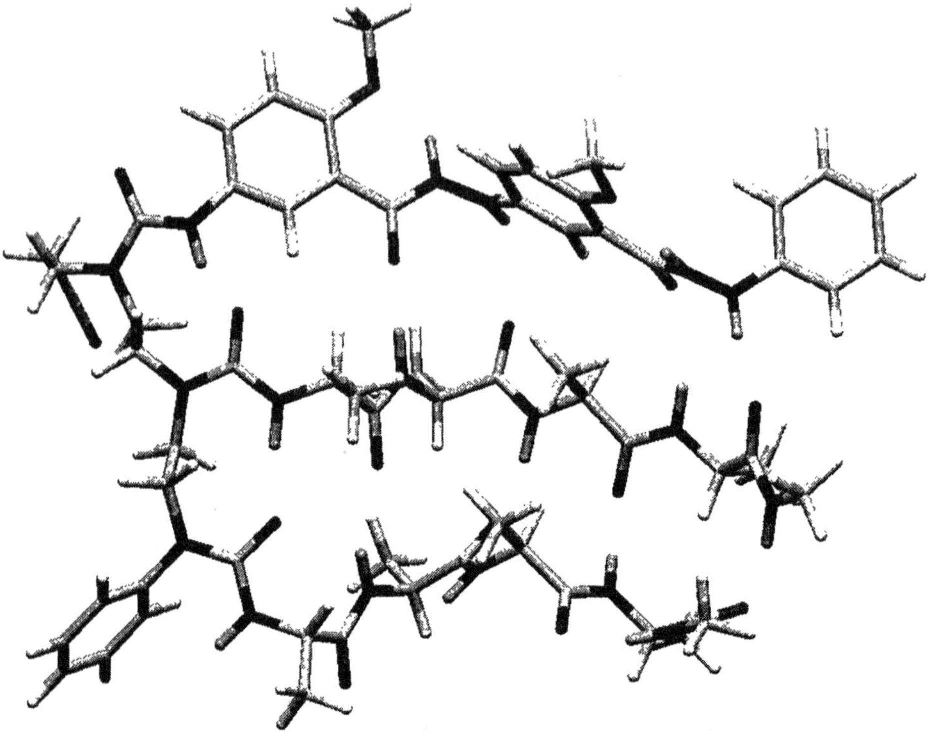

Figure 25. Model for an artificial β-sheet in which a triurea molecular scaffold and an oligomeric 2-methoxy-5-hydrazinobenzoic acid β-strand mimic induce β-sheet structure in two attached peptide strands.

6. Summary and Outlook

This paper has described the development of two peptide structural templates: a molecular scaffold, designed to hold peptide strands in proximity, and a β-strand mimic, designed to induce a β-strand conformation in an adjacent peptide strand. Detailed spectroscopic and crystallographic studies of these templates have provided us with a good understanding of their structures. Each template may be thought of as a means to create structure in one dimension. In combination they provide a means to create structure in two dimensions. By combining these templates with each other and with peptides, we have begun to form small artificial β-sheets.

Thus far, we have focused upon obtaining patterns of hydrogen bonding characteristic of β-sheet structure. For this reason, we have performed studies in chloroform, a solvent that favors hydrogen bonding and resembles, to some extent, the interior regions of proteins. Although hydrogen bonding is not generally thought to be the main driving force in protein folding, the potential to achieve an appropriate pattern of hydrogen bonding is a necessary condition for the formation of secondary structure. As such, we consider chloroform to be an ideal solvent for the studies described herein, as well as for future studies aimed at creating larger artificial β-sheets.

We are also interested in developing molecules that adopt β-sheet structures in water, because naturally occurring proteins fold in water. To achieve structure in water, we anticipate that it will be necessary to buttress hydrogen bonds with hydrophobic interactions. Hydrophobic cluster formation between peptide strands and the templates may provide a means to incorporate such interactions.[8e] The contact seen between the leucine side-chain and the aromatic ring of the β-strand mimic in **51** hints at one way by which such hydrophobic interactions might be attained. A second approach to this goal involves the creation of a sandwich between two amphiphilic β-sheets in which the hydrophobic faces are in contact.[10,12] One way to implement this approach is to employ a template in the third dimension, orthogonal to the molecular scaffold and the β-strand mimic, that holds two artificial β-sheets in contact.

Our current efforts are focused upon the creation of structure. Ultimately, we would like to develop artificial proteins in which β-sheet structure serves as a basis for the creation of function, such as molecular recognition and catalysis. *Intramolecular* recognition is inherent to the structure of β-sheets. If a peptide strand is removed from the middle of a β-sheet, a void remains that has the potential for *intermolecular* recognition of a peptide strand or a peptidelike substrate. The β-sheet may be thought of as a "molecular pegboard" that presents an array of functionality in the form of the amino acid side-chains. We anticipate that it may be possible to develop catalysts in which the peptide side-chains act upon a bound substrate to catalyze various reactions. Future studies are aimed at developing molecular receptors and catalysts based upon these principles.

7. Acknowledgments

This work was supported by National Science Foundation Grant CHE-9118885, National Institutes of Health Grant GM-49076, Zeneca Pharmaceuticals Group, the Upjohn Company, Hoffman-La Roche Inc., and Pfizer Inc. J. S. N. thanks the following agencies for support in the form of awards: the Camille and Henry Dreyfus Foundation (New Faculty Award), the American Cancer Society (Junior Faculty Research Award), the National Science Foundation (Young Investigator Award), and the Arnold and Mabel Beckmann Foundation (Young Investigator Award). A. J. S. thanks the following agencies for support in the form of awards: the National Science Foundation (Presidential Faculty Fellowship), the Alfred P. Sloan Foundation (Fellowship), and the Camille and Henry Dreyfus Foundation (Teacher Scholar Award).

8. References

1. For reviews, see: (a) Kemp, D. S. (1990) Peptidomimetics and the template approach to nucleation of β-sheets and α-helices in peptides, *Trends Biotechnol.* **8**, 249-255. (b) Hölzemann, G. (1991) Peptide conformation mimetics (part 1),

Kontakte 3-12. (c) Hölzemann, G. (1991) Peptide conformation mimetics (part 2), *Kontakte* **1991**, 55-63. (d) Hirschmann, R. (1991) Medicinal chemistry in the golden age of biology: lessons from steroid and peptide research, *Angew. Chem., Int. Ed. Engl.* **30**, 1278-1301. (e) Olson, G. L.; Bolin, D. R.; Bonner, M. P.; Bös, M.; Cook, C. M.; Fry, D. C.; Graves, B. J.; Hatada, M.; Hill, D. E.; Kahn, M.; Madison, V. S.; Rusiecki, V. K; Sarabu, R.; Sepinwall, J.; Vincent, G. P.; Voss, M. E. (1993) Concepts and progress in the development of peptide mimetics, *J. Med. Chem.* **36**, 3039-3049. (f) Wiley, R. A.; Rich, D. H. (1993) Peptidomimetics derived from natural products, *Med. Res. Rev.* **13**, 327-384. (g) Chorev, M.; Goodman, M. (1993) A dozen years of retro-inverso peptidomimetics, *Acc. Chem. Res.* **26**, 266-273. (h) Zuckermann, R. N. (1993) The chemical synthesis of peptidomimetic libraries, *Curr. Opin. Struct. Biol.* **3**, 580-584. (i) Kahn, M. (1993) Peptide secondary structure mimetics: recent advances and future challenges, *Synlett* 821-826. (j) Giannis, A.; Kolter, T. (1993) Peptidomimetics for receptor ligands—discovery, development, and medical perspectives, *Angew. Chem., Int. Ed. Engl.* **32**, 1244-1267. (k) Gante, J. (1994) Peptidomimetics—tailored enzyme inhibitors, *Angew. Chem., Int. Ed. Engl.* **33**, 1699-1720. (l) Liskamp, R. M. J. (1994) Conformationally restricted amino acids and dipeptides, (non)peptidomimetics and secondary structure mimetics, *Rec. Trav. Chim. Pays-Bas* **113**, 1-19. (m) Adang, A. E. P.; Hermkens, P. H. H.; Linders, J. T. M.; Ottenheijm, H. C. J.; van Staveren, C. J. (1994) Case histories of peptidomimetics: progression from peptides to drugs, *Rec. Trav. Chim. Pays-Bas* **113**, 63-78. (n) Beeley, N. (1994) Peptidomimetics and small-molecule drug design: towards improved bioavailability and in vivo stability, *Trends Biotechnol.* **12**, 213-216.

2. (a) Kemp, D. S.; Bowen, B. R. (1988) Synthesis of peptide-functionalized diacylaminoepindolidiones as templates for β-sheet formation, *Tetrahedron Lett.* **29**, 5077-5080. (b) Kemp, D. S.; Bowen, B. R. (1988) Conformational analysis of peptide-functionalized diacylaminoepindolidiones. ^{1}H NMR evidence for β-sheet formation, *Tetrahedron Lett.* **29**, 5081-5082. (c) Kemp, D. S.; Bowen, B. R.; Muendel, C. C. (1990) Synthesis and conformational analysis of epindolidione-derived peptide models for β-sheet formation, *J. Org. Chem.* **55**, 4650-4657.
3. (a) Martin, S. F.; Austin, R. E.; Oalmann, C. J. (1990) Stereoselective synthesis of 1,2,3-trisubstituted cyclopropanes as novel dipeptide isosteres, *Tetrahedron Lett.* **31**, 4731-4734. (b) Martin, S. F.; Austin, R. E.; Oalmann, C. J.; Baker, W. R.; Condon, S. L.; deLara, E.; Rosenberg, S. H.; Spina, K. P.; Stein, H. H.; Cohen, J.; Kleinert, H. D. (1992) 1,2,3-Trisubstituted cyclopropanes as conformationally restricted peptide isosteres: application to the design and synthesis of novel renin inhibitors, *J. Med. Chem.* **35**, 1710-1721. (c) Baker, W. R.; Jae, H.-S.; Martin, S. F.; Condon, S. L.; Stein, H. H.; Cohen, J.; Kleinert, H. D. (1992) Conformationally restricted peptide isosteres. 2. Synthesis and in vitro potency of dipeptide renin inhibitors employing a 2-alkylsulfonyl-3-phenylcyclopropane carboxamide as a P_3 amino acid replacement, *Bioorg. Med. Chem. Lett.* **2**, 1405-1410. (d) Martin, S. F.; Oalmann, C. J.; Liras, S. (1993) Cyclopropanes as conformationally restricted peptide isosteres. Design and synthesis of novel collagenase inhibitors,

Tetrahedron, **49**, 3521-3532.

4. (a) Smith, A. B., III; Keenan, T. P.; Holcomb, R. C.; Sprengeler, P. A.; Guzman, M. C.; Wood, J. L.; Carroll, P. J.; Hirschmann, R. (1992) Design, synthesis, and crystal structure of a pyrrolinone-based peptidomimetic possessing the conformation of a β-strand: potential application to the design of novel inhibitors of proteolytic enzymes, *J. Am. Chem. Soc.* **114**, 10672-10674. (b) Smith, A. B. III; Guzman, M. C.; Sprengeler, P. A.; Keenan, T. P.; Holcomb, R. C.; Wood, J. L.; Carroll, P. J.; Hirschmann, R. (1994) De novo design, synthesis, and X-ray crystal structures of pyrrolinone-based β-strand peptidomimetics, *J. Am. Chem. Soc.* **116**, 9947-9962. (c) Smith, A. B., III; Hirschmann, R.; Pasternak, A.; Akaishi, R.; Guzman, M. C.; Jones, D. R.; Keenan, T. P.; Sprengeler, P. A.; Darke, P. L.; Emini, E. A.; Holloway, M. K.; Schleif, W. A. (1994) Design and synthesis of peptidomimetic inhibitors of HIV-1 protease and renin. Evidence for improved transport, *J. Med. Chem.* **37**, 215-218. (d) Smith, A. B., III; Akaishi, R.; Jones, D. R.; Keenan, T. P.; Guzman, M. C.; Holcomb, R. C.; Sprengeler, P. A.; Wood, J. L.; Hirschmann, R.; Holloway, M. K. (1995) Design and synthesis of nonpeptide peptidomimetic inhibitors of renin, *Biopolymers* **37,** 29-53.
5. Hagihara, M.; Anthony, N. J.: Stout, T. J.; Clardy, J.; Schreiber, S. L. (1992) Vinylogous polypeptides: an alternative to the peptide backbone, *J. Am. Chem. Soc.* **114**, 6568-6570.
6. Kemp, D. S. Studies of template-nucleated β-sheets and α-helices. Presented at the 203rd National Meeting of the American Chemical Society, San Fransisco, April 1992; paper ORGN 151.
7. (a) Feigel, M. (1986) 2,8-Dimethyl-4-(carboxymethyl)-6-(aminomethyl)-phenoxathiin *S*-dioxide: an organic substitute for the β-turn in peptides? *J. Am. Chem. Soc.* **108**, 181-182. (b) Brandmeier, V.; Feigel, M.; Bremer, M. (1989) 2'-Aminomethylbiphenyl-2-carboxylic acid as a component of a cyclopeptide; crystal structure and conformation in solution, *Angew. Chem., Int. Ed. Engl.* **28**, 486-488. (c) Brandmeier, V.; Feigel, M. (1989) A macrocycle containing two biphenyl and two alanine subunits, synthesis and conformation in solution, *Tetrahedron* **45**, 1365-1376. (d) Feigel, M. (1989) Makrocyclen aus Alanin und Phenoxathiin, Synthese und Konformation in Lösung, *Liebigs Ann. Chem.* 459-468. (e) Wagner, G.; Feigel M. (1993) Parallel β-sheet conformation in macrocycles, *Tetrahedron* **49**, 10831-10842. (f) Brandmeier V.; Sauer, W. H. B.; Feigel M. (1994) Antiparallel β-sheet conformation in cyclopeptides containing a pseudo-amino acid with a biphenyl moiety, *Helv. Chim. Acta* **77**, 70-85.
8. (a) Diaz, H.; Kelly, J. W. (1991) The synthesis of dibenzofuran based diacids and amino acids designed to nucleate parallel and antiparallel β-sheet formation, *Tetrahedron Lett.* **32**, 5725-5728. (b) Diaz, H.; Espina, J. R.; Kelly, J. W. (1992) A dibenzofuran-based amino acid designed to nucleate antiparallel β-sheet structure: evidence for intramolecular hydrogen-bond formation, *J. Am. Chem. Soc.* **114**, 8316-8318. (c) Diaz, H.; Tsang, K. Y.; Choo, D.; Espina, J. R.; Kelly, J. W. (1993) Design, synthesis, and partial characterization of water-soluble β-sheets stabilized by a dibenzofuran-based amino acid, *J. Am. Chem. Soc.* **115,** 3790-3791.

(d) Diaz, H.; Tsang, K. Y.; Choo, D.; Kelly, J. W. (1993) The design of water soluble β-sheet structure based on a nucleation strategy, *Tetrahedron* **49**, 3533-3545. (e) Tsang, K. Y.; Diaz, H.; Graciani, N.; Kelly, J. W. (1994) Hydrophobic cluster formation is necessary for dibenzofuran-based amino acids to function as β-sheet nucleators, *J. Am. Chem. Soc.* **116**, 3988-4005.

9. For examples, see: (a) Winningham, M. J.; Sogah, D. Y. (1994) Evidence for intramolecular hydrogen bonding in β-alanine derivatives of 2,8-dimethylphenoxathiin 4,6-dicarboxylic acid. Model studies for nucleation of parallel β-sheets, *J. Am. Chem. Soc.* **116**, 11173-11174. (b) LaBrenz, S. R.; Kelly, J. W. (1995) Peptidomimetic host that binds a peptide guest affording a β-sheet structure that subsequently self-assembles, *J. Am Chem. Soc.* **117**, 1655-1656. (c) Schneider, J. P.; Kelly, J. W. (1995) Synthesis and efficacy of square planar copper complexes designed to nucleate β-sheet structure, *J. Am. Chem. Soc.* **117**, 2533-2546. (d) Gardner, R. R.; Liang, G.-B.; Gellman, S. H. (1995) An achiral dipeptide mimetic that promotes β-hairpin formation, *J. Am. Chem. Soc.* **117**, 3280-3281. (e) Kemp, D. S.; Li, Z. Q. (1995) 2-amino-2'-carboxydiphenylacetylenes as β-turn mimetics. Synthesis and conformational properties, *Tetrahedron Lett.* **36**, 4175-4178. (f) Kemp, D. S.; Li, Z. Q. (1995) A short β-sheet containing proline nucleated by a 2,2'-substituted tolan β-turn mimetic, *Tetrahedron Lett.* **36**, 4179-4180.
10. (a) Mutter, M.; Altmann, E.; Altmann, K.-H.; Hersperger, R.; Koziej, P.; Nebel, K.; Tuchscherer, G.; Vuilleumier, S.; Gremlich, H.-U.; Müller, K. (1988) The construction of new proteins. Part III. Artificial folding units by assembly of amphiphilic secondary structures on a template, *Helv. Chem. Acta* **71**, 835-847. (b) Mutter, M.; Vuilleumier, S. (1989) A chemical approach to protein design—template-assembled synthetic proteins (TASP), *Angew. Chem., Int. Ed. Engl.* **28**, 535-554. (c) Tuchscherer, G.; Dörner, B.; Sila, U.; Kabmer, B.; Mutter, M. (1993) The TASP concept: mimetics of peptide ligands, protein surfaces, and folding units, *Tetrahedron* **49**, 3559-3575.
11. (a) Kim, C. A.; Berg, J. M. (1993) Thermodynamic β-sheet propensities measured using a zinc-finger host peptide, *Nature* **362,** 267-270. (b) Minor, D. L., Jr.; Kim, P. S. (1994) Measurement of the β-sheet-forming propensities of amino acids, *Nature* **367**, 660-663. (c) Smith, C. K.; Withka, J. M.; Regan, L. (1994) A thermodynamic scale for the β-sheet forming tendencies of the amino acids, *Biochemistry* **33**, 5510-5517. (d) Minor, D. L.; Kim, P. S. (1994) Context is a major determinant of β-sheet propensity, *Nature* **371**, 264-267. (e) Otzen, D. E.; Fersht, A. R. Side-chain determinants of β-sheet stability, *Biochemistry* **34**, 5718-5724.
12. (a) Pessi, A.; Bianchi, E.; Crameri, A.; Venturini, S.; Tramontano, A.; Sollazzo, M. (1993) A designed metal-binding protein with a novel fold, *Nature* **362**, 367-369. (b) Yan, Y.; Erickson, B. W. (1994) Engineering of betabellin 14D: disulfide-induced folding of a β-sheet protein, *Protein Science* **3**, 1069-1073. (c) Quinn, T. P.; Tweedy, N. B.; Williams, R. W.; Richardson, J. S.; Richardson, D. C. (1994) Betadoublet: de novo design, synthesis, and characterization of a β-sandwich

protein, *Proc. Natl. Acad. Sci. USA* **91**, 8747-8751.

13. (a) Nowick, J. S.; Powell, N. A.; Martinez, E. J.; Smith, E. M.; Noronha, G. (1992) Molecular scaffolds I: intramolecular hydrogen bonding in a family of di- and triureas, *J. Org. Chem.* **57**, 3763-3765. (b) Nowick, J. S.; Abdi, M.; Bellamo, K. A.; Love, J. A.; Martinez, E. J.; Noronha, G.; Smith, E. M.; Ziller, J. W. (1995) Molecular scaffolds. 2. Intramolecular hydrogen bonding 1,2-diaminoethane diureas, *J. Am. Chem. Soc.* **117**, 89-99.
14. (a) Gellman, S. H.; Adams, B. R. (1989) Intramolecular hydrogen bonding in simple diamides, *Tetrahedron Lett.* **30**, 3381-3384. (b) Dado, G. P.; Desper, J. M.; Gellman, S. H. (1990) Hydrogen bonding in a family of triamides: conformation-directing effects in solution vs the solid state, *J. Am. Chem. Soc.* **112**, 8630-8632. (c) Gellman, S. H.; Dado, G. P.; Liang, G.-B.; Adams, B. R. (1991) Conformation-directing effects of a single intramolecular amide-amide hydrogen bond: variable-temperature NMR and IR studies on a homologous diamide series. *J. Am. Chem. Soc.* **113**, 1164-1173. (d) Liang, G.-B.; Dado, G. P.; Gellman, S. H. (1991) Anatomy of a stable intramolecularly hydrogen bonded folding pattern, *J. Am. Chem. Soc.* **113**, 3994-3995. (e) Dado, G. P.; Desper, J. M.; Holmgren, S. K.; Rito, C. J.; Gellman, S. H. (1992) Effects of covalent modifications on the solid-state folding and packing of *N*-malonylglycine derivatives, *J. Am. Chem. Soc.* **114**, 4834-4843. (f) Liang, G.-B.; Desper, J. M.; Gellman, S. H. (1993) Effects of backbone rigidification on intramolecular hydrogen bonding in a family of diamides, *J. Am. Chem. Soc.* **115**, 925-938. (g) Dado, G. P.; Gellman, S. H. (1993) Structural and thermodynamic characterization of temperature-dependent changes in the folding pattern of a synthetic triamide, *J. Am. Chem. Soc.* **115**, 4228-4245.
15. (a) Ganis, P.; Avitabile, G.; Benedetti, E.; Pedone, C.; Goodman, M. (1970) Crystal and molecular structure of *N,N'*-diethyl-*N,N'*-diphenylurea, *Proc. Natl. Acad. Sci. USA* **67**, 426-433. (b) Itai, A; Toriumi, Y.; Tomioka, N.; Kagechika, H.; Azumaya, I.; Shudo, K. (1989) Stereochemistry of *N*-methylbenzanilide and benzanilide, *Tetrahedron Lett.* **30**, 6177-6180. (c) Yamaguchi, K.; Matsumura, G.; Kagechika, H.; Azumaya, I.; Ito, Y.; Itai, A.; Shudo, K. (1991) Aromatic architecture. Use of the *N*-methylamide structure as a molecular splint, *J. Am. Chem. Soc.* **113**, 5474-5475. (d) Itai, A.; Toriumi, Y.; Saito, S.; Kagechika, H.; Shudo, K. (1992) Preference of *cis*-amide structure in *N*-acyl-*N*-methylanilines, *J. Am. Chem. Soc.* **114**, 10649-10650. (e) Azumaya, I.; Kagechika, H.; Yamaguchi, K.; Shudo, K. (1995) Stereochemistries of aromatic *N*-methylamides in crystal and solution. Temperature-dependent conformational conversion and attracting aromatic-aromatic interactions, *Tetrahedron* **51**, 5277-5290.
16. Nowick, J. S.; Smith, E. M.; Noronha, G. Molecular scaffolds 3. An artificial parallel β-sheet. Manuscript submitted.
17. Goldschmidt, S.; Wick, M. (1952) Über Peptid-Synthesen I, *Liebigs Ann. Chem.* **575**, 217-231.
18. Nowick, J. S.; Powell, N. A.; Nguyen, T. M.; Noronha, G. (1992) An improved method for the synthesis of enantiomerically pure amino acid ester isocyanates, *J. Org. Chem.* **57**, 7364-7366.

19. Holmes, D. L.; Smith, E. M.; Noronha, G.; Nguyen, T. M.; Nowick, J. S. Synthesis of peptide isocyanates. Manuscript in preparation.
20. (a) Bergeron, R. J.; Burton, P. S.; McGovern, K. A.; Kline, S. J. (1981) Reagents for the selective acylation of spermidine, homospermidine, and bis[3-aminopropyl]-amine, *Synthesis*, 732-733. (b) Jasys, V. J.; Kelbaugh, P. R.; Nason, D. M.; Phillips, D.; Rosnack, K. J.; Saccomano, N. A.; Stroh, J. G.; Volkmann, R. A. (1990) Isolation, structure elucidation, and synthesis of novel hydroxylamine-containing polyamines from the venom of the *Agelenopsis aperta* spider, *J. Am. Chem. Soc.* **112**, 6696-6704.
21. (a) McDonald, D. Q.; Still, W. C. (1992) AMBER* torsional parameters for the peptide backbone, *Tetrahedron Lett.* **33**, 7743-7746. (b) McDonald, D. Q.; Still, W. C. (1992) An AMBER* study of Gellman's amides, *Tetrahedron Lett.* **33**, 7747-7750.
22. Dueholm, K. L.; Egholm, M.; Buchardt, O. (1993) An efficient synthesis of Boc-aminoacetaldehyde and its application to the synthesis of *N*-(2-Boc-aminoethyl)glycine esters, *Org. Prep. Proc. Int.* **25**, 457-461.
23. (a) Shintani, R. (1960) The crystal and molecular structure of anhydrous diacetylhydrazine $(-NHCOCH_3)_2$, *Acta Crystallogr.* **13**, 609-618. (b) Urbanczyk-Lipkowska, Z.; Krajewski, J. W.; Gluzinski, P. (1986) The structure of *N,N'*-di-L-phenylalanine hydrazide determined by X-ray diffraction methods, *J. Mol. Struct.* **140**, 151-157. (c) De Ranter, C. J.; Blaton, N. M.; Peeters, O. M. (1979) 1,5-Diphenylcarbonohydrazide (*sym*-diphenylcarbazide), *Acta Crystallogr., Sect. B* **35**, 1295-1297.
24. (a) Tomiie, Y.; Koo, C. H.; Nitta, I. (1958) The crystal and molecular structure of diformylhydrazine, OHC-HN-NH-CHO. I. The X-ray analysis, *Acta Crystallogr.* **11**, 774-781. (b) Ottersen, T. (1974) The structure of diformylhydrazine at 19 °C and -165 °C, *Acta Chem. Scand. A* **28**, 1145-1149. (c) Caira, M. R.; Watson, W. H.; Vögtle, F.; Müller, W. (1984) A 1:2 host-guest complex between 1,4,7,10,13,16-hexaoxacyclooctadecane (18-crown-6) and *N,N'*-diformohydrazide, $C_{12}H_{24}O_6.2C_2H_4N_2O_2$, *Acta Crystallogr., Sect. C* **40**, 136-138.
25. Demers, J. P.; Klaubert, D. H. (1987) Addition of arylmetallics to azodicarboxylates: a novel synthesis of arylhydrazines by aromatic hydrazination, *Tetrahedron Lett.* **28**, 4933-4934.
26. (a) Pauling, L.; Corey, R. B. (1953) Two rippled-sheet configurations of polypeptide chains, and a note about the pleated sheets, *Proc. Nat. Acad. Sci. USA* **39**, 253-256. (b) Ashida, T.; Tanaka, I.; Yamane, T. (1981) β-Pleated sheets in oligopeptide crystals, *Int. J. Peptide Protein Res.* **17**, 322-329.
27. Stott, K.; Stonehouse, J.; Keeler, J.; Hwang, T.-L.; Shaka, A. J. (1995) Excitation sculpting in high-resolution nuclear magnetic resonance spectroscopy: application to selective NOE experiments, *J. Am. Chem. Soc.* **117**, 4199-4200.

ANION RECOGNITION BY NEUTRAL RECEPTORS

Dmitry M. Rudkevich, Jurgen Scheerder, and David N. Reinhoudt*

Laboratory of Organic Chemistry
University of Twente
P.O. box 217
7500 AE Enschede
The Netherlands

Abstract. In this review the design of neutral receptors for anions is described. In these receptors, selective anion recognition takes place either exclusively via hydrogen bonding or by combination of a Lewis acidic UO_2-center and hydrogen bonds. An approach to neutral bifunctional receptors containing both anion and cation complexing sites is described. The results on selective $H_2PO_4^-$ anion transport and simultaneous transport of hydrophilic cations and anions through a supported liquid membrane are presented.

1. Introduction.

The recognition of cations and neutral molecules has had our attention since many years. A remaining challenge is the selective recognition and binding of anions by neutral molecules. In nature, phosphate and sulfate binding proteins are very important for the active transport systems in the cell and organelles [ref 1]. A very high selectivity in binding has been observed in prokariotic, periplasmic phosphate and sulfate binding proteins, which demonstrate $>10^5$ selectivity for complexation phosphate over sulfate and sulfate over phosphate, respectively. In those proteins the specific binding *exclusively* takes place through hydrogen bonding. Native zinc enzymes such as carboxypeptidase A (CPA) bind small inorganic anions like hydrogen phosphate via *combination* of metal-anion

C. S. Wilcox and A. D. Hamilton (eds.), Molecular Design and Bioorganic Catalysis, 137–162.

interactions and hydrogen bonding [ref 1].
The established way to recognize anions is the use of positively charged receptors. The disadvantage of these receptors is the limited selectivity due to the dominant electrostatic interactions that govern the anion recognition [ref 2]. This disadvantage can be overcome by the use of neutral anion receptors. Despite several neutral receptors incorporating Lewis acidic centers have been developed, they lack the possibility of structural variation in order to control the selectivity of the anion binding. To increase the selectivity, we have investigated neutral receptors that have, in addition to a Lewis acidic center, also directional modes of interaction via hydrogen bonds. The proper orientation of hydrogen bond donating and accepting groups enables a more selective anion recognition.

2. Anion recognition exclusively through hydrogen bonding.

In our first attempt, we have developed the tris(aminoethyl)amine based receptors which orient the hydrogen bond donating groups in a tetrahedral fashion (**1** and **2**) [ref 3]. These neutral receptors showed selective recognition of $H_2PO_4^-$ anions over Cl^- and HSO_4^- in MeCN (Table 1).

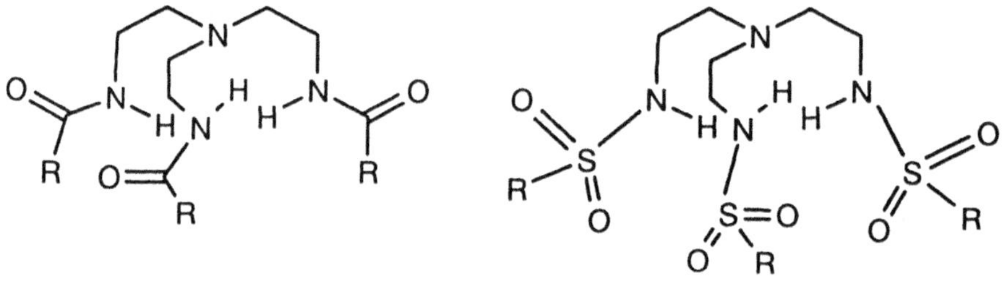

1a: R = CH_2Cl
1b: R = $(CH_2)_4CH_3$
1c: R = C_6H_5
1d: R = 4-$MeOC_6H_4$

2a: R = 4-$MeOC_6H_4$
2b: R = 2-naphthyl

We envisaged that the positioning of hydrogen bond donating groups on a molecular platform would enable further benefit from the directional character of hydrogen bonds and consequently further control the anion complexation. As a molecular platform we have used calix[4]arenes. This versatile building block can be rigidified, and both the upper rim and the lower rim can be used to position hydrogen bond donating groups.

Table 1. *Association constants (K_{ass}, M^{-1}) of receptors **1** and **2** determined by conductometry*

	$H_2PO_4^-$	HSO_4^-	Cl^-
1a	6.1×10^3	1.7×10^2	1.7×10^3
1b	2.8×10^2	3.1×10^1	2.9×10^2
1c	8.7×10^2	5.6×10^1	1.0×10^2
1d	5.1×10^2	7.3×10^1	1.9×10^2
2a	3.5×10^3	7.9×10^1	5.4×10^2
2b	1.4×10^4	3.8×10^1	1.6×10^3

Chlorosulfonylation of calix[4]arenes at the upper rim followed by the addition of a variety of amines gave a class of sulfonamide calix[4]arenes (**3**) [ref 4]. These receptors show a remarkable preference for the complexation of HSO_4^- over $H_2PO_4^-$ and Cl^- in $CHCl_3$ (> 10^2, Table 2). Comparison of the association constants with non-cyclic reference compounds (**4** and **5**, Table 2) shows that the preorganization of the binding sites on calix[4]arene substantially increases the association constants for the anion complexation. Also the lower rim of calix[4]arenes can be used to position binding sites for anions. Functionalization of the phenolic positions with two or four (thio)urea moieties gave a class of

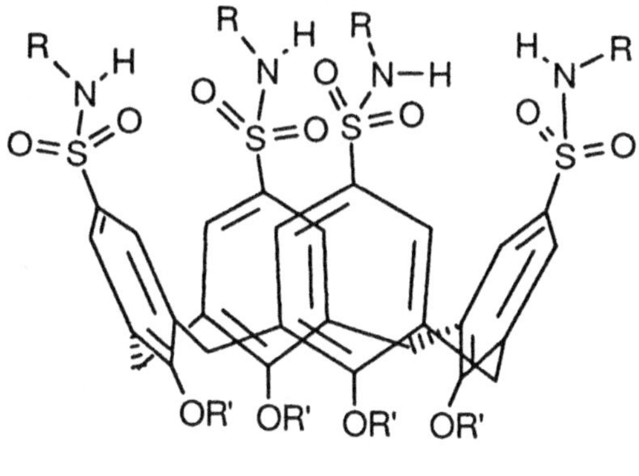

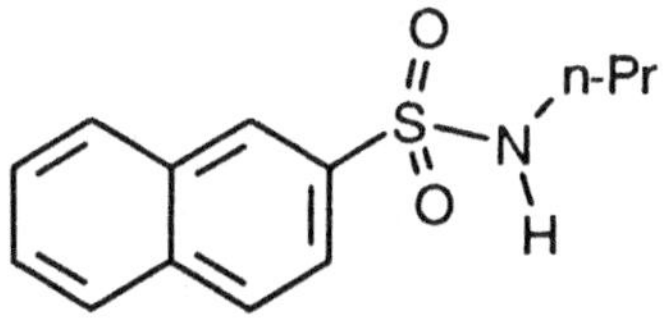

4

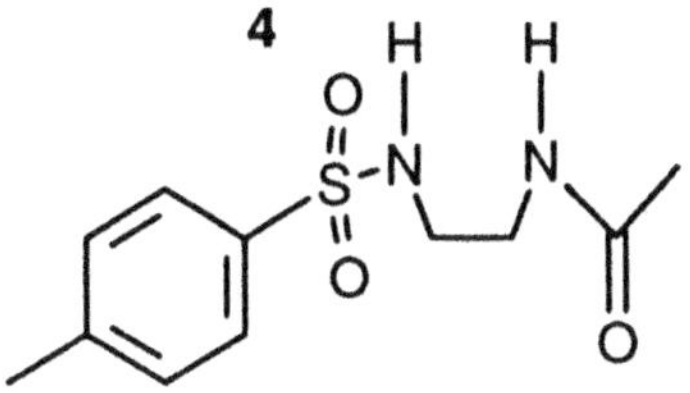

5

R' = $CH_2CH_2OCH_3$
3a: R = H
3b: R = *n*-Pr
3c: R = *tert*-Bu
3d: R = $CH_2CH_2NHC(O)CH_3$

neutral calix[4]arene (thio)urea derivatives (**6-9**) [ref 5]. Despite the flexible butyl spacers, the binding sites are preorganized due to hydrogen bonding between the complementary (thio)urea moieties. The complexation behaviour was studied by ^1H NMR titration experiments and FAB mass

Table 2. *Association constants (K_{ass}, M^{-1}) of receptors **3,4** and **5** determined by ^1H NMR in $CDCl_3$*

	$H_2PO_4^-$	HSO_4^-	Cl^-	NO_3^-
3b	3.5 x 10^2	9.7 x 10^2	3.6 x 10^2	2.4 x 10^2
3c	< 10	1.3 x 10^2	7.2 x 10^1	4.3 x 10^1
3d	-	1.0 x 10^5	1.3 x 10^3	5.1 x 10^2
4	1.4 x 10^1	1.0 x 10^1	1.5 x 10^1	< 10
5	2.6 x 10^2	3.5 x 10^2	3.3 x 10^2	9.9 x 10^1

spectrometry. The receptors showed 1:1 complexation of Cl^-, Br^-, I^-, N_3^-, and SCN^- with a selectivity for Cl^- (Table 3). A preference for hard anions (Cl^-, Br^-, or CN^-) over soft anions (I^-, or SCN^-) is found and spherical anions are bound more strong than linear anions. Despite the fact that the diurea derivative **8** has only two binding sites, the anion complexation is stronger, probably due to less extensive hydrogen bonding in the free ligands. Upon anion complexation these hydrogen bonds must be broken which is more demanding for the (tetrakis)-urea derivatives **6** and **7**. Surprisingly, the thiourea derivatives **7** and **9** show weaker anion complexation than the corresponding urea derivatives (**6e** and **8**, respectively), despite the higher acidity of the thiourea hydrogens.

The same strategy was applied using 1,3,5-trimethoxy-2,4,6-trihydroxycalix[6]arene as a building block [ref 6]. Functionalization of this derivative gave a class of anion receptors (**10** and **11**) which have a C_3 symmetry axis. A receptor with three binding sites arranged in a C_3 symmetry would be able to bind tricarboxylate anions. The anion complexation was studied by ^1H NMR titration experiments, FAB mass spectrometry, and FTIR spectroscopy. The association constants are summarized in Table 4. In all cases a 1:1 stoichiometry was found.

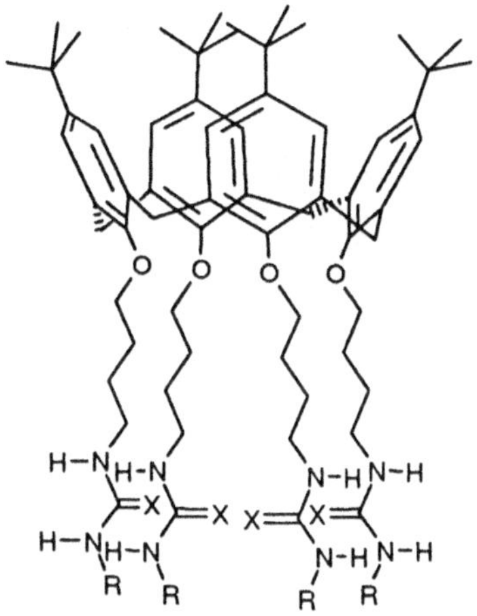

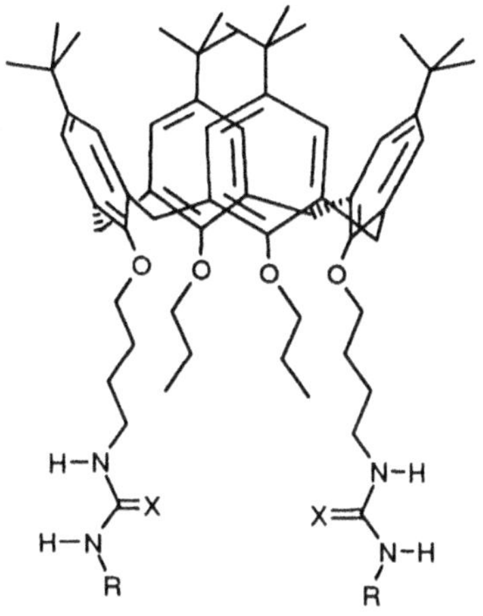

6a: X = O, R = phenyl
6b: X = O, R = *n*-propyl
6c: X = O, R = *n*-octyl
6d: X = O, R = *tert*-butyl
7: X = S, R = phenyl

8: X = O, R = phenyl
9: X = S, R = phenyl

Table 3. *Association constants (K_{ass}, M^{-1}) of receptors **6-9** determined by ^1H NMR in $CDCl_3$*

	Cl^-	Br^-	I^-	CN^-	SCN^-
6a	2.7 x 10^3	1.7 x 10^3	< 25	8.6 x 10^2	< 25
6b	< 25	< 25	< 25	< 25	< 25
6c	2.9 x 10^2	4.5 x 10^2	-	5.5 x 10^2	-
6d	2.1 x 10^3	1.3 x 10^3	-	8.0 x 10^1	-
7	3.4 x 10^2	5.8 x 10^2	< 25	8.6 x 10^2	< 25
8	7.1 x 10^3	2.6 x 10^3	6.1 x 10^2	1.1 x 10^3	< 25
9	1.3 x 10^3	4.9 x 10^2	-	6.7 x 10^2	370

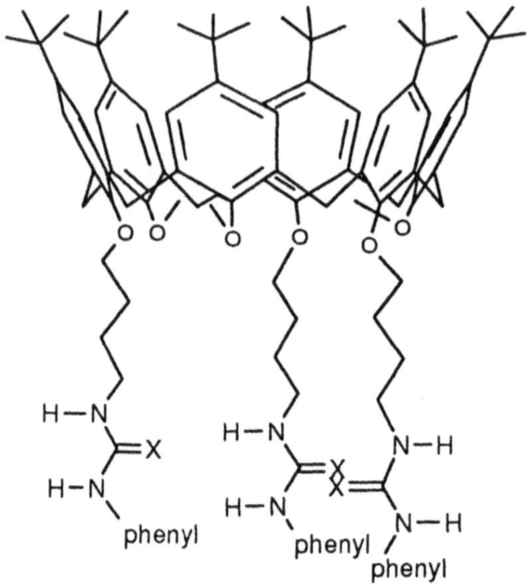

10 X = O **11** X = S

Comparison of the association constants for the complexation of benzoate, isophthalate, and 1,3,5-benzenetricarboxylate revealed cooperativity. The values for the association constants for 1,3,5-benzenetricarboxylate, 1,2,4-benzenetricarboxylate, and 1,2,3-benzenetricarboxylate show a strong preference for the binding of the anion with the same C_3 symmetry as the host. Also, the planar 1,3,5-benzenetricarboxylate is complexed in preference over the non-planar *cis*-1,3,5-cyclohexanetricarboxylate. This is due to the fact that the carboxylate groups of *cis*-1,3,5-benzenetricarboxylate can rotate freely and complexation of this anion will be accompanied with a larger unfavourable entropy term. As for the calix[4]arene (thio)urea derivatives, the thiourea derivatives are less effective to complex anions than the corresponding urea derivatives. To investigate this effect we studied the self-association behaviour of N-(n-butyl)-N'-phenyl-thiourea and N-(n-butyl)-N'-phenyl-urea. ^{1}H NMR dilution experiments showed that the thiourea derivative is more strongly associated than the urea derivative. Upon complexation the self-association must be broken which is more demanding for the thiourea derivative (**11**) than for the urea derivative (**10**) and this will result in a preferred binding of anions by the urea receptor.

Table 4. *Association constants (K_{ass}, M^{-1}) for receptors **10** and **11** determined by ^1H NMR in $CDCl_3$*

	10	11
Benzoate	1.6×10^3	1.4×10^3
Isophthalate	6.9×10^4	6.4×10^3
1,3,5-$(COO^-)_3$-benzene	8.7×10^4	2.9×10^5
1,2,4-$(COO^-)_3$-benzene	2.3×10^4	2.5×10^3
1,2,3-$(COO^-)_3$-benzene	4.7×10^4	1.8×10^4
1,3,5-$(COO^-)_3$-cyclohexane	1.0×10^5	2.9×10^4
Cl^-	4.8×10^2	< 25
Br^-	1.5×10^3	3.5×10^2

However, in case of the 1,3,5-benzenetricarboxylate breaking of the hydrogen bonds is more than compensated by the cooperative binding of three carboxylate groups by the three thiourea groups, and this results in a preferred binding of this anion by the thiourea host **11**. The complexation of Cl^- and Br^- was also investigated. In contrast to the corresponding calix[4]arene (thio)urea derivatives, the Br^- is complexed more strongly than Cl^-. The better complementarity of the Br^- with the cavity formed by the three (thio)urea moieties dominates the larger affinity of the binding site for the harder Cl^-.

3. UO_2-Sal(oph)ens as Anion Receptors

Previously we have reported that metallomacrocycles and clefts containing an immobilized Lewis acidic UO_2-cation, are excellent receptors for the complexation of neutral molecules as the result of coordination of a nucleophilic group (C=O, S=O, =N-) to the uranyl center in addition to H-bond formation and π-π stacking. The uranyl cation complexed in a salophen unit prefers a pentagonal bipyramidal coordination, with the two oxygens at the apical positions and with both the four-coordinating sites of the salophene moiety and a neutral molecule in the equatorial positions. In the same way the presence of both a uranyl Lewis acidic center and additional hydrogen-bonding sites like C(O)-NH fragments in a preorganized receptor molecule should increase the selectivity and efficiency of anion complexation [ref 7].

For the complexation of anions we have first studied the simple UO_2-

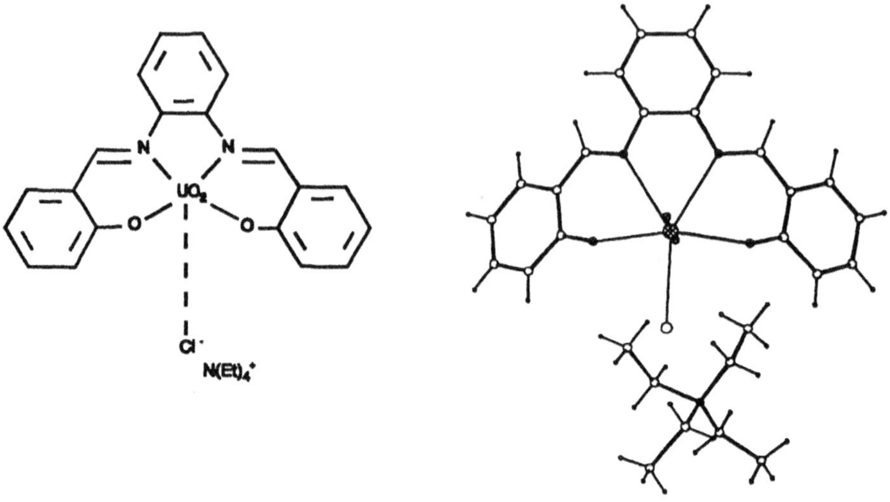

*Fig. 1. X-ray structure of the complex **12a**·Cl⁻*

containing salophen **12a**, the crystal structure of the complex **12a**·$Et_4N^+Cl^-$ is shown in Figure 1. The uranyl cation is coordinated to two oxygen atoms and two nitrogens of the salophen unit and to the *chloride anion* (U···Cl distance 2.76 Å) which clearly demonstrates the tight anion complexation. From ^{1}H NMR titration experiments of **12a** with $Bu_4N^+Cl^-$ in MeCN-d_3 (with 1% of DMSO-d_6) and by conductometry (see also Table 5) high K_{ass} values of 4.2 x 10^2 and 4.0 x 10^2 M^{-1}, respectively, were determined.

Synthesis of Anion Receptors and Anion Complexation in Solid State. Stirring of the uranyl containing salophen **12b** prepared from the corresponding aldehyde, 1,2-benzenediamine and $UO_2(OAc)_2$·$2H_2O$, with tetrabutylammonium dihydrogen phosphate $Bu_4N^+H_2PO_4^-$ in MeCN overnight, followed by evaporation of the solvent gave the corresponding anion complex as an orange powder. The negative FAB mass spectrum of the complex exhibit, in addition to a small peak of the free ligands **12b**, a very intense [**12b** + H_2PO_4]$^-$ signal, while a small [**12b** + H_2PO_4 + Bu_4N]$^-$ peak is also present. The crystal structure of the complex **12b** $Bu_4N^+H_2PO_4^-$ is shown in Figure 2. In this complex the uranium atom has the approximate pentagonal bipyramidal coordination, with the two oxygens in apical positions. In the equatorial plane besides coordination with the two nitrogens and two oxygens of the salophen moiety, the fifth coordination position is occupied by an oxygen atom of the dihydrogen

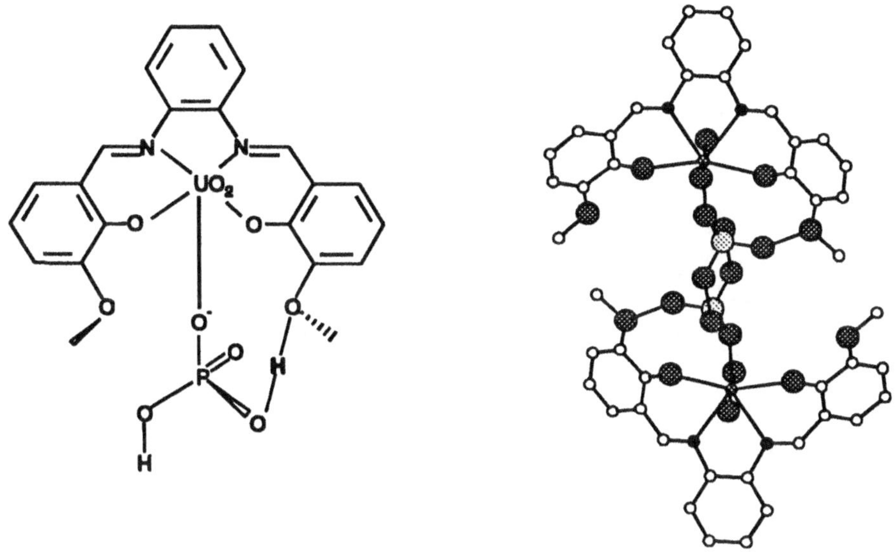

*Fig. 2. X-ray structure of the complex **12b** $H_2PO_4^-$*

phosphate ($H_2PO_4^-$) anion (U···O-P distance 2.28(2) Å). It is evident from Figure 2 that in the solid state the complexes are arranged in centrosymmetric pairs. The "core" of the dimer consists of two $H_2PO_4^-$ anions, which are connected by two short H-bonds (O···O distance 2.52(1) Å). Another H-bond (O···O-P distance 2.68(1) Å) is formed between the $H_2PO_4^-$ anion and a methoxy oxygen of the salophen moiety. This result shows the interesting phenomenon that anion binding in this type of compounds is effected by coordination to the UO_2-cation and is augmented by hydrogen bonding between anion and ligand.
In order to obtain anion receptors which contain a combination of both UO_2-cation and C(O)-NH amido functionalities as additional binding sites, cleft-type molecules **13a-j** and metallomacrocycle **14** containing a polyether bridge were designed [ref 7]. The crystal structure of the complex of ligand **13e** with $Bu_4N^+H_2PO_4^-$ is shown in Figure 3. As in the case of the complex of **12b** with $H_2PO_4^-$ (Figure 2), the $H_2PO_4^-$ anion is tightly coordinated to the UO_2-center (U···O-P distance 2.28(2) Å) in addition to a H-bond formation with the acetoxy oxygen of the salen moiety (O···O-P distance 2.84(2) Å). However, in this case two additional H-bonds between the amido groups of the ligand and the complexed anion are present (N···O-P distance 2.79(2) Å) which clearly shows the

participation of C(O)-NH fragments in anion complexation. As in Figure 2 the $H_2PO_4^-$ complexed to the UO_2-cation forms a H-bonded associate with the second $H_2PO_4^-$ anion (O···O distance 2.48(2) Å) which in this case is not complexed itself by ligand **13e**.

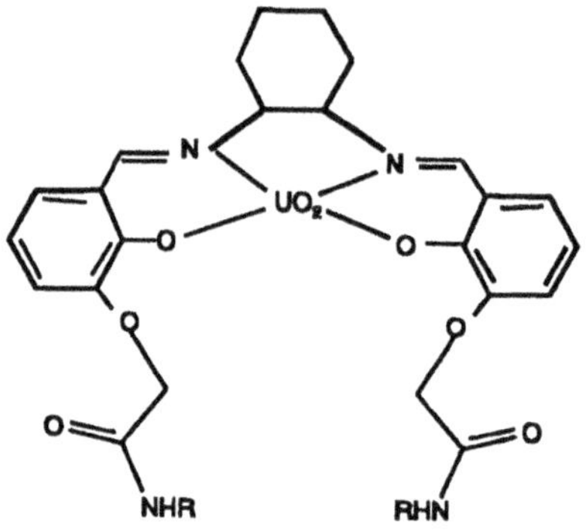

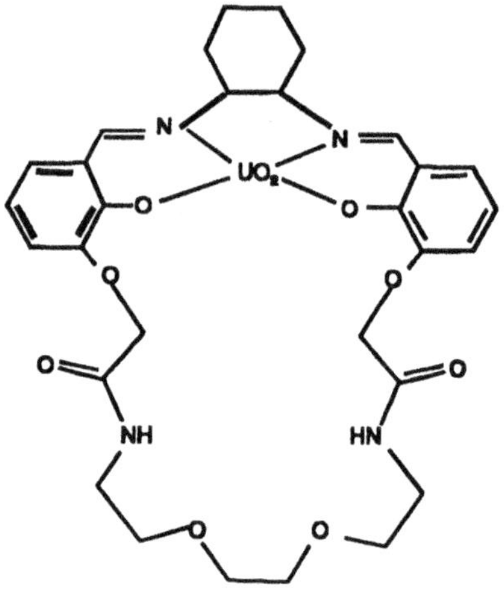

13a: R = H
13b: R = 4-MeC_6H_4
13c: R = n-$C_{18}H_{37}$
13d: R = 4-(n-$C_8H_{17}O$)-C_6H_4
13f: R = C(O)NH-n-$(CH_2)_8O$-C_6H_4-2-NO_2
13j: R = $(CH_2)_2NHSO_2$-4-MeC_6H_4

14

The covalent combination of a UO_2-center and C(O)NH functionalities was also immobilized at the upper rim of a rigid and lipophilic calix[4]arene platform, and consequently calixarene based UO_2-containing anion receptor **15** was prepared [ref 8].
Stirring of a mixture of the ligands **12b**, **13a,b** and **e**, **14**, or **15** and $Bu_4N^+H_2PO_4^-$ in MeCN overnight followed by evaporation of solvent gave the corresponding anion complexes as orange powders. In all cases the negative FAB mass spectra of the solid complexes exhibit, in addition to small peaks of the free ligands, very intense [Ligand + Anion]$^-$ signals, while small [Ligand + Salt]$^-$ peaks are also present. In the ^{1}H NMR spectra of all complexes significant changes of the host were found for the NH amido, the HC=N (except ligand **15**), and the $CH_2C(O)$ signals which clearly indicate the presence of a guest anion in the cavity.
In the ^{31}P NMR spectra of $H_2PO_4^-$ complexes with **12b**, **13a**, **13b**, and **14**

signals of $H_2PO_4^-$ are shifted downfield (1.9-2.3 ppm) in comparison with free $H_2PO_4^-$.

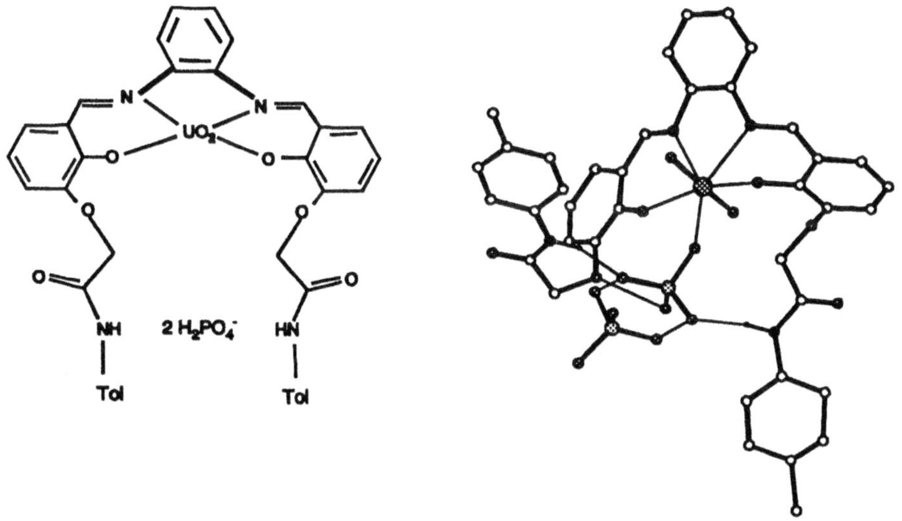

*Fig. 3. X-ray structure of the complex **13e** $2H_2PO_4^-$*

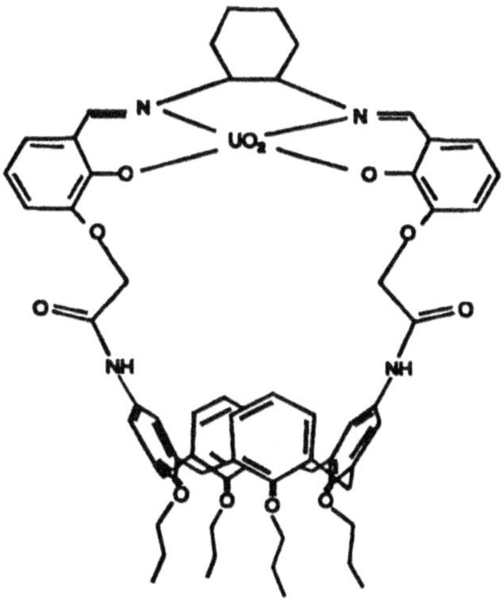

15

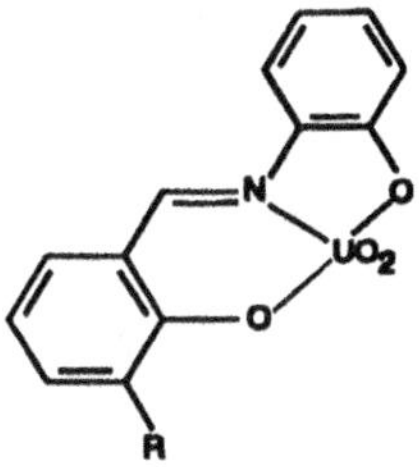

16a: R=H
16b: R=$OCH_2C(O)NH-C_6H_4Me$-4
16c: R=$O(CH_2)_2O-C_6H_4-C(O)NH_2$-2

In receptors **13a-e**, **14**, and **15** the uranyl center contains only one vacant position. The so-called "naked salophens" **16a-c** which have two vacant positions for complexation with guests were also synthesized.
The crystal structure of the complex **16a** · $2Bu_4N^+H_2PO_4^-$ is presented in Figure 4 and clearly shows that binding of two $H_2PO_4^-$ anions takes place (U···O-P distance 2.33 Å).

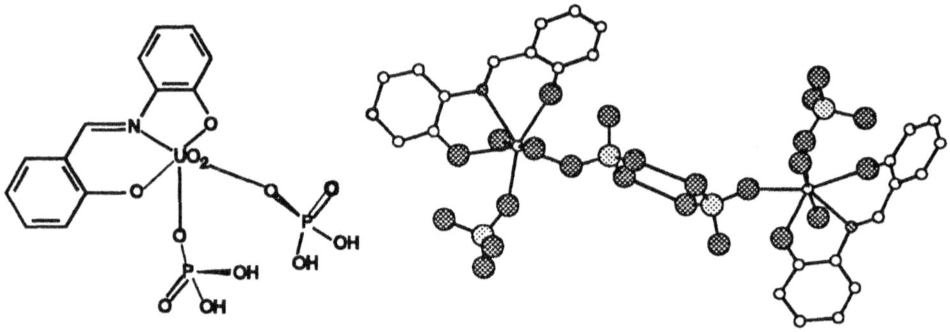

*Fig. 4. X-ray structure of the complex **16a** $2H_2PO_4^-$*

As in previous cases phosphate anions form H-bonded dimers with phosphates complexed by another molecule of **16a** (O···O distance 2.53 Å). At the same time due to the unique fact that two phosphates are complexed by the UO_2-cation, molecules of the complex are organized as a H-bonded ribbon [ref 7].
The first neutral ditopic receptor **17** for adenosine monophosphate (AMP^{2-}) employing the combination of a Lewis acid-anion interaction and complementary base-pairing, has been prepared [ref 9].
In receptor **17** both a UO_2-center and a thymine fragment participate in the nucleotide anion complexation. Solid complex of **17** with the 3'-AMP^{2-} anion exhibits in the negative FAB mass spectrum an intense (ca 55%) signal of [**17** + 3'-AMP + $Bu_4N]^-$ (Figure 5).

Anion Recognition in Solution. The complexation of anions in solution was studied first by conductometry. Already the simple UO_2-salophens **12a,b** show strong binding of different anions (e.g Cl^-, $H_2PO_4^-$, NO_2^-) in MeCN-DMSO, 99:1 solutions (Table 5). In all cases a preference for $H_2PO_4^-$ binding was observed. The binding of anions by metalloclefts **13a,b** and **d** and metallomacrocycle **14**, which contain amido C(O)-NH functionalities, is remarkably strong. The influence of C(O)-NH moieties that are able to form H-bonds with anions complexed is demonstrated (exept $H_2PO_4^-$) by comparing the K_{ass} values of compounds **12b** and **13a,b** and **d** (Table 5).

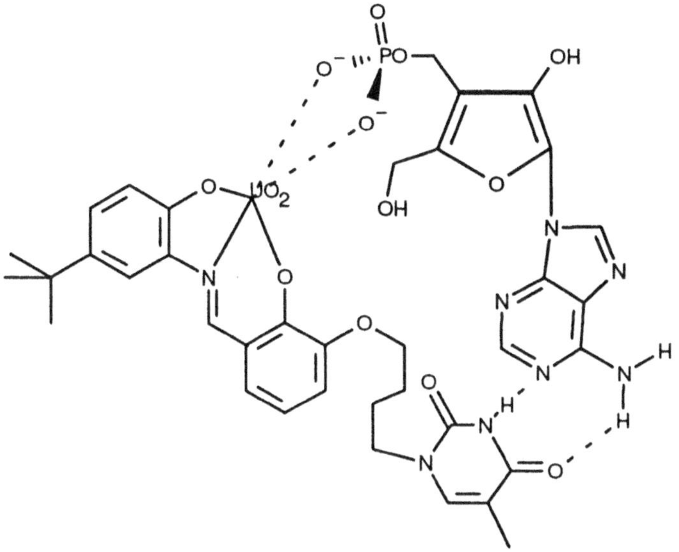

*Fig. 5. Proposed structure of the complex **17**·AMP²⁻*

More preorganized ligands (**13b**, **13d** and **14**) exhibit a very strong ($K_{ass} > 10^5$ M^{-1}) and selective complexation of $H_2PO_4^-$. Compound **13b** shows for $H_2PO_4^-$ selectivities of > 10^2 over Cl^- and > 10^3 over NO_2^-. For HSO_4^- and SCN^- K_{ass} values of < 3 x 10^2 M^{-1} were obtained for all ligands.

The complexation of anions in solution was also studied by NMR spectroscopy. The addition of $Bu_4N^+H_2PO_4^-$ to solutions of **12b**, **13a,b,d,e**, and **14** in DMSO-d$_6$ gave two sets of HC=N, $CH_2C(O)$, and NH signals, both for the free and complexed ligand, indicating that the formation of kinetically stable (on the NMR time scale) complexes with $H_2PO_4^-$ takes place even in pure DMSO-d_6. In case of **15** only the shift of C(O)NH-protons was observed. The contribution of the C(O)NH···$H_2PO_4^-$ hydrogen bond interaction to the anion complexation is clearly reflected in the markedly downfield shifts $\Delta\delta$ 0.4-0.9 ppm of the NH signals for **13a,b,d,e**, **14**, and **15**. From dilution experiments the K_{ass} values of $H_2PO_4^-$ binding were calculated (Table 6). The influence of C(O)NH-moieties that are able to form H-bonds is demonstrated by comparing the K_{ass} values of compounds **12b**, in which binding mainly takes place via electrostatic interaction, and functionalized salens **13a,b,d** and **e** (Table 6).

More preorganized ligands **13b**, **13d**, **13e**, and **14** exhibit a very strong

(K_{ass} > 10^3 M^{-1}) complexation of $H_2PO_4^-$. Under the same conditions, complexation with Cl^-, HSO_4^-, SCN^-, and ClO_4^- was not observed. No changes in the ^1H NMR spectra of ligands **13a-e**, **14**, and **15** were found after addition of tetraalkylammonium salts of these anions which indicates that very *selective* $H_2PO_4^-$ recognition takes place.

Table 5. *Association constants (K_{ass}, M^{-1}) of functionalized sal(oph)ens determined by conductometry in MeCN-DMSO (99:1)*[a]

Anion	$H_2PO_4^-$	Cl^-	NO_2^-
12a	1.5×10^4	4.0×10^2	3.1×10^2
12b	2.0×10^4	< 300	< 300
13a	1.9×10^4	4.0×10^3	8.9×10^2
13b	> 10^5	1.7×10^3	4.5×10^2
13c[b]	8.0×10^3	< 5.0	< 5.0
13d	> 10^5	2.9×10^3	4.7×10^3
14	> 10^5	1.2×10^4	1.5×10^3

[a]Tetrabutylammonium salts were used.
[b]Determined by ^1H NMR in $CDCl_3$-DMSO-d_6 (9:1) due to low solubility in MeCN and DMSO.

Table 6. *Association constants (K_{ass}, M^{-1}) for $H_2PO_4^-$ anion complexation determined with ^1H NMR in DMSO-d_6*[a]

Ligand	K_{ass}, M^{-1}
12b	5.1×10^2
13a	8.4×10^2
13b	8.0×10^3
13c	(8.0×10^3)[b]
13d	2.0×10^3
13e	1.5×10^3
14	1.8×10^3
15	3.5×10^2

[a]Tetrabutylammonium salt was used.
[b]Determined by ^1H NMR in $CDCl_3$-DMSO-d_6 (9:1) due to low solubility in DMSO.

Analogously in the ^{31}P NMR spectra the addition of $Bu_4N^+H_2PO_4^-$ to solutions of **12b**, **13a,b,d,e**, and **14** in DMSO-d_6 gave two signals ($\Delta\delta$ 1.9-2.3 ppm) of $H_2PO_4^-$ both for the free and complexed anion; the phosphate peak of the complex shifting downfield.
Since the "naked" salophens **16a-c** contain two free positions for coordination with guests, the complexation of the dianions of malonate and succinate was investigated. The K_{ass} values for **16a-c** and disodium salts of malonate and succinate dianions were determined by ^1H NMR titration experiments in DMSO-d_6 (Table 7). The HC=N signals of both the free and complexed ligand could be observed separately at 9.50-9.55 ppm and 9.40-9.47 ppm, respectively. Under the same conditions complexation with sodium acetate was not observed. From Table 7 it is clear that in the cases of **16b** and **16c** the contribution of the C(O)NH···dianion hydrogen bond interaction increases the strength of binding significantly. In the ^1H NMR spectra, NH signals of both free ligands **16b** and **16c** and their complexes can be separately observed at 10.50 ppm and 11.34 ppm for **16b**, respectively, and at 7.97, 7.45 ppm and 8.05, 7.85 ppm for **16c**, respectively. The previously described UO_2-salens like **13a,b**, and **14** do not or otherwise very weakly bind dianions as was concluded from the fact that no shifts in the ^1H NMR spectra have been observed after addition of malonate or succinate salts.

Table 7. *Association constants (K_{ass}, M^{-1}) of "naked" receptors determined with ^1H NMR in DMSO*[a]

Dianion	**16a**	**16b**	**16c**
Malonate	8.0×10^1	2.2×10^2	*b*
Succinate	1.7×10^2	4.6×10^2	1.5×10^2

[a]Disodium salts were used.
[b]No visible changes observed in ^1H NMR spectra.

Titration of 3'-AMP^{2-} (as tetrabutylammonium salt) with **17** in DMSO-d_6 exhibited the appearance of an upfield HC=N singlet at 9.34 ppm at the expense of the free host singlet at 9.50 giving association constant of K_{ass} 1.2×10^3 M^{-1} [ref 9].
Dinuclear UO_2-sal(oph)en metallomacrocycles **18** and **19**, containing flexible and rigid cavities, respectively, form strong solution complexes with dicarboxylate anions such as terephthalate, succinate and fumarate in DMSO-d_6; weaker interactions between these hosts and the shorter

malonate and monoanionic benzoate were observed (Table 8) [ref 10].

Table 8. *Association constants (K_{ass}, M^{-1}) of dinuclear receptors **18** and **19** determined with ^1H NMR in DMSO[a]*

Anion	**18**	**19**
Succinate[a]	5.0×10^3	2.7×10^3
Malonate[a]	4.5×10^2	1.7×10^2
Fumarate[a]	$> 10^5$	1.8×10^4
Terephthalate[b]	1.4×10^4	2.0×10^4
Benzoate[b]	2.0×10^2	*c*

[a]Tetrabutylammonium salts were used.
[b]Tetraethylammonium salts were used.
[c]No visible changes observed in the ^1H NMR spectra.

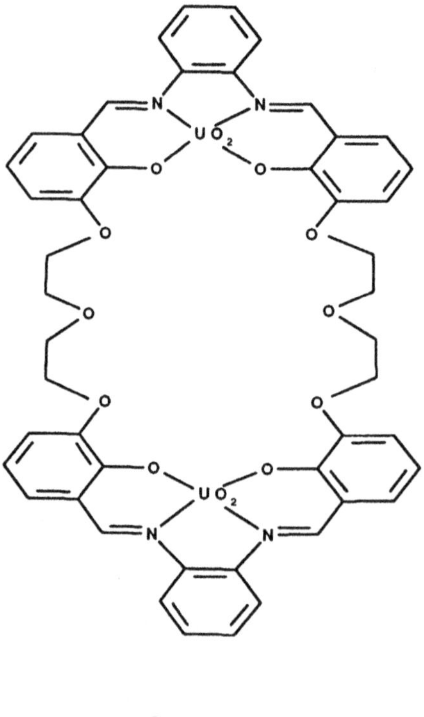

18

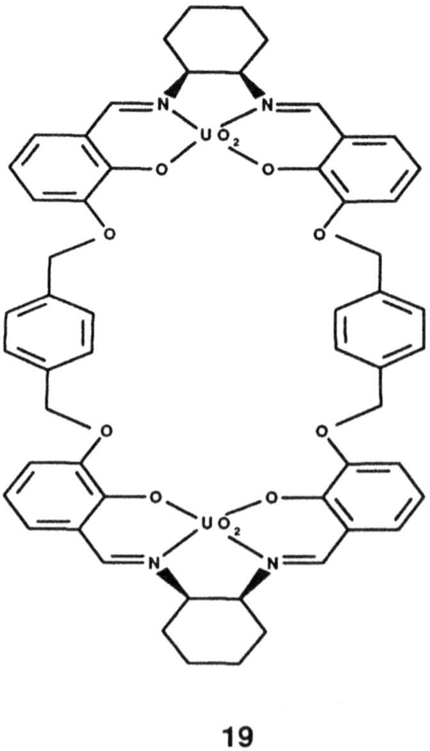

19

4. Bifunctional Receptors

Subsequently we have developed the synthesis of so-called *bifunctional* neutral receptors, which contain both anion and cation binding sites and therefore are able to complex simultaneously anionic and cationic species in apolar solvents.

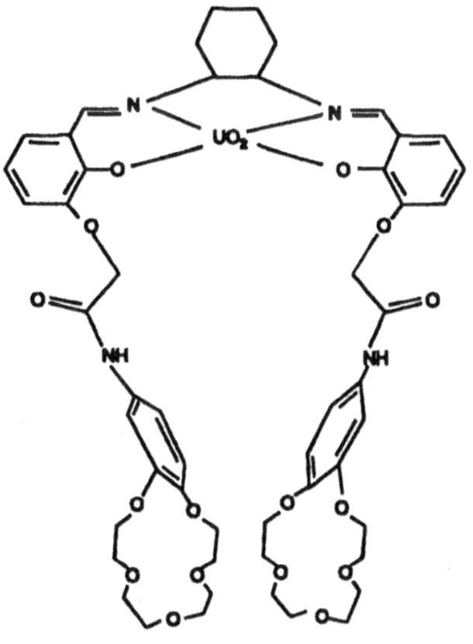

20

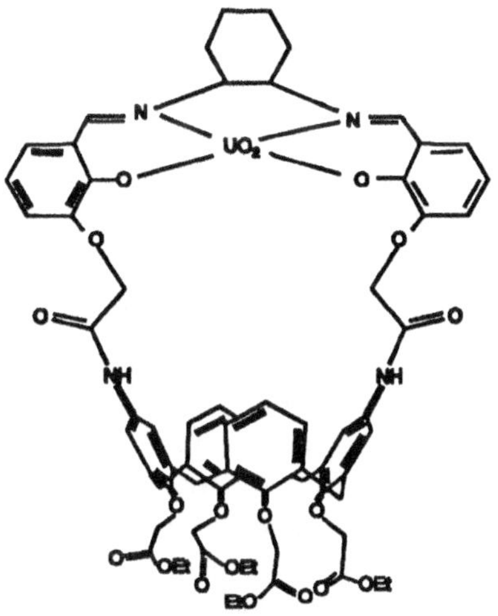

21

Bifunctional Receptor for KH_2PO_4 [ref 11]. In order to find general approaches for the preparation and complexation of bifunctional receptors, first the simple covalent combination of a UO_2-salen with two benzo[15]crown-5 fragments in one neutral molecule **20** was studied. Individually UO_2-containing clefts complex anions, especially $H_2PO_4^-$ (vide supra) and bis(benzo[15]crown-5) ethers bind cations like potassium K^+ with a high degree of selectivity.

The complexation properties of the bifunctional receptor **20** towards cations and anions were first investigated independently by ^1H NMR spectroscopy, cyclic voltammetry, FAB mass spectrometry, and liquid-liquid extraction experiments. In ^1H NMR dilution experiments of **20** with $Bu_4N^+H_2PO_4^-$ in DMSO-d_6 the imino HC=N signals of both the free and complexed ligand can be separately observed at 9.51 and 9.46 ppm, respectively. From this experiment a K_{ass} of 1.1 x 10^3 M^{-1} was calculated. The contribution of the C(O)NH$\cdots H_2PO_4^-$ hydrogen bond interaction to the overall anion complexation can clearly be observed in the ^1H NMR spectrum even in DMSO-d_6 as a solvent, showing the N-H signals of the free and complexed **20** at 10.47 and 11.21 ppm, respectively. Complexation of potassium cation K^+ by receptor **20** was studied by picrate extraction experiments to give in $CHCl_3$ a value of log K_{ex} = 5.3 for the 1:1 **20** K^+Pic^- complex. The complexation of K^+ and $H_2PO_4^-$ ions by **20** has also been investigated by cyclic voltammetry in DMSO with 2% of H_2O using tetrabutylammonium tetraphenylborate as a supporting electrolyte. Addition of $Bu_4N^+H_2PO_4^-$ to a solution of **20** causes a clear shift of both cathodic and anodic peaks toward more negative potentials. This effect is accompanied by a systematic increase of the peak separation, from 85 to 100 mV, and by a decrease of the normalized peak heights. The results suggest that a labile, electroinactive complex with $H_2PO_4^-$ is formed. Assuming 1:1 stoichiometry, a K_{ass} value of 1.3 x 10^3 M^{-1} for $H_2PO_4^-$ was obtained which is in a good agreement with the ^1H NMR measurements. Addition of potassium tetraphenylborate to the DMSO solution of receptor **20** gave no changes in the electrochemical behavior, probably due to the fact that the crown ether moieties in **20** are situated rather far from the electroactive UO_2-center. However, an indirect procedure for the determination of complexation constants, based on competition between Tl^+ and K^+ cations gave a K_{ass} value of 1.0 x 10^2 M^{-1} for K^+. Finally, FAB-MS was used which is an established technique for investigation of non-covalent binding of cationic and anionic complexes. In the positive FAB-MS spectrum (*m*-nitrobenzyl alcohol as a matrix) of the 1:1 complex of receptor **20** and KH_2PO_4, prepared by mixing of the host and guest in MeCN with 10% of water followed by evaporation of solvents, an intense peak corresponding to $[\mathbf{20} + K]^+$ was observed. The corresponding negative FAB-MS spectrum of the same sample showed an intense peak of $[\mathbf{20} + H_2PO_4]^-$, while a signal of the $[\mathbf{20} + H_2PO_4 + K]^-$ was also present, and this clearly proves the complexation of the salt.

Calix[4]arene Based Bifunctional Receptor for NaH_2PO_4 [ref 8]. In this paragraph the first calix[4]arene based bifunctional receptor **21** is described. This compound contains a combination of a UO_2-Lewis acidic center and C(O)NH groups which is known to act as anionic binding site for $H_2PO_4^-$. Besides calixarene **21** contains also four preorganized ester fragments which are known to complex alkali metal cations with a high selectivity for Na^+.

A study of the binding ability of receptor **21** has been carried out using the general strategy we described for the simple bifunctional molecule **20**. In this way it was found that receptor **21** selectively binds $H_2PO_4^-$ anions. From ^1H NMR dilution experiments with $Bu_4N^+H_2PO_4^-$ in

DMSO-d_6 an association constant K_{ass} of 3.9 x 10^2 M^{-1} was calculated. The contribution of the C(O)NH···$H_2PO_4^-$ hydrogen bond interaction to the anion complexation can be seen from a significant downfield shift of the C(O)NH protons of ca 0.4 ppm upon complexation. Only minor shifts were observed upon dilution experiments with tetrabutylammonium salts of Cl^-, HSO_4^- and ClO_4^- anions which indicates their weak binding (K_{ass} < 10 M^{-1}). In the negative FAB mass spectrum of the 1:1 complex of **21** with $Bu_4N^+H_2PO_4^-$, prepared by mixing of host and guest in MeCN, an intense peak corresponding to [**21** + H_2PO_4]$^-$ was observed. Moreover, in the positive FAB mass spectrum of the 1:1 complex of **21** and $Na^+H_2PO_4^-$, prepared by mixing of host and guest in MeCN-H_2O, 10:1, an intense peak corresponding to [**21** + Na]$^+$ was observed, while the corresponding negative FAB mass spectrum of the same sample yielded an intense peak for [**21** + H_2PO_4]$^-$, which proves the complexation of both cation and anion in one bifunctional receptor molecule.

5. Anion Carrier Mediated Membrane Transport of Phosphate; Selectivity of $H_2PO_4^-$ over Cl^-

Since last two decades numerous papers have appeared on the selective transport of salts through supported liquid membranes (SLM) mediated by neutral cation carriers. In transport assisted by neutral cation carriers (CC) such as simple crown ethers, calixarenes, or natural ionophores like valinomycin, anions affect the transport rates because of the different dehydration energies. Therefore, lipophilic anions like ClO_4^-, NO_3^-, or SCN^- are often used as the counterion. Much less lipophilic anions like Cl^- or $H_2PO_4^-$ have only been transported by charged anion carriers via an ion-exchange mechanism, and the transport selectivity follows the Hofmeister series: ClO_4^- > I^- > SCN^- > NO_3^- > Br^- > Cl^- >> CO_3^{2-}, $H_2PO_4^-$, SO_4^{2-}. Here we report the selective transport of $H_2PO_4^-$ over Cl^- through SLM, either exclusively by an anion receptor or by a combination of anion and cation receptors [ref 12]. The receptors exhibit selectivity opposite to the order of dehydration energies of the anions in the Hofmeister series.

Anion receptors **13d,f,j**, described previously, were used in SLM with *o*-nitrophenyl-*n*-octyl ether (NPOE) as the membrane solvent immobilized in an Accurel/1E-PP support (Table 9). Receptors **13d,f,j** are already effective on their own as $H_2PO_4^-$ carriers in the transport of KH_2PO_4, although the fluxes are low. Surprisingly, despite strong $H_2PO_4^-$ binding detected for **13d** by ^{1}H NMR (DMSO) and conductometry (MeCN:DMSO=99:1), transport of KH_2PO_4 through NPOE could not be detected.

Table 9. *Effect of the combination of anion carriers and cation carrier **22** on KH_2PO_4 flux*

Carrier(s)[a]	J (KH_2PO_4)[b], 10^{-8} mol m^{-2} s^{-1}
13d	< 0.2
13f	3.2
13j	5.1
13d + 22	7.2
13f + 22	5.0
13j + 22	12.5

[a][Anion carrier] = [Cation carrier] = 0.02 M.

[b][KH_2PO_4] + [K_2HPO_4] = 0.2 M; pH_s = 6.8

To improve the flux a K^+ selective cation carrier, calix[4]arene crown-5 **22**, was added to the membrane solution for the simultaneous facilitation of the potassium ion transport (Table 9). The flux of KH_2PO_4 mediated by the combination of carriers **13j** and **22** is the highest in the series (12.5×10^{-8} mol m^{-2} s^{-1}) and higher than obtained with any of the anion carriers alone. The absolute concentrations of the 1:1 combination of carriers **13d** and **22** have also been varied. A steady increase of the flux was observed. The fact that receptor **13d** alone was not effective as a carrier for KH_2PO_4, but only in combination with **22**, prompted us to measure the KH_2PO_4 flux as a function of their concentration ratio [**13d**]/[**22**] (Figure 6).

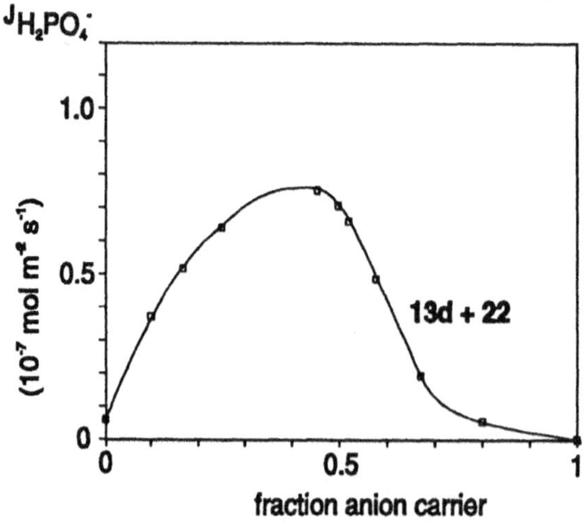

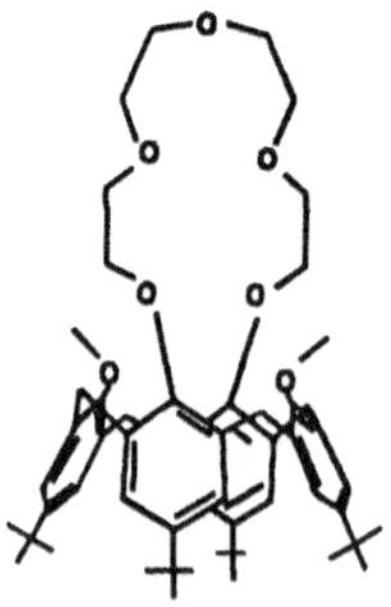

Fig. 6. *Transport of KH_2PO_4 mediated by the combination of carriers **13d** and **22** as a function of the percentage of anion carrier. [**13d**] = [**22**] = 0.04 M; $[KH_2PO_4]_s + [K_2HPO_4]_s$ = = 0.2 M; pH_s = 6.8*

22

The optimum flux is reached at about equal concentrations of **13d** and **22**. Finally, we have investigated the transport selectivity for $H_2PO_4^-$ over Cl^- in competition experiments from a source phase that contained 1 x 10^{-3} M of $KH_2PO_4^-$ and 1 x 10^{-1} M of KCl by the combination of anion receptors **13d,f,j** and carrier **22** (Table 10).

Table 10. *transport selectivity for $H_2PO_4^-$ over Cl^-*

Anion carrier(s)[a]	J ($H_2PO_4^-$)[b], 10^{-8} mol m^{-2} s^{-1}	J (Cl^-)[c], 10^{-8} mol m^{-2} s^{-1}	S
13d	2.19	7.60	29
13f	1.81	4.75	38
13j	7.52	5.25	143

[a][Anion carrier] = [Cation carrier] = 0.02 M.
[b]$[KH_2PO_4]_s$ = 0.001 M, no K_2HPO_4 added.
[c]$[KCl]_s$ = 0.1 M.

The selectivity, *S*, is defined as the ratio of the fluxes divided by the ratio of source phase anion concentrations. All anion carriers show transport selectivity for $H_2PO_4^-$ over Cl^-, which is opposite to the dehydration energy according to the Hofmeister series. The transport selectivity for phosphate over chloride increases as a function of the anion receptor in the order **13d** < **13f** < **13j** up to a value of about 140.

6. Simultaneous Transport of Cations and Anions through a Supported Liquid Membrane.

Previously we have described the neutral bifunctional receptors for the *simultaneous complexation* of hydrophilic anions and cations in organic media. We have also investigated the *simultaneous transport* of cations and anions through SLM assisted by a neutral bifunctional receptor [ref 13].

Our synthetic strategy was based on the attachment of both cation and anion binding sites to the rigid lipophilic calix[4]arene platform The covalent combination of a Lewis acidic UO_2-center and amido C(O)NH moieties provides an excellent receptor site for dihydrogen phosphate $H_2PO_4^-$ and chloride Cl^- anions, and that the calix[4]arene crown-6 (1,3-alternate) fragment is capable of thhe selective complexation of cesium ion Cs^+. Thus, bifunctional receptor **23** has been prepared and used as a carrier to investigate the transport of hydrophilic cesium chloride (CsCl) and the more lipophilic cesium nitrate ($CsNO_3$) [ΔG^0_{tr} (X^-, $H_2O \rightarrow MeCN$)

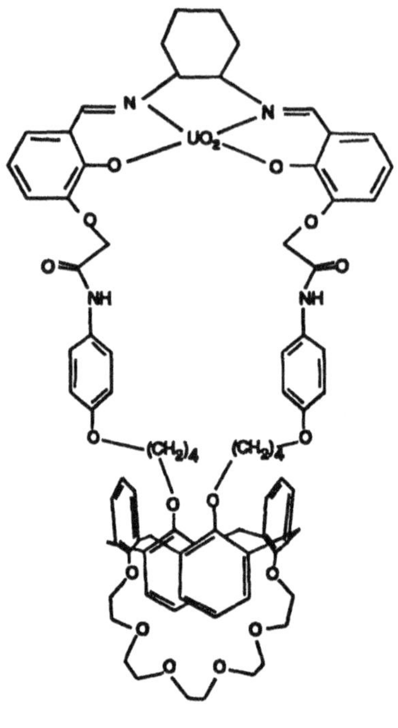

23

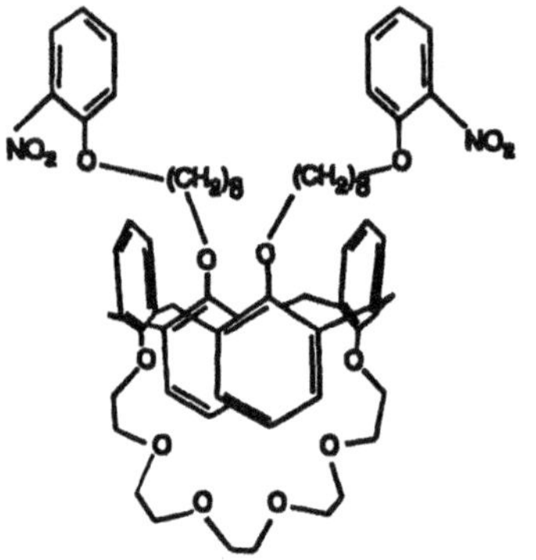

24

42.1 and 21.0 kJ/mol for Cl^- and NO_3^-, respectively] across a supported liquid membrane composed af a porous polymeric support (Accurel) impregnated with *o*-nitrophenyl *n*-octyl ether (NPOE). For comparison, the same expperiment was performed with the receptors **13c** and **24** that have either only anion or cation binding sites, respectively (Table 11). The transport processes for $CsNO_3$ and CsCl are different; NO_3^- is a much more lipophilic than Cl^- and only NO_3^- can easily follow the complexed Cs^+ cation through the hydrophobic membrane, even in the absence of anion carrier. With the cation carrier **24** a high flux of $CsNO_3$ (5.5×10^{-7} mol m^{-2} s^{-1}) was observed but the anion receptor **13c**, which is not selective for NO_3^-, did not transport $CsNO_3$. The flux was very low (0.02×10^{-7} mol m^{-2} s^{-1}) and comparable with the (blank) flux obtained without carrier. It implies that, probably, in case of **24** *only* the cation binding site is responsible for the transport.

Table 11. *Salt fluxes[a-c] trough SLM measured for different carriers[d,e] in NPOE*

Carrier	$CsNO_3$ flux	CsCl flux
13c	0.02	0.07
24	5.50	0.42
23	0.89	1.20

[a]Salt concentration 0.1 M.
[b]Fluxes in 10^{-7} mol m^{-2} s^{-1} after 24 h at 298 K.
[c]Blank fluxes of the salts in NPOE, for CsCl, 0.05 x 10^{-7} mol m^{-2} s^{-1}, and for $CsNO_3$, 0.02 x 10^{-7} mol m^{-2} s^{-1}.
[d]Carrier in membrane, 0.01 M.
[e]No leakage of receptors was observed in blank experiments.

The transport of CsCl by the monofunctional carriers **13c** (anion) and **24** (cation) exhibits low flux values of 0.07 x 10^{-7} mol m^{-2} s^{-1} and 0.42 x 10^{-7} mol m^{-2} s^{-1}, respectively. Obviously, when one of the ionic species is complexed, the uncomplexed counter-ion can not sufficiently penetrate the lipophilic membrane.

However, a significant flux (1.20 x 10^{-7} mol m^{-2} s^{-1}) was observed for bifunctional carrier **23** with CsCl, which is much higher than the corresponding ones for the monofunctional carriers **13c** and **24**. At the same time, carrier **23** showed a surprisingly low flux of $CsNO_3$ (0.89 x 10^{-7} mol m^{-2} s^{-1}) when compared with the flux observed for cation receptor **24** (5.50 x 10^{-7} mol m^{-2} s^{-1}). This proves that: (i) *both* anion and cation binding sites of **23** are involved in the complexation; and (ii) the presence of only an anion or a cation binding site in the receptor molecule is *not sufficient* for effective transport of a *hydrophilic* salt such as CsCl. But more important is that this suggests a *preference* of hydrophilic CsCl over lipophilic $CsNO_3$.

These results indicate the unique feature of receptors in which both binding sites are *covalently* linked.

7. Conclusions.

Approaches to *neutral anion receptors* which are able to *selectively* complex anionic species either by hydrogen bonds or by combination of Lewis acidic UO_2 center and amido functionalities are developed. A new concept has been described of so called *bifunctional* receptors which contain both anion and cation binding sites and therefore are able to complex *simultaneously* anionic and cationic species in apolar solvents. the calix[4]arene platform (cone and 1,3-alternate conformations) and calix[6]arene platform have been successfully applied for the immobilization of (cation and) anion binding sites in the preparation of different types of anion and bifunctional receptors. The results on selective transport of $H_2PO_4^-$ anions through a supported liquid membrane assisted by anion receptors, and simultaneous transport of hydrophilic cations and anions assisted by a neutral bifunctional receptor are described. This is the first example of carrier mediated co-transport, in which the anion and cation of a hydrophilic salt are bound and transported simultaneously through a membrane.

8. Acknowledgements.

The work described in this paper is the result of the efforts of more than 10 coworkers whose names appear in the references. Financial support from the Technology Foundation (STW), Technical Science Branch of the Netherlands Organization for Scientific research (SON/NWO) is gratefully acknowledged.

9. References.

[1] (a) Leucke, H.; Quiocho, F. A. *Nature* **1990**, *347*, 402-406. (b) He, J. J.; Quiocho, F. A. *Science* **1991**, *251*, 1479-1481. (c) Pflugrath, J. W.; Quiocho, F. A. *J. Mol. Biol.* **1988**, *200*, 163-180. (d) Kanyo, Z. F.; Christianson, D. W. *J. Biol. Chem.* **1991**, *266*, 4264-4268. (e) Thatcher, G. R. J.; Cameron, D. R.; Nagelkerke, R.; Schmitke, J. *J. Chem. Soc., Chem. Comun.* **1992**, 386-388. (f) Mangani, S.; Ferraroni, M.; Orioli, P. *Inorg. Chem.* **1994**, *33*, 3421-3423.

[2] For a recent review see: Dietrich, B. *Pure Appl. Chem.* **1993**, *65*, 1457-1464.

[3] Valiyaveettil, S.; Engbersen, J. F. J.; Verboom, W.; Reinhoudt, D. N. *Angew. Chem., Int. Ed. Engl.* **1993**, *32*, 900-901.

[4] Morzherin, Y.; Rudkevich, D. M.; Verboom, W.; Reinhoudt, D. N. *J. Org. Chem.* **1993**, *58*, 7602-7605.

[5] Scheerder, J.; Fochi, M.; Engbersen, J. F. J.; Reinhoudt, D. N. *J. Org. Chem.* **1994**, *59*, 7815-7820.

[6] Scheerder, J.; Engbersen, J. F. J.; Casnati, A.; Ungaro, R.; Reinhoudt, D. N., submitted for publication.

[7] (a) Rudkevich, D. M.; Stauthamer, W. P. R. V.; Verboom, W.; Engbersen, J. F. J.; Harkema, S.; Reinhoudt, D. N. *J. Am. Chem. Soc.* **1992**, *114*, 9671-9673. (b) Rudkevich, D. M.; Verboom, W.; Brzozka, Z.; Palys, M. J.; Stauthamer, W. P. R. V.; van Hummel, G. J.; Franken, S. M.; Harkema, S.; Engbersen, J. F. J.; Reinhoudt, D. N. *J. Am. Chem. Soc.* **1994**, *116*, 4341-4351.

[8] Rudkevich, D. M.; Verboom, W.; Reinhoudt, D. N. *J. Org. Chem.* **1994**, *59*, 3683-3686.

[9] Lacy, S. M.; Rudkevich, D. M.; Verboom, W.; Reinhoudt, D. N. *Tetrahedron Lett.* **1994**, *35*, 5953-5956.

[10] Lacy, S. M.; Rudkevich, D. M.; Verboom, W.; Reinhoudt, D. N. *J Chem. Soc., Perkin Trans. 2* **1995**, 135-139.

[11] Rudkevich, D. M.; Brzozka, Z.; Palys, M.; Visser, H. C.; Verboom, W.; Reinhoudt, D. N. *Angew. Chem., Int. Ed. Engl.* **1994**, *33*, 467-468.

[12] Visser, H. C.; Rudkevich, D. M.; Verboom, W.; de Jong, F.; Reinhoudt, D. N. *J. Am. Chem. Soc.* **1994**, *116*, 11554-11555.

[13] Rudkevich, D. M.; Mercer-Chalmers, J. D.; Verboom, W.; Ungaro, R.; de Jong, F.; Reinhoudt, D. N. *J. Am. Chem. Soc.* **1995**, *117*, 6124-6125.

ENZYMES IN THE CONTINUOUS GAS PHASE BIOREACTOR

Fangxiao Yang and Alan J. Russell
Center for Biotechnology and Bioengineering
Department of Chemical Engineering
University of Pittsburgh
300 Technology Drive
Pittsburgh, Pennsylvania, 15219, USA

Enzyme catalyzed gas phase reactions may offer advantages over many conventional solution-based chemical processes. For example, in the analytical field, enzyme-based gas sensors may be more sensitive, faster and accurate than conventional electric or chemical probes. One of the critical parameters in enzyme gas phase substrate reactions is the water content of the gas mixture. The study of water adsorption isotherms for enzymes could provide insight into how water interacts with enzymes and the relationship between enzyme activity and hydration. Alcohol dehydrogenase from baker's yeast (YADH) was used as model enzyme to study the effect of hydration on catalytic performance. As expected, enzyme activity increases and stability decreases with increasing hydration. It has been found that for YADH stability is significantly enhanced in the presence of gaseous substrates and the cofactor can be efficiently regenerated in the system. Indeed, the cofactor turnover number is higher in the gas reactor than in water or organic solvents. A temperature independent critical hydration level is necessary in order to "switch on" both catalytic activity and protein stability. Nevertheless, the level of enzyme hydration has no effect on the enzymatic reaction mechanism or apparent activation energy.

1. Introduction

Since the discovery that enzymes catalyze unique reactions in nonaqueous media [18], our understanding of enzymes in aqueous/nonaqueous environments has considerably increased. Enzymes are particularly stable at low hydration levels when suspended in organic solvents [9, 42]. The introduction of enzymes into gas phase systems broadens the potential industrial applications of biocatalysts, and also presents a fascinating system in which to study the role of hydration in enzyme activity. Transformation of gaseous reactants by dehydrated enzymes or intact cells is emerging as a new technology with applications in chemical analysis gas-solid bioreactors [1, 2, 12, 13, 24, 26, 27, 38-40]. There are many advantages for carrying out enzymic reactions with gas phase substrates

C. S. Wilcox and A. D. Hamilton (eds.), Molecular Design and Bioorganic Catalysis, 163–190.

[1, 20], including the following:

(1) Many gaseous substrates have low solubility in all liquid solvents. The concentration of these substrates in the gas phase can be readily increased, however, by manipulating the operation pressure.
(2) Enzymes in a dehydrated form are usually more resistant to thermoinactivation.
(3) The rate of mass transfer for gaseous substances is much higher than in liquids, therefore, reactions are less likely to be limited by external diffusion.
(4) Enzyme preparation for gas reactors is straightforward because immobilization is either unnecessary or much simplified. Recovery of the biocatalyst form either batch or continuous gas phase processes is also straightforward.
(5) No solvent is needed, therefore, process operations cost can be minimized.
(6) Products or unconverted substrates can be separated and recovered with high purities by facilitated fractional condensation.
(7) Volatile, toxic or inhibitory products may be removed continuously.

There are two main concerns for applying enzymes in gas phase systems: (a) volatilizing the substrate and product, and (b) optimizing the catalytic performance (activity and stability). Substrate concentration in the gas phase depends on the reactants vapor pressure, and the temperature. Either raising temperature or reducing pressure would result in increased substrate concentration. Enzyme activity and stability in the gas phase reactions are dependent on the degree of enzyme hydration and reaction temperature. Thus, optimization of the system design and operating conditions would enable to improve enzyme performance.

While the application of enzymes with gaseous substrates is a new area of research, there are already a number of published investigations which provide insight into the relevant underlying processes.

2. Enzyme Hydration in the Continuous Gas Phase

The persistence of catalytic activity and enhanced stability of enzymes at low degree of hydration calls into question the role of water plays in enzyme catalysis [2, 27, 29, 36, 42]. The study of enzyme structure-function in aqueous solution does not enable a careful study of the structural role of water, since the concentration of the water is invariable. Studies of enzyme water adsorption behavior at low levels of protein hydration can, however, more clearly define the mechanisms by which water participates in enzyme catalysis [23].

As we have stated before [39] "The ideal water adsorption study will address a number of relevant questions: (1) How much water does an enzyme adsorb at equilibrium under given conditions? (2) How strong is the interaction between the enzyme and water molecules? (3) What is an enzyme's integrity in the gas phase? i.e. Does enzyme structure and shape change during water adsorption process? (4) If an enzyme structure changed during a water adsorption, does it relate to alternations in enzyme activity and stability? (5) Can the degree of hydration of the enzyme be affected by the presence of substrates?"

The intermolecular structure of liquid water deposited on large biomolecule surface was the subject of significant research in the 60s and 70s [3, 10, 17]. The sorption of H_2O vapor by dried proteins is generally assumed to involve the binding of H_2O molecules to specific hydrophilic sites at lower relative humidities followed by condensation or multimolecular adsorption as the humidity increases [3]. Isotherm data for the sorption of H_2O and D_2O vapors by dry (lyophilized) lysozyme at different temperatures suggested that the protein undergoes structural changes as the temperature of adsorption is increased. These changes are not reversible until the protein has adsorbed relatively large amounts of H_2O and D_2O [10].

Enzyme performance in nonaqueous media very much depends on the water content and activity in the reaction system [2, 9, 36, 43]. In fact, enzyme activity and stability are directly related to the amount of water associated with the enzyme. Therefore, a model that can be used to quantify the amount of water adsorbed by the enzyme is desired. Proteins differ from inorganic materials not only in chemical structure but also in physical properties such as less rigidity of proteins, and as such cautious must be used when applying isotherm models based on rigid inorganic materials to water-protein systems. Several thermodynamic models which predict water adsorption isotherms for nearly anhydrous proteins hydrated from the vapor phase have been proposed. The differences between the models are primarily in the number of adjustable parameters, or in the physical interpretation of these parameters. A dual-mode sorption model was the simplest model capable of adequately describing experimental data. McMinn and co-workers [23] used a linear combination of the Langmuir sorption and Henry's law expressions to predict water sorption by proteins suspended in nearly anhydrous organic solvents. Two parameters of the model, the binding equilibrium constant and Henry's constant, were obtained by fitting experimental data to the model. A comprehensive study of H_2O and D_2O vapor sorption by methemoglobin A, methemoglobin S, and metmyoglobin was conducted by Killion and co-workers [17]. For these heme-proteins, the BET model was used to describe adsorption behavior, and the amount of water necessary for monolayer coverage was compared between protein derivatives with H_2O or D_2O.

Much of the existing protein-water adsorption research has been performed using the "weighing bottle" method [10, 23] or a McBain sorption system [17], in batch mode. Since water adsorption is a thermodynamic process, equilibrium should be reached before the measuring the adsorbed water. In a batch process with no agitation applied to enzyme particles, equilibrium would take a long time (24 hours to 5 days) to establish [17]. Applying agitation to easy deformable protein powder particles will have significant effect on water adsorption behavior. In a continuous flow gas phase system, however, diffusion limitation can be minimized and the shape of enzyme particles can be maintained if the packed enzyme bed is not too high.

To develop our understanding of how enzyme hydration affects enzyme function we have selected baker's yeast alcohol dehydrogenase (YADH) as a model enzyme. Alcohol dehydrogenase is an enzyme which requires a cofactor to be catalytically active. Alcohol dehydrogenase is active with gas phase substrates [26. 27], and our laboratory has had particular interest in the use YADH in extreme environments for some time [36, 37]. In the following section we present a detailed investigation of the hydration and dehydration of YADH in continuous flow gas phase system.

2.1 EXPERIMENTAL METHOD

Materials: Alcohol dehydrogenase (EC 1.1.1.1) from baker's yeast (YADH, once crystallized and lyophilized) and β-nicotinamide adenine dinucleotide (NAD^+) were obtained from Sigma Chemical Co. (St. Louis, MO). 3-methyl-2-buten-1-ol (UOL), 3-methyl-2-butenal (UAL), acetone and isopropanol were purchased from Aldrich Chemical Co. (Milwaukee, WI). All chemicals used were of analytical grade.

Enzyme preparation: Dry enzymatic preparations were prepared as described previously [36]. In a typical experiment, YADH (50 mg) and NAD^+ (0.5 mg) were dissolved in 20 ml, 0.01 M phosphate buffer (pH 7.8) and lyophilized for 48 hours. The water content of lyophilized enzyme/NAD^+ mixture powder was measured by Karl Fisher titration. Typically, 50 mg lyophilized enzyme powder were dissolved in 5 ml of dimethyl sulfoxide (DMSO), and the water contents of the neat DMSO and the enzyme-DMSO solution were measured. The difference in water concentration between the enzyme solution and neat DMSO was taken as the water content of enzyme. The water content of the lyophilized enzyme powder was 20% (w/w).

Experimental system setup: A schematic of experimental setup is shown in Figure 1. The reactions catalyzed by YADH involve alcohol as substrate and ketone as co-substrate. Helium was used as the carrier gas and its flow rate was controlled and measured with a float flowmeter (F1). Saturators contained buffer (S1, S4)), alcohol (S2), and acetone (S3). The make-up gas flow rate was controlled by a flowmeter (F4). The concentrations of alcohol and acetone were controlled by varying flow rates through F2 and F3. The humidity of the gas mixture was adjusted by varying the make-up flow rate. The relative humidities of the gas mixture before and after the reactor were determined with two humidity probes. The system was placed in an incubated box which temperature was controlled by two heat guns and a programmable temperature controller. The operation pressure was fixed at 14.7 psi throughout the study. The reactor was a cylindrical glass column (15 cm × 1 cm) in which enzyme powder was supported by glass wool.

Composition analysis: A Hewlett-Packard Series II 5890 GC was used to determine the vapor phase concentration of alcohol and acetone. A packed column (6 ft × 1/8 in., stainless steel, 60/80 Carbopark™ B/1% SP™ - 100, Supelco, Inc., Belletone, PA) and a FID detector were used. The temperatures for oven, injector, and detector were 180, 200 and 200 °C, respectively. The carrier gas was helium, with a flow rate of 30 ml/min.

Water adsorption and desorption measurements: At a given temperature, same batch of enzyme was used for the complete set of hydration experiments. Experiments were initiated at low relative humidity, then humidity was increased until the inlet gas reached 100%. After each re-equilibration, the gas stream was bypassed away from reactor until the humidity stabilized at the new desired level. The same batch enzyme was used to perform water desorption study. In desorption experiments, inlet gas relative humidity was decreased from 100%. For experiments designed to assess the effect of substrates

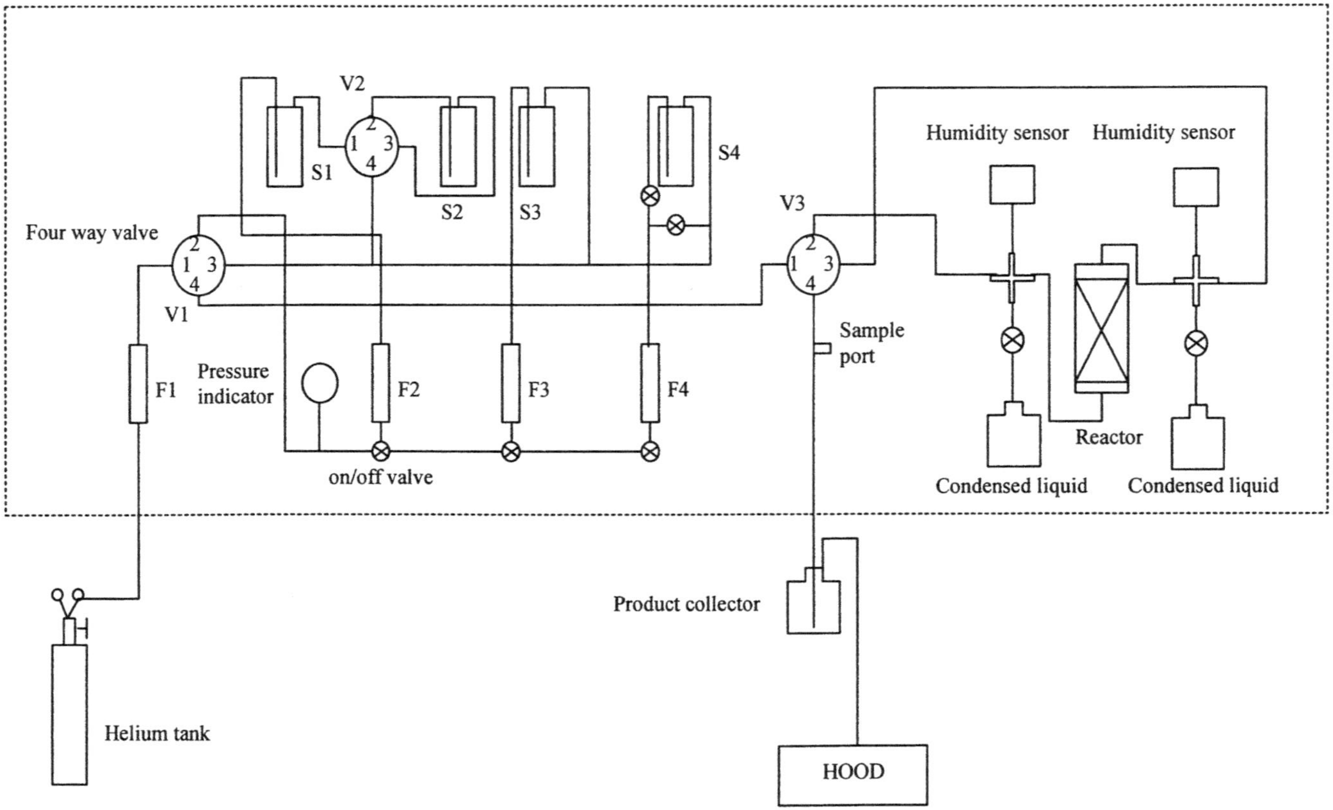

Figure 1. Schematic of gas phase reactor

on water adsorption, the desired humidity level and substrate concentration were obtained by adjusting the flow rates of various streams. The experiments were performed as described above.

2.2 WATER ADSORPTION CALCULATION

The amount of water adsorbed by enzyme can be determined by measuring the change in relative humidity of gas stream at the inlet and outlet of the reactor. The calculation of the total amount of water adsorbed assumes that (1) the ideal gas law applies, (2) the total pressure change due to water absorption can be neglected.

Absolute humidity ($g_{water}/g_{dry\ gas}$) can be defined as follows:

$$H_a = \frac{M_w \, p_w}{M_{He} \, (P - p_w)} \tag{1}$$

where P is the total system pressure and p_w is the partial pressure of water vapor in the gas mixture. M_w and M_{He} are water and helium molecular weights, respectively.

If $H_{a,i}$ and $H_{a,o}$ represent the reactor inlet and outlet absolute humidities, respectively, then the change in the humidity between inlet and outlet can be expressed as:

$$\begin{aligned} \Delta H_a &= H_{a,i} - H_{a,o} \\ &= \frac{M_w}{M_{He}} \left(\frac{p_{w,i}}{1 - p_{w,i}} - \frac{p_{w,o}}{1 - p_{w,o}} \right) \end{aligned} \tag{2}$$

where $p_{w,i}$ and $p_{w,o}$ are the partial pressure of water vapor of gas stream at the inlet and outlet, respectively.

The change in the absolute humidity of gas mixture between inlet and outlet is the amount of water that has been adsorbed by, or desorbed from, the enzyme in the reactor. The humidity sensors used in the experiment, however, only report a percentage relative humidity, therefore, it is necessary to express the change of absolute humidity in terms of the change in relative humidity. The definition of relative humidity is given by:

$$H_R = \frac{p_w}{p_w^o} \times 100\ \% \tag{3}$$

where p^o_w is the saturated partial pressure of water at the operating temperature.

Substituting Equation 3 into Equation 2, we have

$$\Delta H_a = \frac{M_w}{M_{He}} \left(\frac{H_{R,i}}{100 / p_w^o - H_{R,i}} - \frac{H_{R,o}}{100 / p_w^o - H_{R,o}} \right) \quad (4)$$

where $H_{R,i}$ and $H_{R,o}$ are relative humidity at the inlet and outlet of the reactor, respectively.

If the carrier gas helium volumetric flow rate is V (ml/min), the mass flow rate (F_{He}) can be expressed as:

$$F_{He} = \frac{M_{He} P V}{R T} \quad (5)$$

where P is the helium pressure in the system, T represents the operating temperature and R is the universal gas constant.

Since the rate of water absorption can be expressed as,

$$G = \Delta H_a \times F_{He} \quad (6)$$

the amount of water absorbed during the process can be calculated from:

$$W = \int_0^t G\,dt = \int_0^t \frac{M_w P V}{R T} \left(\frac{H_{R,i}}{100 / p_w^o - H_{R,i}} - \frac{H_{R,o}}{100 / p_w^o - H_{R,o}} \right) dt \quad (7)$$

2.3 EFFECT OF RELATIVE HUMIDITY AND TEMPERATURE ON HYDRATION

Since the water associated with the enzyme in the continuous gas reactor will be at equilibrium with the gas mixture, initially, the enzyme will absorb or desorb water until equilibrium is established. For example, if the humidity of the inlet gas is higher than the equilibrium level, the enzyme will adsorb water from the gas stream during the hydration process [39]. The amount of water adsorbed or desorbed by the enzyme in the reactor then can be quantified by numerical integration of difference between the inlet and outlet gas humidity curves according to Equation 7.

Lyophilized enzyme powder contains 20% (w/w) water, and the final water content of the enzyme for a given gas mixture is the sum of water associated with the enzyme powder and amount of water gained or lost during the absorption or desorption process.

The effect of humidity on enzyme water content (w/w) at four temperatures, 22 °C, 30 °C, 40 °C, and 50 °C, is shown in Figure 2. The closed symbols and solid lines represent the adsorption process while open symbols and dashed lines are for desorption isotherms. The data shown are averages of two experiments at each point (the data for each pair of points take approximately six hours to generate).

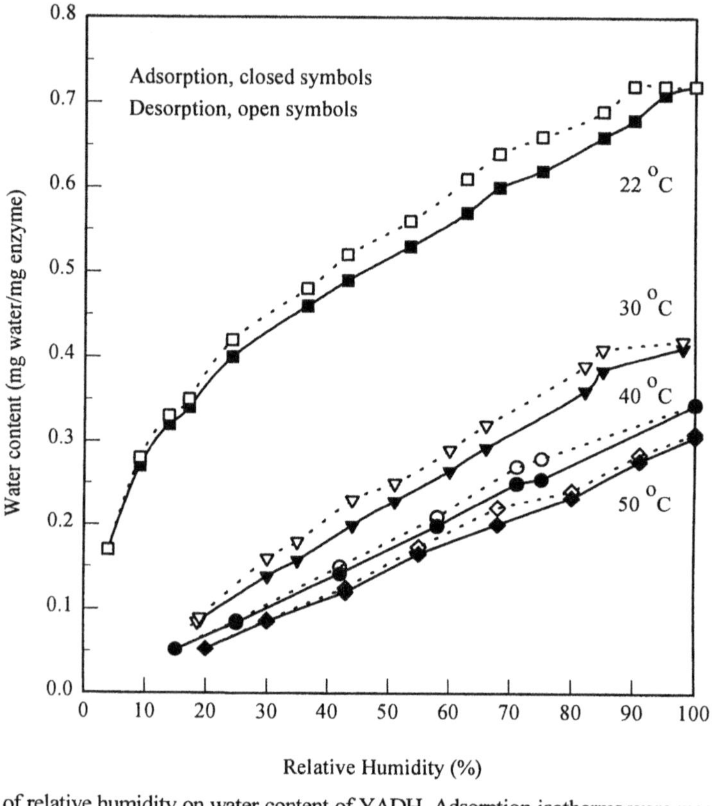

Figure 2. Effect of relative humidity on water content of YADH. Adsorption isotherms were measured at 22 °C, 30 °C, 40 °C and 50 °C. Open symbols refer to the equivalent temperature for the desorption process.

At 22 °C, the amount of water retained by the enzyme powder increases with relative humidity at low hydration levels. At intermediate and high hydration levels, the slope of the sorption isotherm decreases and the adsorption is almost proportional to relative humidity. At higher temperatures, the adsorption curves become less concave. These isotherms also show an almost proportional relationship between adsorption and relative humidity at intermediate and high hydration level. The changing shape of adsorption curves with temperature has been observed previously with the sorption of H_2O and D_2O vapors by lysozyme [10]. The changing slope of the sorption isotherms with hydration levels also implies that the mechanism of water binding by enzyme is dependent on the hydration level. As expected, YADH retains less water at higher temperatures, and the effect of temperature on water retention by YADH is so evident that the enzyme retains only 50% of the water at 30 °C than at 22 °C. Calorimetric measurements of the isosteric heat of sorption of water to protein indicate that the first 5% of water sorbed by lysozyme (which lyophilized powder contains 17% w/w water) is bound tightly [3].

Water adsorption-desorption isotherms should coincide if the properties of adsorbate do not change during the adsorption-desorption process. Figure 2 indicates that YADH hydration exhibits hysteresis at all four temperatures and the magnitude of the hysteresis decreases with increasing temperature. As we have stated previously "Hysteresis during the hydration has been observed previously, and it is thought to occur because intramolecule and/or intermolecule of protein molecules are altered as hydration proceeds [10, 23]. Such a change in protein structure would be expected to alter the catalytic behavior of YADH. In a previous paper we reported that pre-incubation of YADH in a solvent significantly activates the enzyme [36]. We hypothesized that the solvent activation phenomena was related to "slow morphological changes in the enzyme particle during pre-incubation, which in turn could affect the activity/stability of the catalyst and/or the mass transfer characteristics of the catalyst". The data reported here substantiate the notion that exposure of a dry enzyme preparation to a solvent (water) can result in an irreversible alteration in protein structure. Hysteresis has also been explained with an "ink-bottle" analogy [22]. In many porous materials large cavities are accessible to system constituents through small channels or orifices. In such a system, as the relative humidity of vapor is gradually increased, the condensation of water will begin at the narrowest points and will then extent to wider areas as the relative humidity is increased. When the vapor is almost saturated the pore will be completely filled. Upon a subsequent decrease of the relative humidity, however, in general no evaporation will occur from this particular pore until the relative humidity has fallen to the level necessary to achieve evaporation from the largest orifices or passage. Such a set of assumptions has been used to explain why the curve representing the amount of water retained during dehydration will lie above that for progressive hydration [22]."

2.4 EFFECT OF ORGANIC MOLECULES ON ENZYME HYDRATION

In enzyme-catalyzed reactions with gas phase substrates, the enzyme particles will be in contact with substrate(s), carrier gas and water molecules. Substrate molecules could be competitive adsorbates for water thus altering the properties of the protein. In our experiments, 3-methyl-2-buten-1-ol (UOL) and acetone interact with YADH and, therefore, the effect of these molecules on water adsorption must be assessed in order to determine the actual amount of water adsorbed by the enzyme during catalysis. The significance of the effect of substrates on hydration may also depend on substrate concentration in the gas mixture. Therefore, measuring water adsorption in the presence of substrate, at a range of actual reaction concentrations, has significant interest. The vapor pressure of UOL, acetone and water at a given temperature were obtained using Antoine's equation (Table 1) and simple calculations were used to predict the necessary flow rates through each saturator to achieve a given humidity and substrate concentration. The actual concentrations of all components were also measured experimentally before and after the reactor to verify the appropriateness of theses predictive calculations. We observed that the effect of UOL and acetone on the humidity probes was negligible.

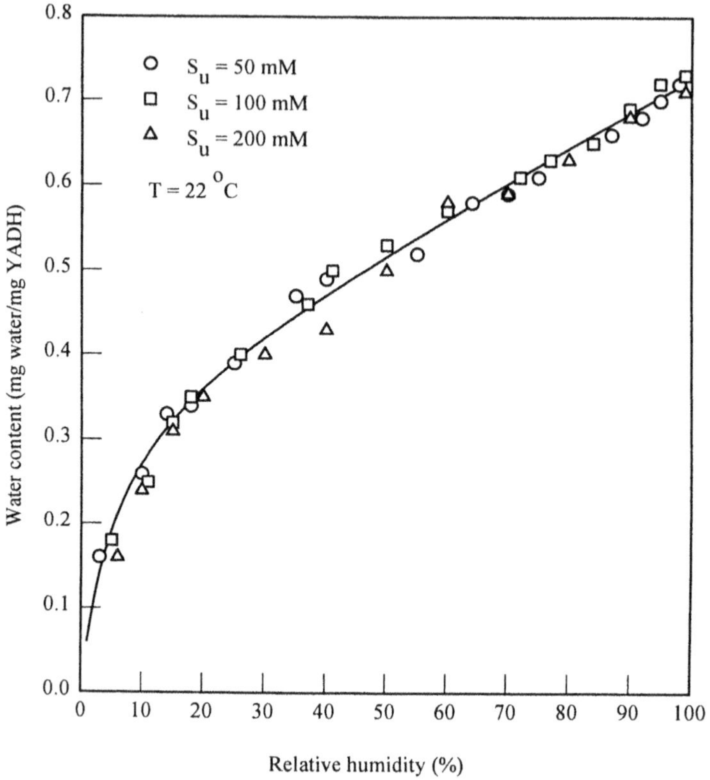

Figure 3. Effect of 3-methyl-2-buten-1-ol on YADH hydration at 22 °C. The measurement was carried out at various concentration of UOL in the gas mixture. Solid line represents the sorption isotherm obtained in the absence of the alcohol.

Figure 3 describes enzyme hydration in the presence of various concentrations of UOL at 22 °C, the solid line representing YADH water adsorption in the absence of substrates. The data indicate that the presence of UOL in the gas mixture, within the experimental concentration range, has no effect on YADH hydration for the whole range of relative humidity. We suggest that, since these organic molecules will not competitively bind to the enzyme surface, nonpolar organic molecules do not affect the interaction between enzyme molecules and hydrophilic molecules such as water. This also implies that enzyme molecules at low hydration levels are preferentially surrounded by water molecules. Other molecules such as substrate(s) have to penetrate this water layer to reach enzyme active sites and product molecules have to diffuse out.

When UOL is replaced by acetone, some suppression in water adsorption is observed as shown in Figure 4. At all acetone concentrations used, the measured water adsorption isotherm in the presence of acetone followed the acetone-free adsorption isotherm up to 20% relative humidity. Above this humidity, however, the amount of water adsorbed by

YADH is suppressed by acetone, compared to the amount of water adsorbed from the acetone-free gas mixture. The suppression in water adsorption increases with increasing acetone concentration and reaches maximum at $S_a = 300\ \mu M$.

Table 1. Antoine Constants

Substance	A	B	C
Water	8.1	1750.3	235.0
UOL	7.1	1435.2	202.0
Acetone	7.0	1161.0	224.0

Constants are obtained from Lange's Handbook of Chemistry [6].
Constants for water are only valid for the range of 0 - 60 °C.

2.5 PREDICTION OF ENZYME HYDRATION

As mentioned above, a model that predicts quantitatively the amount of water adsorbed by YADH in the gas reactor would be helpful. Currently there are two major adsorption theories, monolayer and multilayer. YADH has a molecular weight of about 140,000 daltons and specific volume 0.769 ml/mg [34], which gives an approximate molecular radius of 36 Å [32]. The surface area of this size molecule is 16,300 Å^2, assuming the molecule is a solid sphere ball. The projection area of a water molecule is about 10 Å^2 [30], thus we can quantitatively predict that the monolayer coverage of a YADH molecule will be approximately 1,630 water molecules (0.21 mg water/mg YADH).

The actual water content for most YADH isotherms is above 0.21 mg/mg (Figure 2), suggesting that a monolayer adsorption isotherm model may not be suitable for our situation [21]. Multilayer adsorption theories assume that adsorbate molecules can cover an absorbent surface with more than one layer. Among these theories, the B.E.T. model [4] has been used most extensively.

In its most general form, for the first layer, the adsorption equilibrium equation can be expressed as:

$$a_1 p(\chi - n_1) = b_1(n_1 - \gamma n_2) e^{-E_1/RT} \quad \textbf{(8)}$$

where p is the adsorbate pressure, E_1 is the heat of adsorption in the first layer, and a_1, b_1, and γ are constants. χ represents the maximum number of molecules that is adsorbed on the first layer and, n_1, n_2 are the number of molecules in the first and the second layer, respectively. The right-hand side of this equation signifies that the rate of evaporation of molecules on the first layer is proportional to the total number of molecules in that layer minus a certain effect produced by those molecules that are

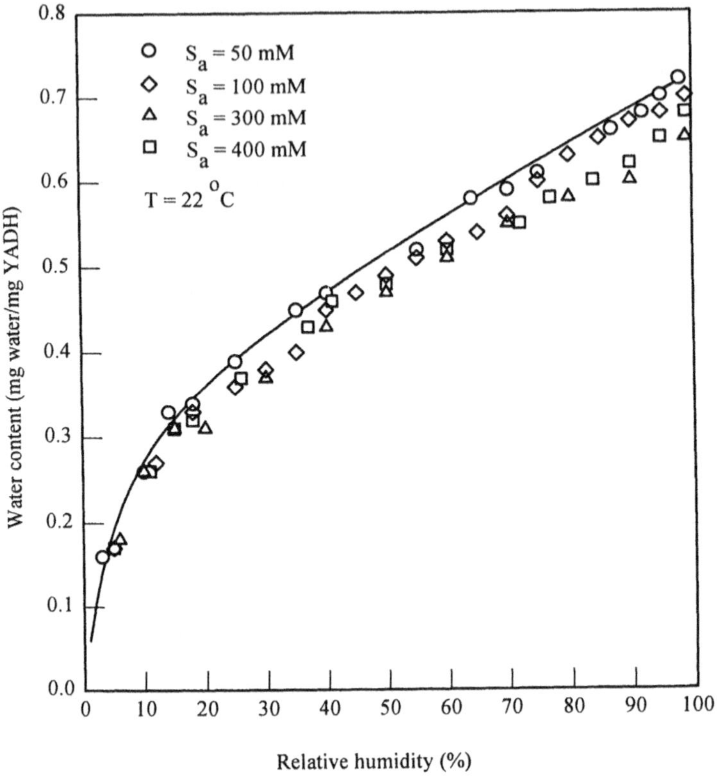

Figure 4. Effect of acetone on YADH hydration at 22 °C. The measurement was carried out at various concentration of acetone in the gas mixture. Solid line represents the sorption isotherm obtained in the absence of the acetone.

covered by one or more higher layers. The B.E.T. theory assumes that this effect is powerful enough to prevent evaporation of first-layer molecules so covered ($\gamma = 1$), while the theory of Huttig [11] uses the other extreme viewpoint, namely, that molecules covered by second and higher layers still play their role in evaporation as freely as if the higher layer were not present ($\gamma = 0$). Indeed, this is the only point of difference between these two theories [28].

Huttig's mechanism for equilibrium on each layer gives:

$$a_1 p(\chi - n_1) = b_1 n_1 e^{-E_1/RT} \tag{9}$$

$$a_i p(n_{i-1} - n_i) = b_i n_i e^{-E_i/RT}, \ ect. \tag{10}$$

where n_i is the molecule number in the ith layer and, $E_i = E_L$ ($i \neq 1$) is the heat of condensation of pure water.

Straightforward algebra leads to:

$$\frac{V}{V_m} = \left(\frac{cp_w / p_w^o}{1 + cp_w / p_w^o} \right) (1 + p_w / p_w^o) \tag{11}$$

where V_m is the minimum volume of absorbed molecules necessary to form the first layer, and c is a temperature-dependent constant.

The isotherm equation can also be expressed in term of absorbed weight and the relative humidity:

$$W = \frac{W_m}{100} \left(\frac{cH_R}{100 + cH_R} \right) (100 + H_R) \tag{12}$$

where W_m is minimum weight of adsorbed molecules in the first layer.

This equation can be linearized to:

$$\frac{H_R}{W} (100 + H_R) = \frac{10^4}{W_m c} + \frac{100}{W_m} H_R \tag{13}$$

For comparison the classical B.E.T. equation in a similar form is:

$$\frac{W}{W_m} = \frac{100cH_R}{(100 - H_R)[100 + (c - 1)H_R]} \tag{14}$$

According to the B.E.T. equation, adsorption will be infinite when relative humidity approaches 100%. Since our experiments demonstrate that hydration reaches a limiting value when H_R approaches 100%, YADH hydration process does not follow the B.E.T. isotherm.

Figure 5 indicates that the Huttig isotherm, however, can be used to describe YADH-water adsorption behavior with remarkably good accuracy. The results which can be calculated from the data are summarized in Table 2. The parallel lines indicate that the amount of water for monolayer coverage of YADH molecules is independent of temperature, assuming that the shape and size of YADH does not change between 22 °C and 50 °C.

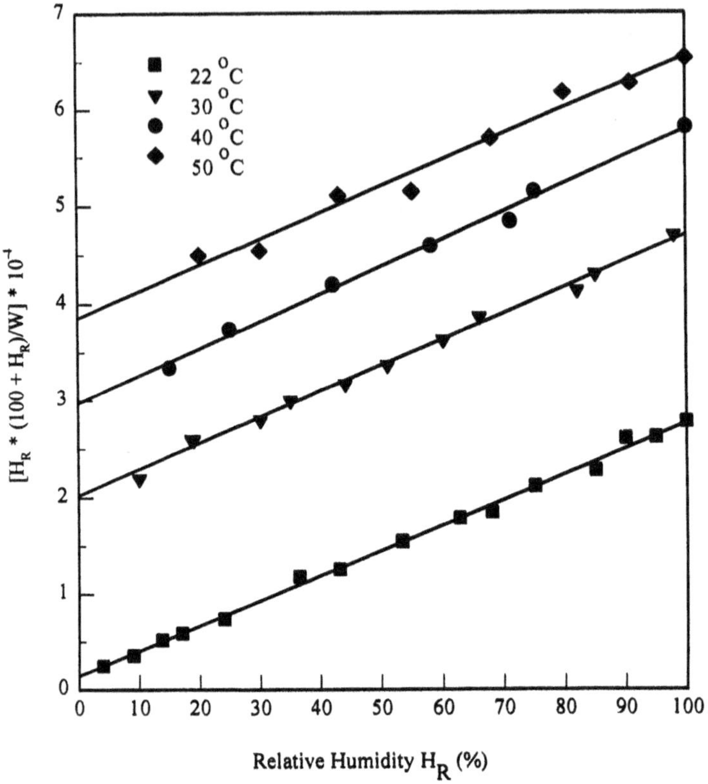

Figure 5. Temperature and humidity dependence on enzyme hydration as predicted by Huttig's isotherm. Water adsorption isotherms were measured at different temperatures as shown in the figure. Data points are calculated based on Equation 13. Solid lines are the linear regression of these data points.

Table 2. Experimental Determinations of Parameters in the Isotherm Model

Parameters	Calculated values	Experimental values			
		22 °C	30 °C	40 °C	50 °C
W_m, mg/mg	0.21	0.38	0.37	0.36	0.37
c	-	1.83	1.33	0.95	0.70

The temperature dependent constant, c, in Huttig's isotherm model is related to the heat of adsorption of water on the protein by following equation.

$$c = Ae^{(E_1 - E_L)/RT} \tag{15}$$

or

$$\ln c = \frac{(E_1 - E_L)}{RT} + lnA \tag{16}$$

where E_1 is the heat of adsorption for the first layer, E_L is the heat of condensation of water (-10.42 kcal/mol), and A is a temperature independent constant.

The temperature dependence of "c", which fits Huttig's model can be used to calculate E_1 (-16.88 kcal/mol). Clearly, the first layer of water molecules adsorbed to the enzyme is very tightly bound. Similar results have been observed by Hnojewyj and Reyerson [10], who found that the isosteric heat of adsorption at low levels of hydration is as much as three times the heat of condensation of water. The heat of adsorption decreases at higher hydration levels and approaches a value characteristic of the heat of condensation of pure water at approximately 20% (w/w) water adsorbed. Naturally, the value of the heat of adsorption will be protein dependent. When lyophilized preparations of β-Lactoglobulin, bovine serum albumin, and calfskin collagen sorbed at least 0.18 gm H_2O per gram of dried protein, the heat of vaporization of the adsorbed water was 1.4 - 2.3 kcal/mol more than the heat of vaporization of liquid water [3].

3. Enzyme Activity with Gas Phase Substrates

One of the advantages of applying enzymes to gas phase substrates is the ease with which reactions can be carried out in a continuous manner. However, there are few studies on continuous gas phase reactions [12,24,26,27].

Alcohol dehydrogenases (ADH) have important applications for flavor aldehyde production. Although they have already been used in aqueous solution [25], there are serious limitations to their widespread application. These include operational instability of ADH's and cofactor, product inhibition of ADH's, lack of stability of some substrates and products, and the insolubility of most substrates and products in aqueous solution. Therefore, organic solvents have to be used to solubilize substrates and products. In this case, the limitations are instability of enzyme in the presence of organic solvents and the slow rates of mass transfer to the insoluble enzyme in organic solvents [5, 7, 16, 33, 36]. An ADH solid-gas bioreactor has been proposed as an alternative to overcome the problems. Alcohol dehydrogenases from horse liver (HLADH) and *Salfolobus solfataricus* (SSADH) were coimmobilized with NAD^+ (or NADH) onto albumin glutaraldehyde porous particles in either batch or continuous feed column reactors for alcohol and/or aldehyde production. Reduction of an aldehyde was coupled to a second alcohol oxidation in order to regenerate the cofactor [26, 27]. Although these studies have demonstrated the feasibility of application of cofactor dependent alcohol dehydrogenases with gas phase substrates, more detailed study is necessary to offer insight on how enzyme hydration and reaction temperature affect the enzyme performance in this kind of environment. As described above, we now have a unique insight into the hydration of YADH in such a system.

The YADH-catalyzed model reaction we have studied [36] uses 3-methyl-2-buten-1-ol and acetone as substrate and cosubstrate. In the reaction, YADH converts UOL to UAL, simultaneously reducing NAD^+ to NADH. The reduced NADH is oxidized back to NAD^+ by the conversion of acetone going to isopropanol, thus regenerating the cofactor.

Dry enzymatic preparations were prepared as described above and in a typical experiment, lyophilized enzyme powder (50 mg YADH and 0.5 mg NAD^+) was packed in a cylindrical glass reactor with dimension of 1 cm × 15 cm. The typical packed bed height was 6 cm, and the voidage fraction was 0.46. The reaction was initiated by introducing the reactant gas mixture to the bioreactor. The composition of the gas mixture was adjusted by varying the carrier gas flow rate through the saturators. Periodically, concentrations of UOL, UAL, acetone and isopropanol were determined gas chromatographically by withdrawing 0.5 ml gas sample with a gas-tight syringe.

3.1 INFLUENCE OF TEMPERATURE ON ACTIVITY

In order to investigate the effect of temperature on enzyme activity and stability with gas phase substrates, a series of reactions with water saturated (H_R = 100%) feed gas mixtures at various temperatures were performed. The reaction progress curves are shown in Figure 6.

A period with constant conversion, or pseudo-steady state, was observed for all reactions, although the length of this pseudo-steady state varied with reaction temperature. In the hydration experiments discussed above, the hydration of the enzyme takes approximately three hours to equilibrate (outlet gas humidity equals inlet gas humidity), and the higher the temperature, the faster the equilibration at any relative humidity of the feed gas. Enzyme activity (the level of conversion in a continuous process) gradually increases during the hydration process until reaching a maximal value corresponding to pseudo-steady state. Because of the short time period for hydration relative to the scale of Figure 6, it is difficult to analyze activity data during hydration.

As observed for conventional non-protein catalyzed reactions, as reaction temperature increases, reaction rate increases. The increased activity is accompanied by decreased stability or half life. The half life of enzyme can be used to evaluate stability, whereas the pseudo-steady state is related to activity of YADH with gas phase substrates. Figure 6 shows that YADH stability is greatly enhanced in the gas phase reactor. The half life of the enzyme is more than 20 days at 22 °C and as long as 8 days even at temperatures as high as 50 °C. The steady state can be maintained for 15 days at 22 °C, and 4 days at 50 °C. As we have previously stated "In the presence of a solvent the enzyme deactivates in a classical first order mode, losing activity from the moment of exposure to the solvent. The unusual conversion versus time profiles that we see in the gas reactor could be characteristic of the combination of substrate and/or product poisoning and the presence of multiple steady states [35]. Multiple steady states would be present if individual enzyme molecules had distinct activities and specificities. The existence of multiple steady states also indicates that the steady state conversion is, as with all examples of immobilized enzyme preparations, the sum of the activities of the enzyme forms which are represented. One must, of course exercise caution when interpreting changes in such average properties." [40].

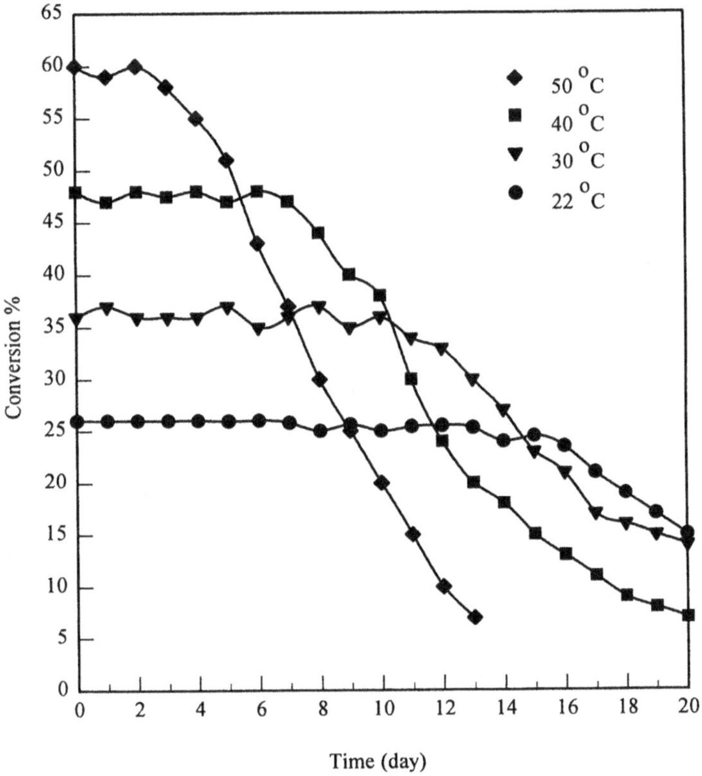

Figure 6. Effect of temperature on YADH activity and stability in the gas phase. Reactions were carried out at inlet gas humidity at 100% and the gas phase system temperature was controlled at 22 °C, 30 °C, 40 °C and 50 °C as indicated in the figure. Substrate UOL concentration was maintained at 100 μM, and inlet gas flow rate was 25 ml/min.

Cofactor turnover number (TON), defined as moles of product over moles of NAD^+ placed in the reactor, can be calculated from the ratio of total moles of product synthesized over initial moles of cofactor present. The results indicate that cofactor can be regenerated in the gas phase reaction system. The TON values are generally higher than those obtained for YADH in heptane [36]. The lower stability of YADH in organic solvents could result from (a) the direct effects of organic solvents and, (b) higher product concentration.

3.2 INFLUENCE OF HYDRATION ON ACTIVITY

To optimize enzyme performance with gas phase substrates, it is important to understand the role of water in retaining enzyme activity and stability. In the system we describe, enzyme hydration is directly related to inlet gas humidity and temperature as quantified in Equation 12 and Figure 2. The conversion shown in Figure 6 was obtained at an inlet gas relative humidity of 100%, where enzyme hydration is maximal. Decreasing inlet gas humidity results in significant changes in enzyme activity as shown in Figure 7A. In order to establish the effect of enzyme hydration on YADH's activity in gas phase reactor, it was necessary assure that enzyme inactivation was negligible during the activity measurements. Therefore, pseudo-steady state conversions were used throughout the study to assess activity. At 22 °C, enzyme activity in the gas phase increases six to ten times as inlet gas humidity increases from 3% to 100%. At relatively low temperatures such as 22 °C and 30 °C, enzyme activity with the gas phase substrates exhibits an optimum conversion with inlet gas humidity. As discussed above, in the gas-phase system, the temperature independent monolayer water content for YADH is ~0.4 mg H_2O/mg YADH. At 22 °C and 30 °C, the gas mixture relative humidities necessary to achieve a monolayer water coverage are 20% and 82% respectively. Above 30 °C there is always less than a monolayer of H_2O associated with enzyme at all humidities. Figure 7 demonstrates that the reaction can still be catalyzed with significantly less than monolayer water coverage. This observation is consistent with results for alcohol oxidase in the batch oxidation of ethanol supplied in the gaseous phase [2]. Rupley and co-workers [29, 30] have studied molecular events occurring during the sequential hydration of dry lysozyme using several experimental techniques and concluded that the enzyme is active below monolayer coverage, and that enzymatic activity increases with increasing hydration level.

As relative humidity increases from 80% to 100%, the reduction in enzyme activity which is observed at 22 °C and 30 °C (Fig. 7A) could result from (a) mass-transfer limitations due to enzyme aggregation at high levels of hydration, and thus an increasing thickness of the adsorbed water layer around enzyme particle surface, or (b) dilution of the adsorbed substrate. A similar response to water activity was observed for the ethanol oxidation by alcohol oxidase in gas phase [2]. It was hypothesized that at water activities higher than 0.9, ethanol would be dissolved in pools of condensed water, reducing its availability for the enzyme. As reported in the previous section, for YADH at 22 °C, the maximal adsorption of water on YADH is 0.8 mg water/mg YADH, or a bilayer coverage, thus reducing the possibility of significant pooling. At 30 °C, since water adsorption only results in monolayer coverage at relative humidity > 85%, and it is at this point that conversion starts to decline, we hypothesize that the main factor, which causes the reduced reaction rate at high humidity and low temperature, is increased mass transfer caused by enzyme aggregation and morphological changes.

Our results demonstrate that enzymatic activity is highly dependent on the inlet gas mixture relative humidity. Only when the humidity is above a temperature dependent value can the enzyme function effectively. Studying humidity effects does not, however, reveal an important function of hydration. As shown in Figure 7B, when we plot activity

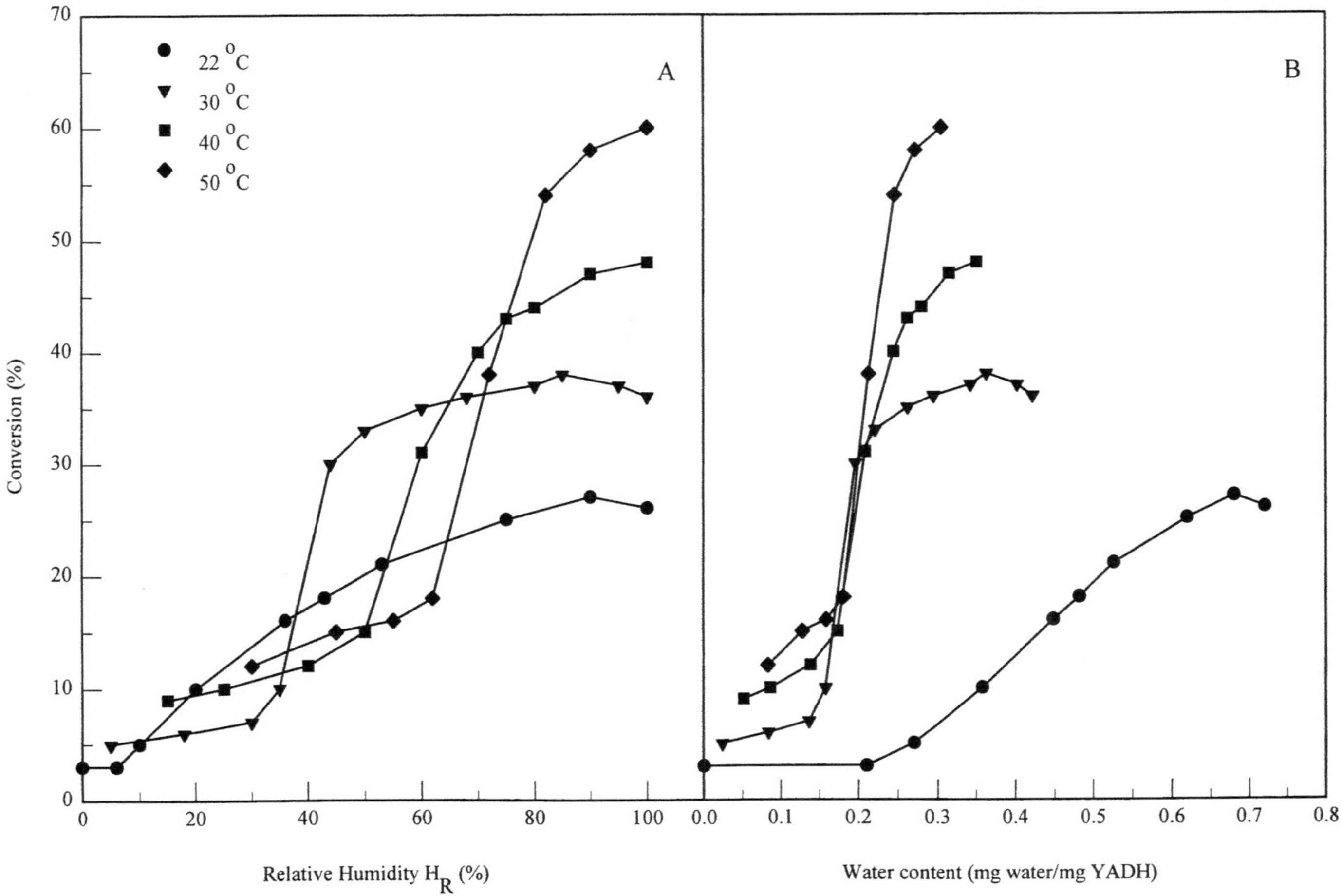

Figure 7. Effect of inlet gas relative humidity (A) and YADH hydration (B) on enzyme activity at various temperatures.

versus the enzyme water content, we find that the water content needed for activation is temperature *independent*. The critical water content (CWC) is 0.2 mg water/mg YADH, which is equivalent to 1555 molecules of water per molecule of enzyme. Since we have also observed that YADH water adsorption-desorption isotherms are not superimposable, it is of importance to determine if the changes in water binding affect enzyme function. Data for activity of YADH during hydration and dehydration (Figure 8) indicate that the hysteresis in water adsorption is mirrored by equivalent changes in activity.

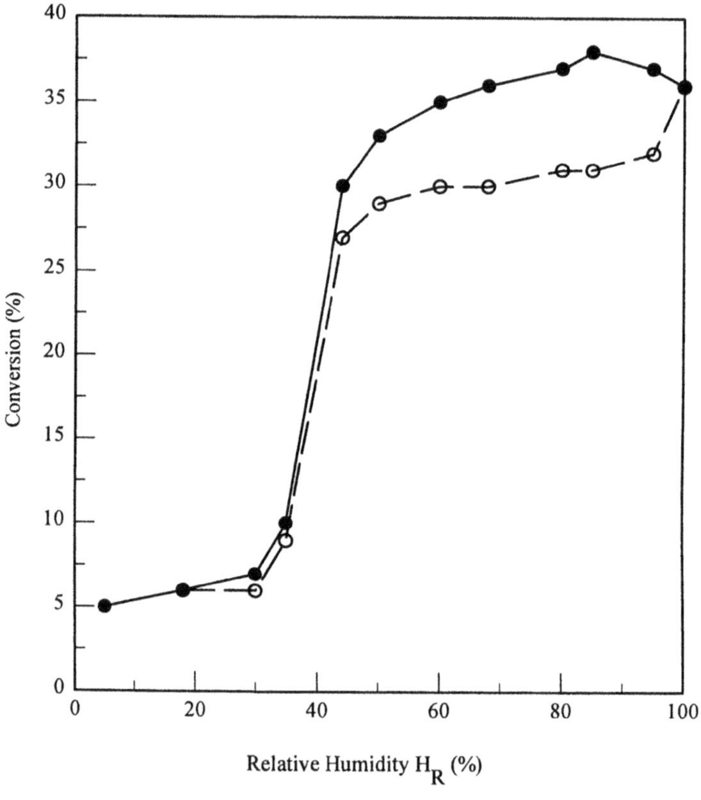

Figure 8. Effect of hydration-dehydration cycle on YADH activity. Reactions were measured at 30 °C with enzyme hydrated at a given gas humidity with simultaneously reaction (closed cycle) and enzyme hydrated at 100% relative humidity then dehydrated to the desired water content then started reaction (open cycle). Other experimental conditions were as described for Figure 6.

4. Enzyme Stability with Gas Phase Substrates

4.1 EFFECT OF TEMPERATURE

Our results (Figure 6) indicate that YADH is more stable than immobilized alcohol oxidase [12] in the continuous gas phase reactor we describe. A number of mechanisms are presumably simultaneously causing the eventual reduction in activity. During the reaction, and over a period of days, the initially white lyophilized enzyme powder becomes yellow as result of reaction with the product 3-methyl-1-butenal (UAL). Indeed, the solubility of enzyme particles which are exposed to UAL for long periods is significantly decreased [36]. Aldehydes are known to react with free amino groups on the protein molecules, and such a reaction may result in irreversible non-competitive product inhibition. In the gas reactor, as temperature increases, concentration of product at steady state will also increase, and not surprisingly the rate of discoloration and deactivation also increase.

Increasing the temperature of the reactor also alters the morphology of the enzyme bed. Significant long term (days) shrinkage of the bed volume was observed at raised temperatures. This shrinkage undoubtedly resulted in channeling of the substrate through the bed, thus reducing the effective catalyst concentration. Finally, the effect of temperature on intrinsic protein stability is well established, and in the absence of either product inhibition or channeling one would expect the half life of the enzyme to be reduced by an increase in temperature.

4.2 EFFECT OF HYDRATION

The stability of YADH in organic solvents is significantly increased relative to that in aqueous systems [36]. Given the increased stability at low water content, it is perhaps not surprising that YADH in the gas reactor is extremely stable. The stability of the enzyme (enzyme halflife) in the continuous flow gas phase reaction system is temperature and inlet humidity dependent (Figure 9A). YADH stability decreases substantially as temperature increases. The change in stability as a function of inlet gas relative humidity is more significant at ambient temperatures. Figure 9A demonstrates again that there is a transition point at all temperature where YADH becomes stable. The dramatic stabilization of the enzyme in the gas phase is clear in Table 3, which is a comparison of stability of YADH in different media at 30 °C.

Table 3. YADH stability in different media

Reaction Media	Half-life	Conditions	Reference
Gas phase	16 days	$H_R = 100\%$	This study
Organic solvent	50 hours	Heptane, 0.25% (v/v) water	36
Reverse micelles	2 hours	AOT/Cyclohexane, $W_o = 15$	37
	5 min	AOT/Isooctane	32
Aqueous	30 min	pH = 8.8	32

All the data shown in this table were obtained at temperature of 30 °C.

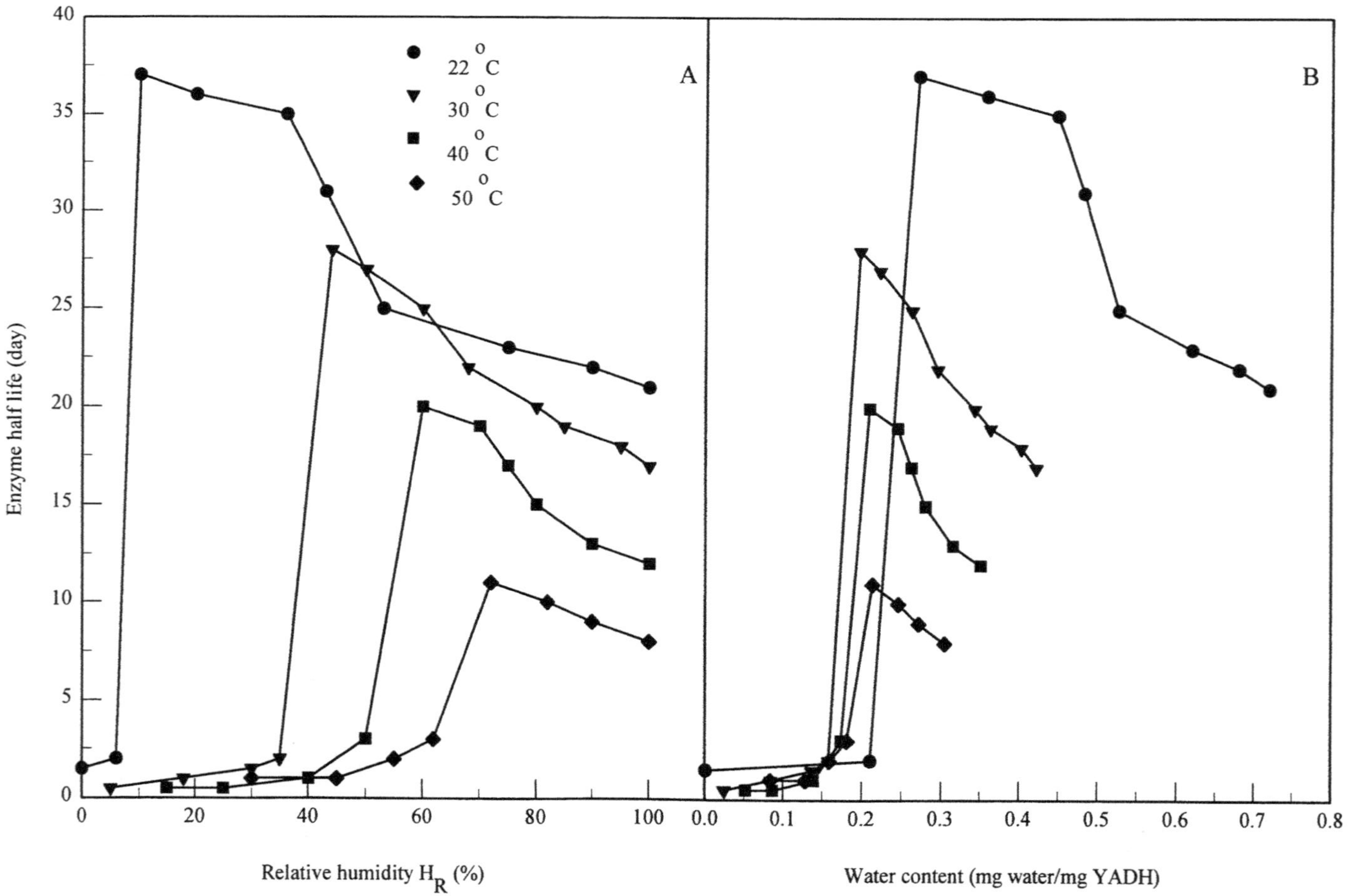

Figure 9. Effect of inlet gas relative humidity (A) and YADH hydration (B) on enzyme stability at various temperatures

Remarkably, when the stability data is plotted versus enzyme water content (Figure 9B), the same temperature-independent amount of water needed for activity is required for stabililization. This water content is equivalent to ~46% of a monolayer. Comparing Figure 7 and Figure 9, one can clearly see that enzyme hydration has an opposite effect on enzyme activity and stability. This phenomena was also observed for alcohol oxidase with gas phase substrates [2] and YADH in other water restricted media [35]

5. Enzyme Kinetics in the Gas Phase

Typical reaction progress curves (Figure 2) indicated that YADH catalyzed reactions are at a steady state for 4 to 16 days before activity begins to decrease. The activity data at steady state was used to study the kinetics of YADH-catalyzed reactions in the gas phase reactor. The enzyme follows an Ordered Bi Bi mechanism in aqueous media [8, 25]. Since the concentration of the cosubstrate (NAD^+) is fixed, the initial rate expression can be simplified to a Michaelis-Menten equation. The Michaelis constants for ethanol and propanol are 21 mM and 256 mM respectively [8]. In the gas phase, UOL concentration was maintained at about 100 μM through out our study, and thus the Michaelis-Menten equation can be further simplified to a first order reaction rate expression:

$$v = \frac{dC_s}{dt} = \frac{V_{\max} C_s}{K_m} = kC_s \qquad \textbf{(17)}$$

where C_s is substrate concentration in the gas phase and k is the apparent V_{max}/K_m for the enzyme.

In a packed bed reactor, substrate concentration C_s will decrease along with bed from inlet to outlet. Assuming F, V_R, ε and X_s represent superficial flow rate, fixed bed volume, bed voidage fraction and UOL conversion, respectively, integration of Equation 17 yields:

$$-\ln(1 - X_s) = \frac{kV_R\varepsilon}{F} \qquad \textbf{(18)}$$

At constant temperature, inlet gas humidity and at the pseudo-steady state, a plot of (-ln(1 - X_s)) versus 1/F will be straight line with slope of $kV_R\varepsilon$, from which the kinetic constant k can be obtained. Figure 10 demonstrates the linearity of a plot of conversion versus flow rate at 30 °C, thus substantiating the proposed mechanistic scheme.

Figure 10 also implies that the enzyme mechanism is independent of hydration level (the hydration level used in Figure 10 ranges from below the CWC to water saturated gas). As expected, the specificity constant (V_{max}/K_m) increases with increasing humidity (at 100%, 44% and 30% relative humidity the V_{max}/K_m's are 5.27, 4.23 and 0.98 min^{-1}, respectively).

We have already discussed the effect of temperature on enzyme activity and stability (Figure 6), but the pseudo-steady state data enable a continuous reactor to be assessed for the validity of Arrhenius equation:

$$-\ln(1 - X_s) = \frac{V_R \varepsilon}{F} k_o \exp(-\frac{E_a}{RT}) \tag{19}$$

where E_a is the activation energy. Equation 19 can also be expressed as

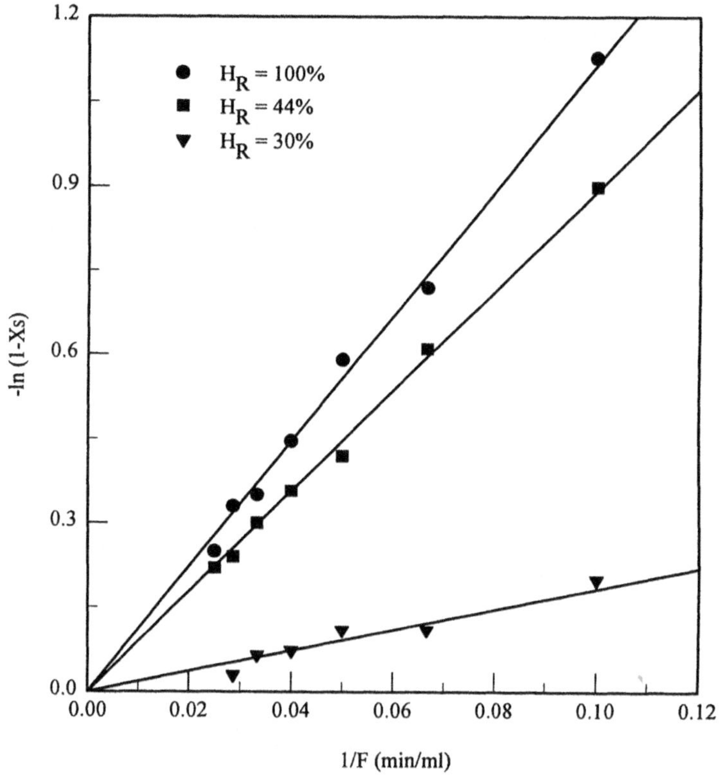

Figure 10. Effect of flow rate on conversion. Conversions were measured at 30 °C and various gas relative humidity. Other experimental conditions were as described for Figure 6 except flow rate. For details see Equation 18 and text.

$$\ln(-\ln(1 - X_s)) = \ln\frac{V_R \varepsilon k_o}{F} - \frac{E_a}{RT} \tag{20}$$

Once again, the linearity of the data in Figure 11 substantiates the efficacy of the equations.

Interestingly, the calculated activation energy (7.47 kcal/mol) appears to be independent of the hydration of YADH (the lines in Figure 11 are parallel). Barzana and co-workers [2] have observed that apparent activation energy of alcohol oxidase catalyzed ethanol oxidation reaction is a function of water activity in a batch process. It has been shown that apparent activation energy decreases as water activity increases, approaching the value founded in aqueous solution. For YADH, only the frequency factor (k_0) is dependent on the enzyme hydration level (at relative humidity 100%, 90% and 80%, k_0 is 1.3×10^6, 1.1×10^6 and 1.0×10^6 min^{-1}, respectively). The activation energy is about half of that in aqueous solution (14 kcal/mol) [34], and similar reductions in activation energy have also been observed for organophosphorus hydrolase catalyzed pesticide hydrolysis in a continuous gas phase reaction [38]. The decreased activation energy in the gas phase system could be caused by either ground-state destabilization or transition-state stabilization [14, 31, 40]. There is no method available yet to determine why the activation energy for YADH catalyzed oxidation reduction reactions is decreased significantly by removal of solvent.

6. CONCLUSIONS

We have shown in this study that alcohol dehydrogenase from baker's yeast can catalyze oxidation-reduction reactions with gas phase substrates. Enzyme stability is greatly enhanced in this water limited environment, the cofactor NAD^+ can be regenerated more effectively than in either water or other nonaqueous media. The study has revealed that enzyme activity and stability require a temperature independent level of hydration (surface coverage of water) which is closely related to the proportion of the protein [15] which is hydrophilic (as defined by Kyle and Doolittle [19]). YADH deactivates and destablizes rapidly if the enzyme loses this critical water content. There is no optimal gas humidity for enzyme activity if reaction temperature is higher than 30 °C.

YADH catalyzed reactions in the gas phase exhibit kinetics which fit to a pseudo first order mechanism. The enzyme hydration level apparently has no influence on the putative mechanism. Our results have also shown that the activation energy for YADH catalyzed reactions with gas phase substrates is independent of the enzyme hydration level (7.47 kcal/mol). The optimum reaction temperature, based on enzyme overall performance, for the continuous bioreactor is 30 °C. With the reactor system described, it is now possible to control the hydration of enzyme while determining catalytic function. As such we are in a position to elucidate the role of water, and other solvents, in catalysis.

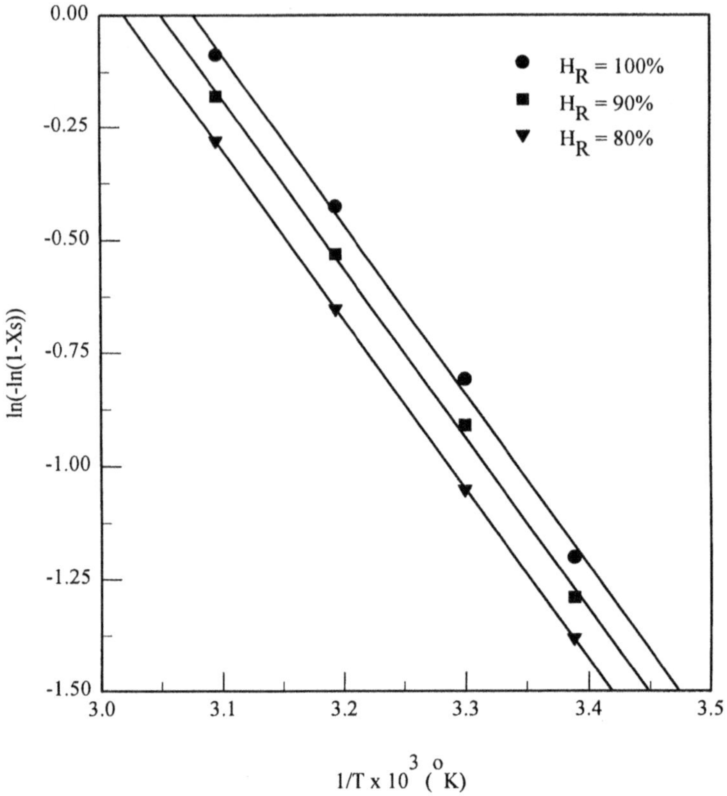

Figure 11. Arrhenius plot for YADH catalyzed reactions with gas phase substrates. Data were taken at various gas relative humidity as shown in the figure. Experimental conditions were as described for Figure 6. For details see Equation 20 and text.

REFERENCES

1. Barzana, E., Klibanov, A. M. and Karel, M. (1987) Enzyme-Catalyzed Gas-Phase Reactions. *Appl. Biochem. Biotechnol.* 15, 25 - 34.
2. Barzana, E., Karel, M. and Klibanov, A. M. (1989) Enzymatic oxidation of ethanol in the gaseous phase. *Biotechnol. Bioeng.* 34, 1178 - 1185.
3. Berlin, E., Kliman, P. G. and Pallansch, M. J. (1970) Changes in state of water in proteinaceous systems. *J. Colloid Interface Sci.* 34, 488 - 494.
4. Brunauer, S., Emmett, P. H. and Teller, E. (1938) Adsorption of gases in multimolecular layers. *J. Am. Chem. Soc.* 60, 309 - 319.
5. Carta, G., Gainer, J. L. and Benton, A. H. (1991) Enzymatic Synthesis of Ester Using an Immobilized Lipase. Biotechnol. Bioeng. 37: 1004 - 1009.
6. Dean, J. A. (1979) *Lange's Handbook of Chemistry*. 12th ed., McGraw-Hill, New York, pp 10 - 54.
7. Deetz, J. S. and Rozzell, J. D. (1991) Chapter 9, Catalysis by Alcohol Dehydrogenase in organic Solvents, in *"Biocatalysts for Industry"* (edited by Dordick, J. S.), Plenum Press, New York.
8. Dickinson, F. M. and Monger, G. P. (1973) A Study of the Kinetics and Mechanism of Yeast Alcohol Dehydrogenase with Variety of Substrates. Biochem. J. 131: 261 - 270.

9. Dordick, J. S. (1989) Enzymatic Catalysis in Monophasic Organic Solvents. *Enzyme Microb. Technol.* 11, 194 - 211.
10. Hnojewyj, W. S. and Reyerson, L. H. (1961) Further studies on the sorption of H_2O and D_2O vapors by lysozyme and the deuterium-hydrogen exchange effect. *J. Phys. Chem.* 65, 1694 - 1698.
11. Huttig, G. F. (1948) Zur Auswertung der Adsorptions-Isothermen. *Monatshefte fur Chemie* 78, 177 - 184.
12. Hwang, S. O., Trantolo, D. J., Wise, D. L. (1993) Gas Phase Acetaldehyde Production in a Continuous Bioreactor. *Biotechnol Bioeng.* 42, 667 - 673.
13. Hwang, S. O. and Park, Y. H. (1994) Ethyl acetate production in the gas phase. *Biotechnol. Lett.* 16, 379 - 384.
14. Janssen, A. E. M. and Halling, P. J. (1994) Specificities of enzymes corrected for solvation depend on the choice of the standard state. *J. Am. Chem. Sco.* 116, 9827 - 9830.
15. Jornall, H. (1997) The Primary Structure of Yeast Alcohol dehydrogenase. *Eur. J. Biochem.* 72, 425 - 442.
16. Kery, V., Haplova, J. and Tihlarik, K. (1990) Factors Influencing the Activity and Thermostability of Immobilized Porcine Pancreatic Lipase. *J. Chem. Tech. Biotechnol.* 48: 201 - 207
17. Killion, P. J., Reyerson, L. H. and Cameron, B. F. (1970) Sorption of H_2O and D_2O Vapor by Human Methemoglobin A and Methemoglobin S Derivatives, and Sperm Whale Metmyoglobin. *J. Colloid Interface Sci.* 34, 495 - 504.
18. Klibanov, A. M. (1986) Enzymes That Work in Organic Solvents. *Chemtech* 16, 354 - 359.
19. Kyle, J. and Doolittle, R. F. (1982) A simple method for displaying the hydrooathic character of a protein. *J. Mol. Biol.* 157, 105 - 132.
20. Lamare, S. and Legoy, M. D. (1993) Biocatalysis in the gas phase. *Trends in Biotech.* 11, 413 - 418.
21. Langmuir, I. (1916) The constitution and fundamental properties of solids and liquids, part I. solids. *J. Am. Chem. Soc.* 38, 2221 - 2295.
22. McBain, J. W. (1935) An explanation of hysteresis in the hydration and dehydration of gels. *J. Am. Chem. Soc.* 57, 699 - 700.
23. McMinn, J. H., Sowa, M. J., Charnic, S. B. and Paulaitis, M. E. (1993) The hydration of proteins in nearly anhydrous organic solvents. *Biopolymers* 33, 1213 - 1224.
24. Parvaresh, F., Robert, H., Thomas, D. and Legoy, M. D. (1992) Gas phase transesterification reactions catalyzed by lipolytic enzymes. *Biotechnol. Bioeng.* 39, 467 - 473.
25. Plapp, B. V., Ganzhorn, A. J., Gould, R. M., Green, D. W., Jacobi, T., Warth, E. and Kratzer, D. (1991) Catalysis by yeast alcohol dehydrogenase. p. 241 - 251. In: *Enzymology and molecular biology of carbonyl metabolism 3.* Edited by Weiner, H., Wermuth, B. and Crabb, D. W. Plenum Press, New York.
26. Pulvin, S., Legov, M. D., Lorde, R., Pensa, M. and Thomas, D. (1986) Enzyme technology and gas phase catalysis: alcohol dehydrogenase example. *Biotechnol. Lett.* 8, 783 - 784.
27. Pulvin, S., Parvaresh, F., Thomas, D. and Legoy, M. (1988) Solid-Gas Reactors. A Comparison between the Horse Liver and the Thermostable *Sulfolobus solfataricus* ADH. *Ann. N. Y. Acad. Sci.* 542, 434 - 439.
28. Ross, S. (1949) A comparison of two theories of multilayer adsorption. *J. Phys. Chem.* 53, 383 - 391.
29. Rupley, J. A. and Careri, G. (1988) Enzyme hydration and function. p. 223 - 234 In: *The enzyme catalysis process: energetics, mechanism, and dynamics.* (edited by Cooper, A., Houben, J. L. and Chien, L. C.), Plenum Press, New York and London.
30. Rupley, J. A. Gratton, E. and Careri, G. (1983) Water and globular proteins. Trends Biochem. Sci. 8: 18 - 22.
31. Ryu, K.and Dordick, J. S. (1992) How Do Organic Solvents Affect Peroxidase Structure and Function? *Biochemistry* 31, 2588 - 2598.
32. Sarcar, S., Jian, T. K. and Maitra, A. (1992) Activity and Stability of Yeast Alcohol Dehydrogenase (YADH) Entrapped in Aerosol OT Reverse Micelles. *Biotechnol. Bioeng.* 39, 474 - 478.
33. Snijder-Lambers, A. M., Vultson, E. N. and Doddema, H. J. (1991) Optimization of Alcohol Dehydrogenase Activity and NAD(H) Regeneration in Organic Solvents. *Red. Trav. Chim. Pays-Bas* 110, 226 - 230.
34. Sund, H. and Theorell, H. (1963) Chapter 2. Alcohol Dehydrogenase. p. 25 - 83 In: P. Boyer (ed.), *The Enzymes*, vol. 7. 2nd edition. Academic Press, New York.
35. Vasudevan, P. T. and Thakur, D. S. (1994) Soluble and Immobilized Catalase - Effect of Pressure and Inhibition on Kinetics and Deactivation. *Appl. Biochem. Biotechnol.* 49, 173 - 189.

36. Yang, F. X., and Russell, A. J. (1993) Optimization of Bakers Yeast Alcohol Dehydrogenase Activity in an Organic Solvent. *Biotechnol. Prog.* 9, 234 - 241.
37. Yang, F. X. and Russell, A. J. (1994) Two-step biocatalytic conversion of an ester to an aldehyde in reverse micelles. *Biotechnol. Bioeng.* 43, 232 - 241.
38. Yang, F. X. and Russell, A. J. (1995) Nonaqueous Biocatalytic Degradation of A Nerve Gas Mimic. Biotechnol. Prog. in press.
39. Yang, F. X. and Russell, A. J. (1995) Water adsorption by Alcohol Dehydrogenase in a Continuous Gas Phase Reactor. Submitted to *Biotechnol. Bioeng.*
40. Yang, F. X. and Russell, A. J. (1995) The Role of Hydration in Enzyme Activity and Stability. Submitted to *Proc. Natl. Acad. Sci.*
41. Yang, Z., Zacherl, D. and Russell, A. J. (1993) pH dependence and activity of subtilisin in organic solvents. *J. Am. Chem. Soc.* 115; 26, 12251 - 12257.
42. Zaks, A. and Klibanov, A. M. (1984) Enzymatic Catalysis in Organic Media at 100°C. *Science* 224, 1249 - 1251.
43. Zaks, A. and Klibanov, A. M. (1988) The effect of water on enzyme action in organic media. *J. Biol. Chem.* 263, 8017 - 2021.

FOLDABLE ANION HOSTS BASED ON BICYCLIC GUANIDINIUM ANCHOR GROUPS

FRANZ P. SCHMIDTCHEN†*, MICHAEL BERGER†; AXEL METZGER†; KARSTEN GLOE‡ and HOLGER STEPHAN‡

† *Institut für Organische Chemie und Biochemie, TU München, D-85747 Garching, Germany;* ‡ *Institut für Anorganische Chemie, TU Dresden, D-01062 Dresden,* Germany

1. General Concept

Among the three basically different approaches to bind anions non-covalently to an artificial host structure as exemplified in Figure 1, the interaction of unlike charges constitutes the farthest reaching and energetically strongest alternative. This is also the type present in almost every natural receptor molecule which is to complex an anionic moiety in a substrate.

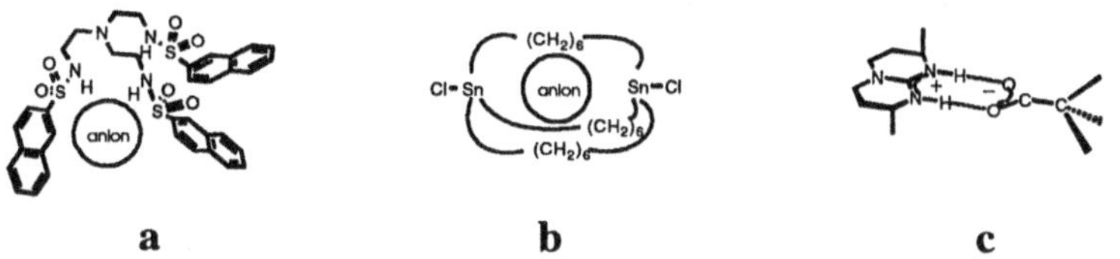

a **b** **c**

Figure 1: Representative examples of the three principally different types of anion interaction with molecular host compounds: a) ion-dipole interaction, *cf.* [1]; b) coordination to poly Lewis-acids [2]; c) dedicated ion pairing [3]

C. S. Wilcox and A. D. Hamilton (eds.), Molecular Design and Bioorganic Catalysis, 191–210.

Many X-ray crystal structures of e.g. enzyme-substrate complexes reveal that the guanidinium side chain of an arginine residue is the most prominent positively charged primary anchor group in direct contact with the anionic site in the guest molecule [4]. Backed by site-directed mutagenesis experiments, which document the extraordinary stability of guanidinium-oxoanion pairing - this host-guest arrangement was found to be ~5 kcal/mol more stable than an analogous ion pair using a prim. ammonium group as the cationic partner [5] - one can adapt this natural binding motive for incorporation into an artificial host compound. For this to be achieved one must allow only one unique and unambiguous host-guest binding mode. In proteins the guanidinium side chain of arginine is part of an extensive H-bonding network which places and orients this anchor group in space.

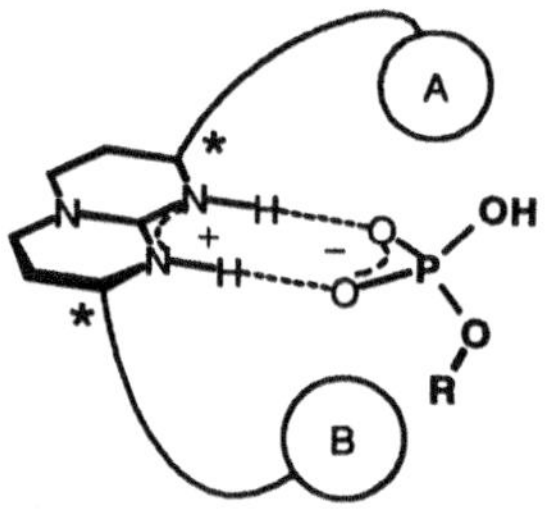

Figure 2: Conceptual design of a guanidinium anchor group for oxoanions.

We considered bicyclic guanidinium salts of the type shown in Fig. 2 to model this directing effect rather nicely. The guanidinium structure is imbedded into an almost strainfree bicyclic framework which leaves only one side open for oxoanion binding. If the anionic guest is complexed by virtue of the charge interaction complemented by two parallel hydrogen bonds its location is well defined. Another beneficial feature of the design would emerge, if the bicycle could be functionalized at the α, α'-positions in chiral fashion. This would pave the way to the regioselective introduction of more anchor functions (e.g. A and B) capable to interact with additional substructures of the guest and thus contribute to the overall attractive association.

2. Synthesis and Use of Ditopic Guanidinium Receptors

The synthesis of substituted bicyclic guanidinium salts has been elaborated early on [6], but the introduction of chirality was realized much later [7]. Of the two alternative strategies depicted in Fig. 3 the lower route proved more profitable and is detailed in Fig. 4:

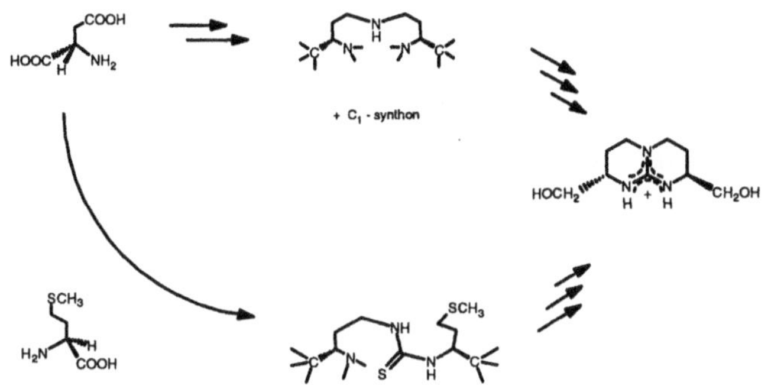

Figure 3: Strategies for the construction of chiral bicyclic guanidines.

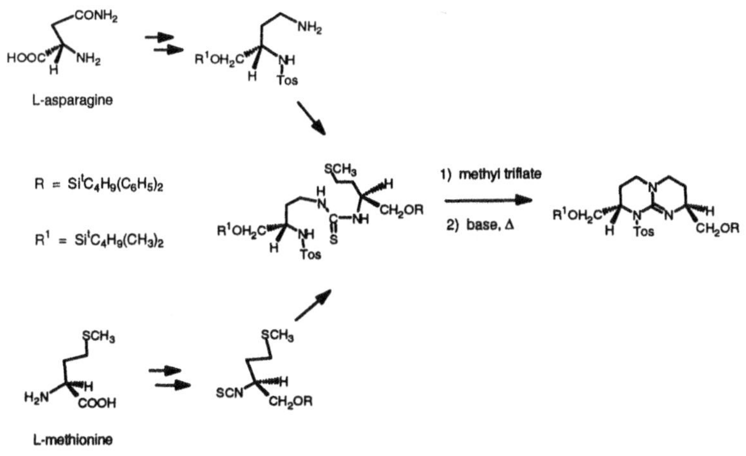

Figure 4: Synthetic route to chiral functionalized guanidinium anchor groups.

Starting from readily available components of the natural chiral pool two amino acids, asparagine and methionine, were transformed into a primary amine and an isothiocyanate, respectively. Since all of the other functional groups were blocked, reaction of these compounds with each other gave an almost quantitative yield of the thiourea. A four step one-pot reaction concludes the sequence: Methylation successively converted the sulfur functions into sulfonium and isothiuronium moieties. On subsequent addition of a Hünig tert. amine base and gentle warming the twofold cyclisation commenced, giving the fully but differentially protected bicyclic guanidine in over 80% isolated yield. Chemistry was developed to manipulate any of the protected functions independently [8], so that this compound served as the parent building block for elaboration into polytopic artificial receptor molecules.

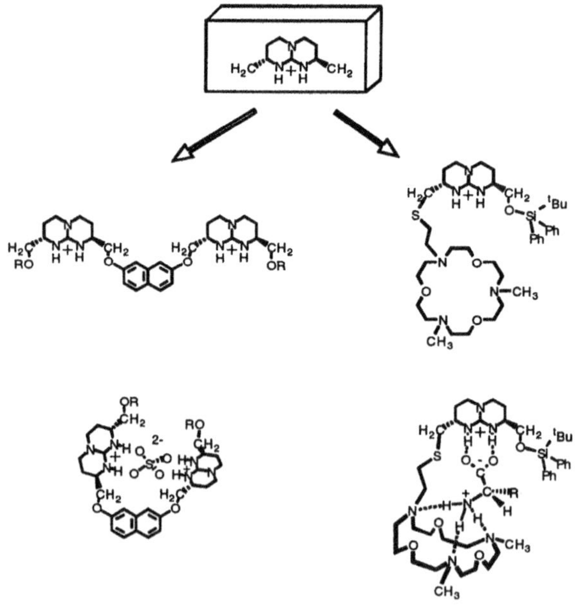

Figure 5: Foldable ditopic hosts for anions and amino acid zwitterions, respectively.

The initial step towards our target of flexible hosts that would adopt defined 3 dimensional structures by a folding process [9] thereby imitating the natural analogs, is the connection of two anchor moieties via a spacer unit. Following this basic idea ditopic hosts for binding anions or amino acids under physiological solvent conditions as shown in Fig. 5 were of prime interest. In both cases the molecular guests are to alter the equilibrium conformation of the uncomplexed host and organize it into a novel and unique "tertiary" structure (induced fit).

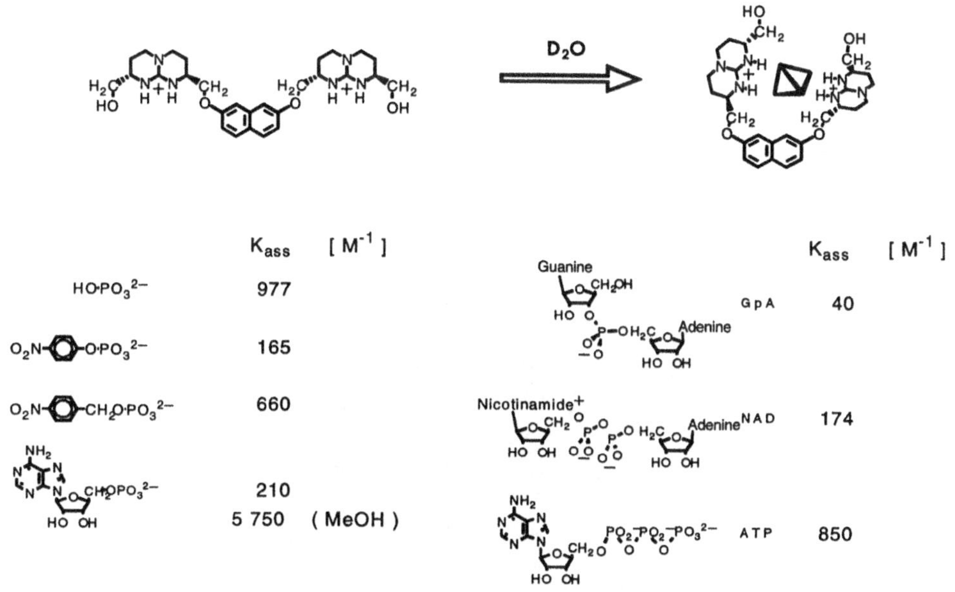

Figure 6: Association constants K_{ass} as determind by ^{1}H-NMR in the host-guest complexation of biologically relevant phosphate anions with an artificial ditopic guanidinium host in water.

Bridging two guanidinium anchor groups with a planar 2,7-disubstituted naphthalene unit gave, after desilylation, a water soluble dication which showed extraordinary host-guest binding properties (Fig.6): In addition to unexpected selectivities in the complexation of dicarboxylates [10] this host design serves to bind strongly solvated and biologically important phosphates in water [11]. One reason that certainly adds to the stability of the complexes is the peculiar folding pattern accessible to this host. In the process of guest binding the guanidinium substructures must move like lobsters´ claws to pinch the guest in between them. Owing to the chirality of the cations along with the flatness of the spacer the guanidinium bicycles end up with their main planes arranged perpendicular to each other in a perfect set to ease hydrogen bonding towards a tetrahedral oxoanion. However, more subtle influences govern the binding features as well. Thus, it was originally expected that the attachment of organic residues to the phosphate anion would boost binding due to additional favorable interaction modes. The data given in Fig. 6, however, indicate that inorganic hydrogenphosphate is the best dianionic substrate for this host in spite of putatively beneficial moieties in phosphate esters like p-nitrophenyl- or p-nitrobenzylphosphate, which could stack to the aromatic spacer. The higher K_{ass} value for the latter guest suggests the involvement of steric restrictions for complexation which is amazing viewing the flexible nature of the host.

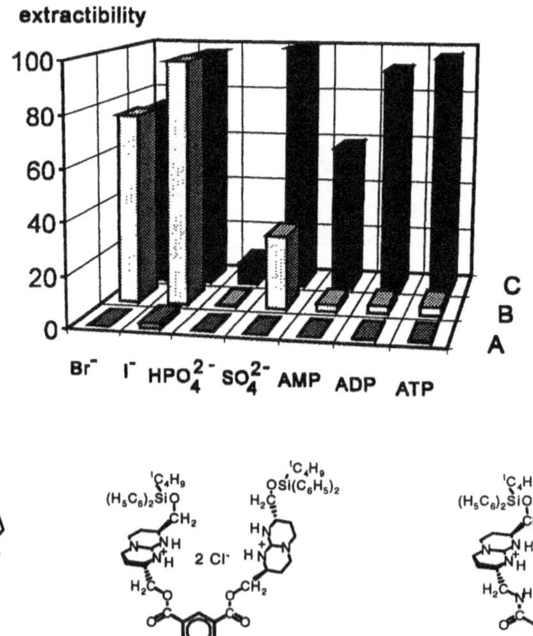

Figure 7: Extraction of various anions from water into chloroform by the ditopic host compounds A, B and C monitored by liquid scintillation counting of appropriate radioisotopes. Receptors: 1*10^{-3} M in $CHCl_3$; anions: 1*10^{-4} M in water buffered at pH 7.4

A possible application of abiotic phosphate receptors could lie in their use as sensor materials in ion-selective electrodes. The mandatory requirement for an appropriate host compound to qualify in this sense besides anion selectivity is high hydrophobicity in order to avoid leaking from the hydrophobic membrane phase. As an initial model system the three ditopic guanidinium compounds shown in Fig. 7 were studied in their ability to transfer anions from a very dilute buffered aqueous phase to a bulk organic phase (chloroform). The hosts all contained the same lipophilic isophthaloyl spacer and an identical set of anchor groups, but differed in the distal substituent (hydroxygroup in A versus bulky silylether functions in B and C) and the junction connecting the modular array (ester functions in A and B versus a secondary amide in C). The phase transfer of a number of anions was unambiguously monitored by radiotracer methods (Fig. 7) [12]. Host A failed to extract any anion from the aqueous phase. Apparently the pronounced amphiphilicity of this compound led to its concentration in the liquid interphase boundary so that it functions as a block rather than a carrier for anion transfer. Replacement of the alcoholic groups in A by lipophilic silyl ethers in B gave efficient anion extraction, but only for the less hydrated halide anions. The more heavily solvated oxoanions phosphate, sulfate and some adenine nucleotides required host C for phase transfer which is the most potent transfer agent for these anions known to date. The outstanding efficiency of host C for oxoanion extraction presumably derives from two additional hydrogen bonds to the amide functions which can form on complexation. In all cases, however, there is a counter transport of chloride ions which eventually may hamper an ion sensing process. Thus, it would be desirable to have an electroneutral anion host which still retains the favorable binding properties and hydrophobicity of the ditopic guanidinium receptors but lacks counter anions with complete motional freedom. Towards this end we constructed a zwitterionic host containing an icosahedral borane cluster as an anionic site (Fig. 8). The *closo*-$B_{12}H_{11}SH$ cluster is a chemically rather stable dianionic yet hydrophobic compound with negligible H-bond donor-acceptor properties. We expected this anion to be almost non-coordinating to the guanidinium site leaving it "naked" and eager for binding oxoanions. Since one top the thiol derivative could be easily incorporated into the host structure via a straightforward nucleophilic substitution, this borocaptate cluster seemed first choice as an anionic building block. When the resulting zwitterionic receptor was compared to a dicationic host of analogous design (Fig. 8) in host-guest complexation with squarate and sulfate in homogeneous DMSO solution hardly any difference in binding strength was observed. The absence of competing counter ions in the electroneutral receptor apparently is balanced by the reduction in overall charge leaving a marginal advantage to the positively charged receptor. More detailed investigation will tell whether or not a peculiar ground state conformation of the zwitterionic compound is responsible for the compensatory effect observed.

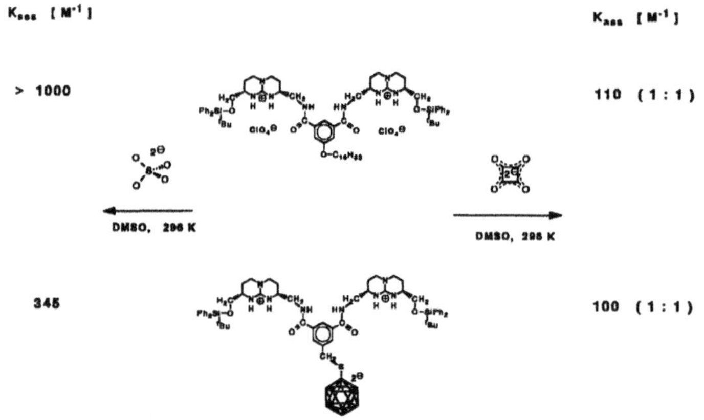

Figure 8: Comparsion of dicationic versus zwitterionic guanidinium host in the binding of squarate and sulfate anions in homogeneous DMSO-solution. Analysis was done by ^{1}H-NMR at ambient temperature.

3. Evaluation of Host Flexibility

Folding on open-chain host molecule into a defined tertiary structure in the host guest complex requires restriction of internal degrees of rotational freedom. Numerous estimates on the size of this entropically costly process have been published [13]. However, there is little doubt about the qualitative statement advanced by Cram [14] claiming more stable host-guest association the more preorganized the host is i.e. the less rotors within a chain molecule have to be restricted on complexation. In an effort to obtain an experimental basis on this effect we prepared an ensemble of ditopic receptors which were identical in every respect exept for the spacer unit. This was to be altered in its flexibility in an incremental fashion allowing to correlate the association constant with a common anionic guest with spacer rigidity. The idea underlying this approach was based on the scheme of factorising the total free energy of complexation into a number of components as advanced by Williams [13] (Fig. 9).

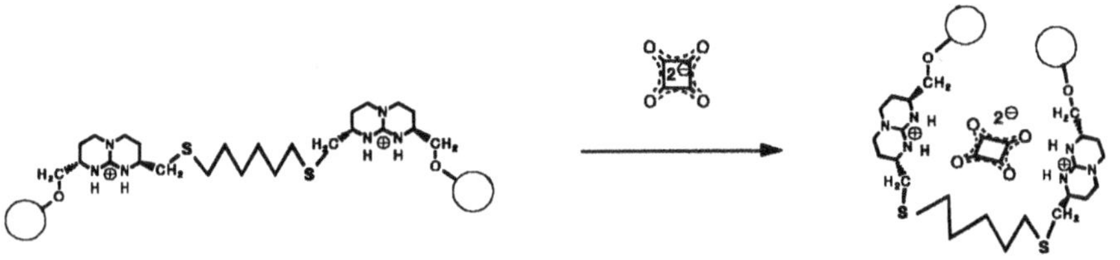

$$\Delta G_{total} = \Delta G_{tra+rot} + \Delta G_{solv} + \Delta G_{rot} + \sum_i \Delta G_{i,pol}$$

Figure 9: Strategy to correlate guest binding to spacer rigidity. In a host series differing in spacer flexibility only, K_{ass} (ΔG_{total}) should yield a measure of $\Delta G_{solv} + \Delta G_{rot}$.

Keeping the guest species and the anchor groups identical within the set of hosts we assumed that the free energy of host-guest binding attributable to changes in translation and rotation of the participating species $\Delta G_{tra+rot}$ and all polar interactions (Σ $\Delta G_{i,polar}$) would be the same. Any differences in K_{ass} should directly relate to changes in solvation ΔG_{solv} and the internal restriction on rotors ΔG_{rot}. Moreover, if the differences in solvation between free host and host-guest complex would stay constant for any ditopic host the K_{ass}-value would directly reflect the rigidity of the spacer. This condition was thought to be met most likely if spacer groups of identical size and chemical nature could be found. For this reason we developed C-6 spacer units starting from D-mannitol, all of which contained acetal functionality only, but differed in the amount of backbone rigidification. Fig. 10 sketches the preparation of the spacers which reacted with the thiol derivatives of the guanidinium anchor groups [8] to give a series of 4 ditopic hosts, in which zero to 3 backbone rotors were arrested.

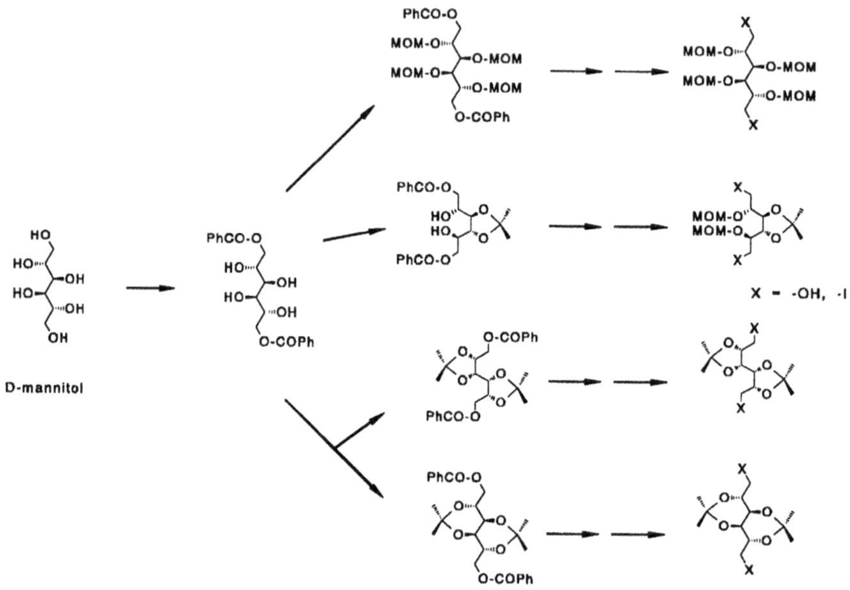

Figure 10: Synthetic pathways to a set of C-6 spacer di-iodides of variable rigidity, usable for attachment of guanidinium anchor groups by thioether formation.

The determination of the host-guest binding constants with squarate in DMSO yielded surprising results (Fig. 11). There was no apparent correlation of the binding constant with host rigidity. One out of the quartet of hosts possessing intermediate rigidity exhibited much better binding of squarate than all the others. The precise K_{ass}-value could not be determined in this case by the ^{1}H-NMR-method employed. The evaluation of thermodynamic parameters from linear van´t Hoff plots revealed that contrary to our expectation host-guest binding was exclusively entropy driven. The enthalpies of association fell within 1.2 kcal/mol around zero confirming our initial premise to use isenthalpic spacer elements. Positive entropies of host-guest interactions are conceivable only if massive solvation effects outmatch the inherently negative entropy of combining host and guest into one complex and sacrifice on top internal degrees of rotational freedom. We are driven to conclude that at least in strongly solvating environments preorganisation and complementarity in host-guest binding which so far have been the main points of emphasis in the design of artificial receptors may be easily screened or even turned to the contrary by the huge free energies available from changes in solvation. These are, however, even much harder to assess or predict than direct host-guest interactions.

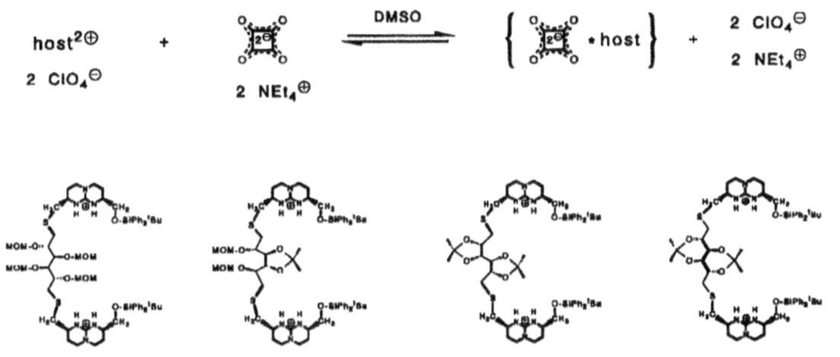

ΔG_0 [kcal]	- 2.95	- 2.57	> - 4.5	- 2.74
ΔH_0 [kcal]	+ 0.72	+ 1.1		- 0.1
ΔS_0 [e.u.]	+ 12.3	+ 12.3		+ 8.7

Figure 11: Thermodynamic parameters of the complexation of squarate in DMSO by ditopic guanidinium hosts of differential flexibility.

4. An Artificial Host for Amino Acid Extraction

In recent years several host systems have been published which would extract amino acids (aa) into an organic phase [15]. The main challenge in this task is to construct a set of anchor groups that would mask the highly hydrophilic α-aminocarboxylate substructure with a lipophilic moiety. The absolute amount of extraction of the amino acid existing preferentially in zwitterionic form in water thus depends on the stability of host-guest complex formation at this partial structure as well as on the free energy of transfer of the complex from water to the organic phase. If no binding function to accommodate the side chain is provided by the host one expects extraction to be modulated according to aa hydrophobicity.

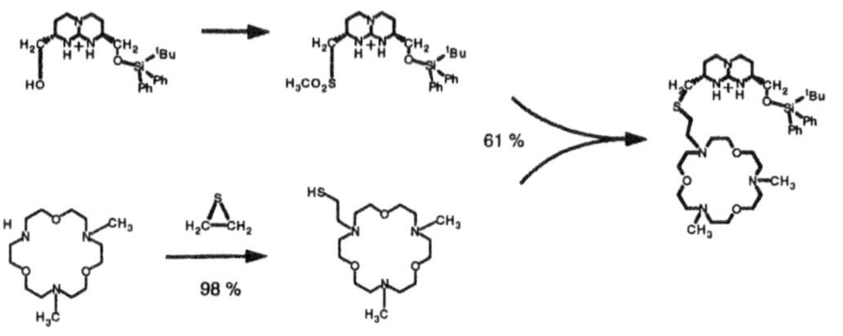

Figure 12: Synthesis of a ditopic host molecule for zwitterionic amino acids.

A useful combination of receptor groups to meet the requirements for aa extraction was thought to result from the covalent connection of bicyclic guanidinium cation and a triazatrioxa crown ether (Fig. 12) either of which is of proven utility in binding carboxylate and prim. ammonium functions, respectively [16]. The functionalized building blocks of this modular host were readily derived from known precursors [17] and were assembled in satifactory yield into the synthetic target (Fig. 12).

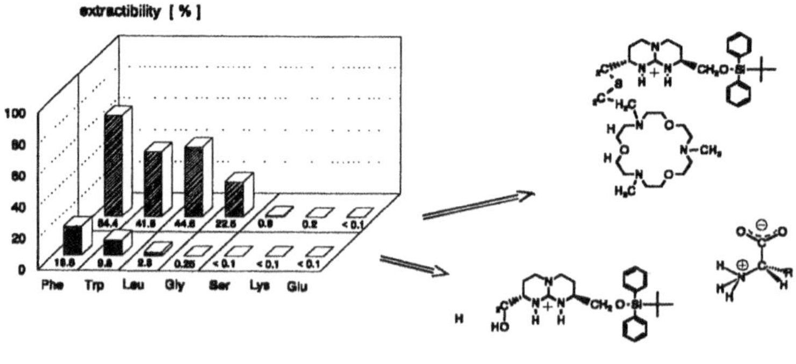

Figure 13: Two-phase extraction of amino acids by a ditopic host and a monotopic guanidinium salt for comparsion in a system aqueous buffer, pH 8.9 / dichloromethane. [host] = $1*10^{-3}$ M in CH_2Cl_2; [aa] = $1*10^{-4}$ M in water.

Amino acid extraction was probed in a two-phase water/CH_2Cl_2 system using radiotracer methods. The comparsion of the newly constructed host to a guanidinium salt locking a second binding function clearly demonstrated the superiovity of the ditopic design (Fig. 13). In any case studied the extractability was greater than with the monotopic standard and the differences were more pronounced with the less hydrophobic amino acids. So glycine and even serine which showed no detectable extraction with the simple guanidinium salt were transfered to dichloromethane by ditopic host. From the concentration dependence which revealed clean 1:1 extraction stoichiometry the extraction equilibrium constants K_{eq} were calculated. When compared to literature values [15] our host is a better extracting agent than its predecessors by factors ranging from 450 to 3000 depending on the amino acid studied [18]. The participation of both anchor groups of the ditopic host and the charge state of the quest species extracted is apparent from the pH-profile of the transfer process (Fig. 14):

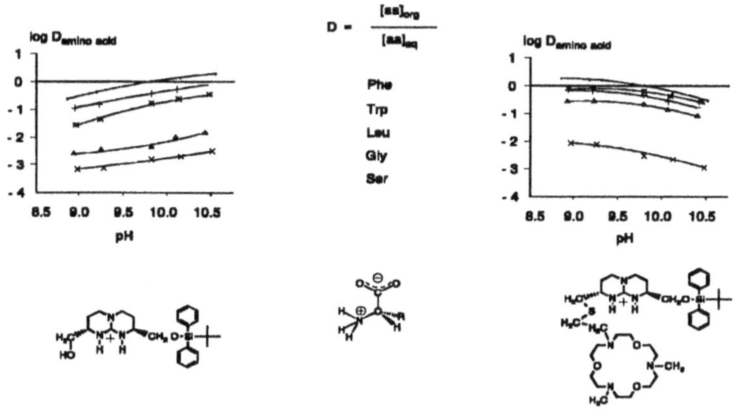

Figure 14: pH-profile of amino acid extractions. The exp. conditions are given in the legend to fig. 13.

With the monotopic guanidinium salt extractability (given as log D_{aa}) increases with higher pH-value. This is indicative of ion pairing in which the amino acid is transfered as the anionic partner in a complex with the guanidinium cation. In contrast extraction of amino acids with the ditopic host has a maximum around pH 9 and diminishes with increasing pH value. In addition to the guanidinium-carboxylate interaction which does not participate in protonation equilibria in the pH range under investigation here, there must be another dissociable proton involved in host-guest binding. In view of the rather similar pK_a-values of all the amino functions in the host and the prim. amino group of the quest it appears a matter of taste where to place this proton. It is definitely part of the hydrogen bonding network errected in the azacrown ether-amino group interaction, but may be easily relocated between all available nitrogen sites in rapid transprotonation equilibria.

5. A Model for Asymmetric Transamination

An intriguing goal of supramolecular chemistry is the rational design of non-covalent interactions that would direct the pathway of reacting systems. Shining examples of the potential power of this approach are provided by the enzymes which may select one out of several possible reaction routes by precise positioning of reactive functionality and predisposition of molecular conformations in the substrate. In Vitamin B_6-dependent enzymes, for instance, chemoselectivity in the further transformation of a reactive intermediate (Fig. 15), itself accessible from a reaction of pyridoxal and an α-aminocarboxylic acid, is supposedly controlled by the conformation of the C_α-nitrogen single bond.

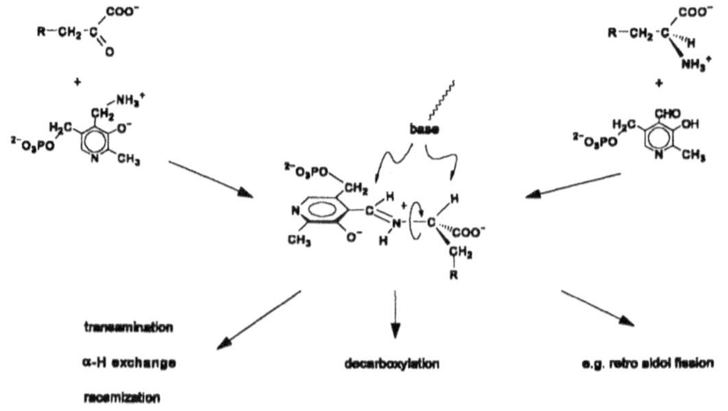

Figure 15: General reaction pattern of vitamin B_6-dependent enzyme reactions.

According to a hypothesis of Dunathan [19] the role of the enzyme protein is to fix this rotor to the effect that the bond at C_α oriented perpendicular to the extended π-electron system is broken and thus may give decarboxylation, aldol fission or the products resulting from C_α-H bond cleavage (racemisation, transamination etc.). Quite a number of enzyme models have been elaborated in the past based on this idea [20], but in most cases the models involved binding of the amino acid side chain and consequently were not applicable to amino acids as a structural class. With the bicyclic guanidinium moiety as an effective anchor group for carboxylates at hand an enzyme mimick for the great transaminases subgroup seemed to be in reach. This model would consist of a coenzymatically active pyridoxamine derivative covalently attached to one side of a chiral disubstituted guanidinium unit, the other side of which would hold a general base (Fig. 16).

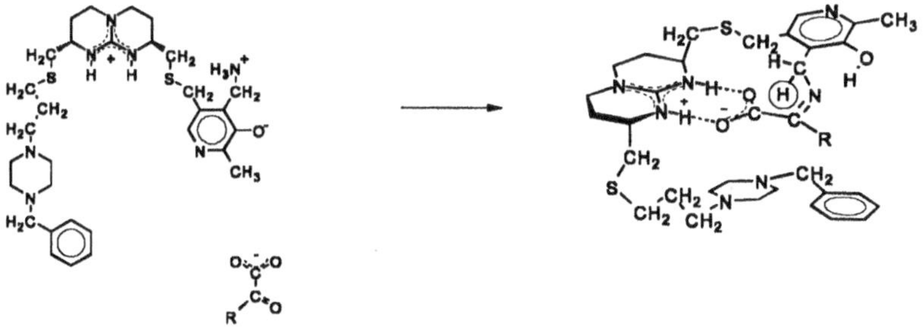

Figure 16: Design and concept of an open-chain transaminase model.

The idea underlying this open-chain array of functionality was built on the asumption that in analogy to the enzymic system the pyridoxamine could react with an α-ketocarboxylic acid to form an azomethine very rapidly (this is one step of the natural transamination sequence interconverting α-oxo- and α-amino acids).The flexibility of the resulting linear molecule in combination with the host-guest relationship of its guanidinium and carboxylate functions would force the chain into a folded conformation, which by virtue of the chirality at the guanidinium site would resemble a chiral helix. As a corollary the base can reach only one face of the prochiral azaallyl system to promote proton transfer converting the initially formed ketimine into an aldimine structure with creation of a novel asymmetric center at the C_{α}-carbon atom. The aldimine obtained could be hydrolyzed to an α-amino acid having predictable chirality which is based on the enantiomeric type of the guanidinium anchor group. According to this rational one expects to observe optical induction and the formation of *S*-amino acids, if the chiral host is administered as the *S*, *S*-enantiomer.

Since it is very unusual that a flexible linear organic molecule folds into a defined secondary structure, experimental advice was sought from an analogous but non-reacting system.

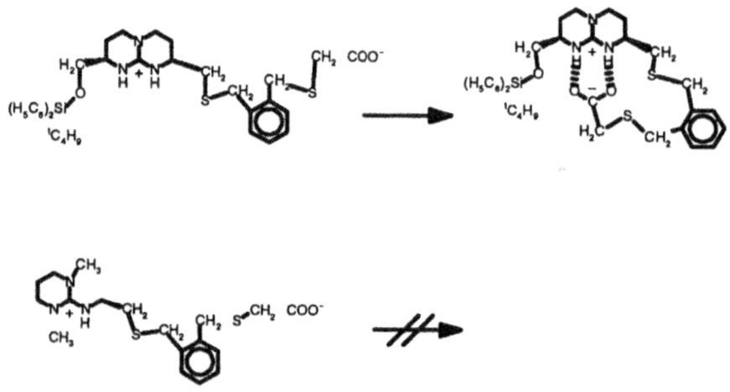

evidence for folding

NH chemical shift

→ **value ($CDCl_3$, DMSO, CD_3OH)**

→ **concentration independence**

→ **inverse solvent shift ($CDCl_3$ - DMSO)**

→ **temperature coeff. < 0.002 ppm/K**

→ **diastereotopic splitting of the S-CH_2-C=O signal**

Figure 17: NMR-results on folding (i.e. formation of intramolecular hydrogen bonds) of non-reacting analogues of the transaminase model.

The NMR-investigation of the thioether analogue of the reactive transamination intermediate (Fig. 17) revealed the presence of intramolecular guanidinium-carboxylate hydrogen bonds on the basis of chemical shift, dilution-, solvent replacement and temperature dependence data. None of these effects were reproduced using the analog B

(Fig. 17) incapable to form the peculiar binding motif. This result confirmed the special conformation features of linear molecules having guanidinium and carboxylate moieties in an appropriate spacing justified embarking on the synthesis of the enzyme model. Its preparation followed straightforward chemistry and furnished three target compounds differing in the base substituent in about 50% overall yield (Fig. 18).

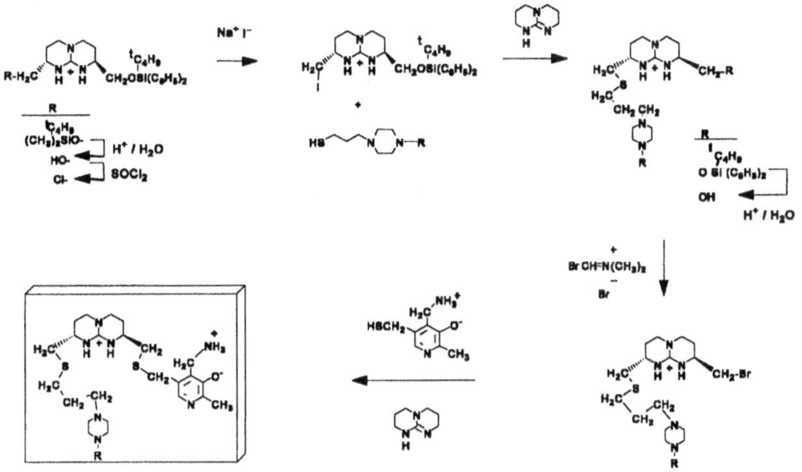

Figure 18: Synthetic scheme for the preparation of linear transaminase models.

The kinetic evaluation of the transaminase models was done under single turnover conditions in methanol and was monitored by UV-spectroscopy. Mixing the enzyme models with an α-ketocarboxylate was accompanied by a burst reaction indicating the formation of the intermediate ketimine. The subsequent slower UV-active reaction followed clean first order kinetics over at least 4 half-lives and corresponded to the rate limiting proton transposition interconverting the ketimine and aldimine structures. At the end addition of aqueous acid released the amino acids, which were then analysed for chirality. In Fig. 19 the results employing ketoleucine as the α-keto acid with enzyme models having various basic substituents are shown. In every case we observed optical induction of the predicted enantiomer of the α-amino acid albeit to only a marginal extent if no base was present in the chiral host. When an anionic or hydrophobic benzyl substituent at the piperazine assisted in forming the desired folding pattern the ee-values were boosted to 38% ee but do not reach preparatively useful levels. Nevertheless the optical induction found is significant and is one of the very rare cases of remote steric influence in an open-chain flexible molecule. The experimental observation of an **1,9 asymmetric induction** of the predicted stereochemistry (the center receiving chirality is separated by 8 heavy atoms from the next chiral center) forces one to conclude that a peculiar supramolecular arrangement is in operation in this open-chain intermediate which has been purposely designed into the system. Destruction of the essential guanidinium-carboxylate host-guest motif by changing to the more competitive solvent water, addition of transition metal ions like Zn^{2+}, which complexes to the carboxylate, or converting the carboxylate to an ester all anihilate the optical induction in this system. Taken together there is ample evidence that this model provides an example for reaction engineering. We are in the position to deliberately design the stereochemical outcome of suitable reacting systems using artificial supramolecular interactions.

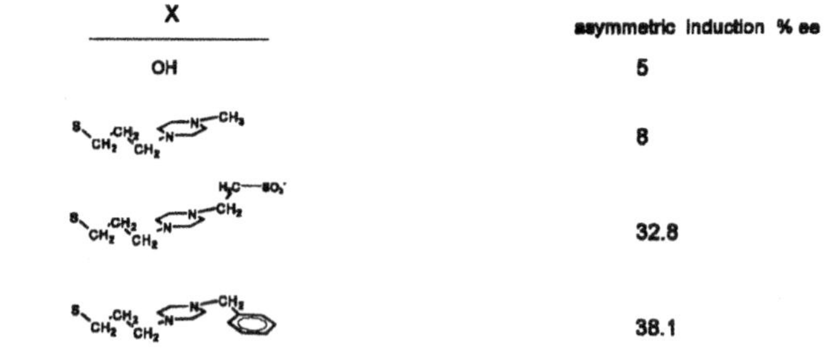

Figure 19: Transaminase model reaction under single turnover conditions in methanol. The model with *S*, *S*-chirality at the guanidinium anchor group yield assymmetric induction to the *S*-amino acid as predicted.

6. References

1. a) Valiyaveettil, S., Engbersen, J.F.J., Verboom, W., and Reinhoudt, D.N. (1993) Synthese und Komplexierungsverhalten ungeladener Anionen-Rezeptoren, *Angew. Chemie* **105**, 942-944; b) Raposo, C., Pérez, N., Almaraz, M., Mussons, M.L., Caballero, M.C., and Morán, J.R. (1995) A cyclohexane spacer for phosphate receptors, *Tetrahedron Lett.* **36**, 3255-3258;

2. Newcomb, M., Horner, J.H., Blanda, M.T., and Squattrito, P.J. (1989) Macrocycles containing tin. Solid complexes of anions encrypted in macrobicyclic Lewis acidic hosts, *J.Am.Chem.Soc.* **111**, 6294-6301; cf. Yang, X., Knobler, C.B. Zheng, Z., and Hawthorne, M.F. (1994) Host-guest Chemistry of a new class of macrocyclic multidentate Lewis acids comprised of carborane-supported electrophilic mercury centers, *J.Am.Chem.Soc.* **116**, 7142-7159;

3. Schmidtchen, F.P., Gleich, A., and Schummer, A. (1989) Selective molecular hosts for anions, *Pure Appl.Chem.* **61**, 1535-1546;

4. Anslyn, E.V., and Hannon, C.L. (1993) The guanidinium group. Its biological role and synthetic analogs, in H.Dugas and F.P.Schmidtchen (eds.), Bioorganic Chemistry Frontiers, Vol. 3, Springer Verlag, Berlin, p. 193-255;

5. Inoue, Y., Kuramitsu, S., Inoue, K., Kagamiyama, H., Hivomi, K., Tanase, S., and Morino, Y. (1989) Substitution of a lysyl residue for Arginine 386 of Escherichia coli aspartate aminotransferase, *J.Biol.Chem.* **264**, 9673-9681;

6. Schmidtchen, F.P. (1980) Synthese symmetrisch substituierter bicyclischer Guanidine, *Chem.Ber.* **113**, 2175-2182;

7. a) Echavarren, A., Galán, A., de Mendoza, J., Salmerón, A., and Lehn, J.-M. (1988) Anion receptor molecules: Synthesis of a chiral and functionalized binding subunit, a bicyclic guanidinium group derived from *L*- or *D*-asparagine, *Helv.Chim.Acta* **71**, 685-693; b) Schmidtchen, F.P., and Kurzmeier, H. (1990) Abiotic anion receptor functions. A facile and dependable access to chiral guanidinium anchor groups, *J.Am.Chem.* **55**, 3749-3755; c) Schmidtchen, F.P., Oswald, H., and Schummer, A. (1991) Chiral anchor groups for Oxoanions: Asymmetric synthesis of tetrasubstituted bicyclic guanidines, *Liebigs Ann.Chem.* **1991**, 539-543;

8. Peschke, W., Schiessl, P., Schmidtchen, F.P., Bissinger, P., and Schier, A. (1995) Building blocks for artificial anion receptors: Derivatives of chiral bicyclic guanidines, *J.Org.Chem.* **60**, 1039-1043;

9. a) Schmidtchen, F.P. (1986) Probing the design of a novel ditopic anion receptor, *J.Am.Chem.Soc.* **108**, 8249-8255; b) Schmidtchen, F.P. (1987) Synthese und Bindungsverhalten eines künstlichen Rezeptors für biogene Amine, *Z.Naturforsch.* **42C**, 476-485;

10. Schiessl, P., and Schmidtchen, F.P., (1993) Abiotic molecular recognition of dicarboxylic anions in methanol, *Tetrahedron Lett.* **34**, 2449-2452;

11. Schiessl, P., and Schmidtchen, F.P. (1994) Binding of phosphates to abiotic hosts in water, *J.Org.Chem.* **59**, 509-511;

12. Stephan, H., Gloe, K., Schiessl, P., and Schmidtchen, F.P. (1995) Lipophilic ditopic guanidinium receptors: Selective extractants for tetrahedral oxoanions, *Supramolec.Chem.* in press;

13. For a general discussion see Searle, M.S., Westwell, M.S., and Williams, D.H. (1995) Application of a binding co-operativity and weak associations in solution, *J.Chem.Soc.Perkin Trans. 2*, **1995**, 141-151;

14. Cram, D.J. (1988) Von molekularen Wirten und Gästen sowie ihren Komplexen, *Angew.Chem.* **100**, 1041-1052;

15. a) Galán, A., Andreu, D., Echavarren, A.M., Prados, P., and de Mendoza, J. (1992) A receptor for the enantioselective recognition of phenylalanine and tryptophane under neutral conditions, *J.Am.Chem.Soc.* **114**, 1511-1512; b) Reetz, M.T., Huff, J., Rudolph, J., Töllner, K., Deege, A., and Goddard, R. (1994) Highly efficient transport of amino acids through liquid membranes via three-component supramolecules, *J.Am.Chem.Soc.* **116**, 11588-11589; c) Aoyama, Y., Asakawa, M., Yamagishi, A., Toi, H., and Ogoski, H. (1990) Simultaneous hydrogen bonding and metal coordination interactions in the two-point fixation of amino acids with a bifunctional metalloporphyrin receptor, *J.Am.Chem.Soc.* **112**, 3145-3151;

16. a) Lehn, J.-M., and Vierling, P. (1980) The [18]-N_2O_3 aza-oxamacrocycle: A selective receptor unit for primary ammonium cations, *Tetrahedron Lett.* **21**, 1323-1326; b) Müller, G., Riede, J., and Schmidtchen F.P. (1988) Host-guest bonding of oxoanions to guanidinium anchor groups, *Angew.Chem.Int.Ed.Engl.* **27**, 1516-1518; c) Gleich, A., Schmidtchen, F.P., Mikulcik, P., and Müller, G. (1980) Enantiodifferentiation of carboxylates by chiral building blocks for abiotic anion receptors, *J.Chem.Soc.Chem.Commun.* **1990**, 55-57;

17. Schmidtchen, F.P. (1986) Tetazac: A novel artificial receptor for binding ω-amino carboxylates, *J.Org.Chem.* **51**, 5161-5168;

18. Metzger, A., Gloe, K., Stephan, H., and Schmidtchen, F.P. Molecular recognition and transfer of underivatized amino acids by a foldable artificial host, *J.Org.Chem.*, submitted;

19. Dunathan, H.C. (1966) Conformation and reaction specificity in pyridoxal phosphate enzymes, *Proc.Natl.Acad.Sci.USA* **55**, 712-716;

20. a)Tabushi, I., Kuroda, Y., Yamada, M., Higashimura, H., and Breslow, R. (1985) A-(modified B_6)-B-[ω-amino(ethylamino)]-β-cyclodextrin as an artificial B_6 enzyme for chiral amino transfer reaction, *J.Am.Chem.Soc.* **107**, 5545-5546; b) Kuzuhara, H., Komatsu, T., and Emoto, S. (1978) Synthesis of a chiral pyridoxamine, *Tetrahedron Lett.* **1978**, 3563-3566; c) Martell, A.E. (1989) Vitamin B_6 catalyzed reactions of α-amino and α-ketoacids: Model systems, *Acc.Chem.Res.* **22**, 115-124.

TRANSITION METAL ION-BASED SUPRAMOLECULAR CATALYSTS

P. SCRIMIN[a], P. TECILLA[b], U. TONELLATO[b]
AND A. VERONESE[b]
[a]*Universita' di Trieste, Dipartimento di Scienze Chimiche,*
Via L. Giorgieri, 1 - 34127 Trieste, Italy
[b]*Universita' di Padova, Dipartimento di Chimica Organica and*
Centro CNR Meccanismi di Reazioni Organiche
Via Marzolo, 1 - 35131 Padova, Italy

ABSTRACT. Transition metal ions may take part in the modification of conformation of molecules as well as in the modification of reaction paths and rates. Examples of simple supramolecular systems exploiting these abilities of transition metal ions are reported.

1. Introduction

Transition metal ions in biological systems may organize molecular structures, orient substrates and activate small molecules.[1] Similar roles can be predicted for synthetic molecules. For instance one can synthesize a rigid macrocycle where co-ordination sites for transition metal ions are properly located for the interaction with a suitable substrate (Figure 1).

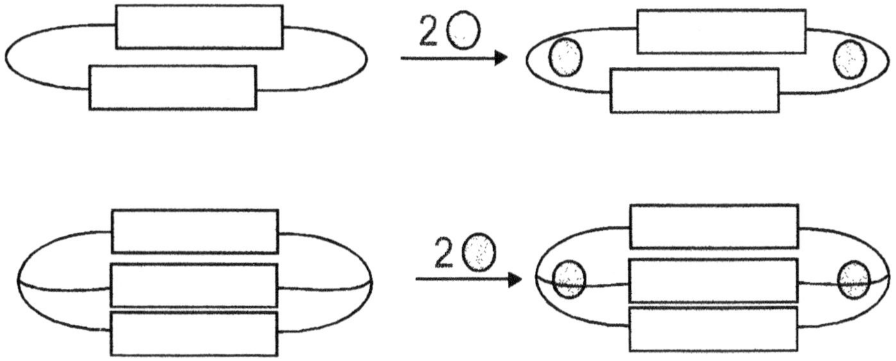

Figure 1. Properly designed macrocyclic molecules may bind transition ions placing them at a well defined distance.

C. S. Wilcox and A. D. Hamilton (eds.), Molecular Design and Bioorganic Catalysis, 211–226.

The metal ions may hence be involved in properly directing the substrate within the cavity of the macrocycle and trigger a reaction on a functional group present in the substrate. In such a system the synthetic work has already set the basis for the space definition of the resulting supramolecular structure. However, one may also conceive a molecule where the metal ion preorganizes the structure before the interaction with the substrate. In this case, the painful step of macrocyclization is avoided and still one may end up with supramolecular structures whose geometrical organization in the space is very similar to the macrocycles mentioned before (Figure 2).

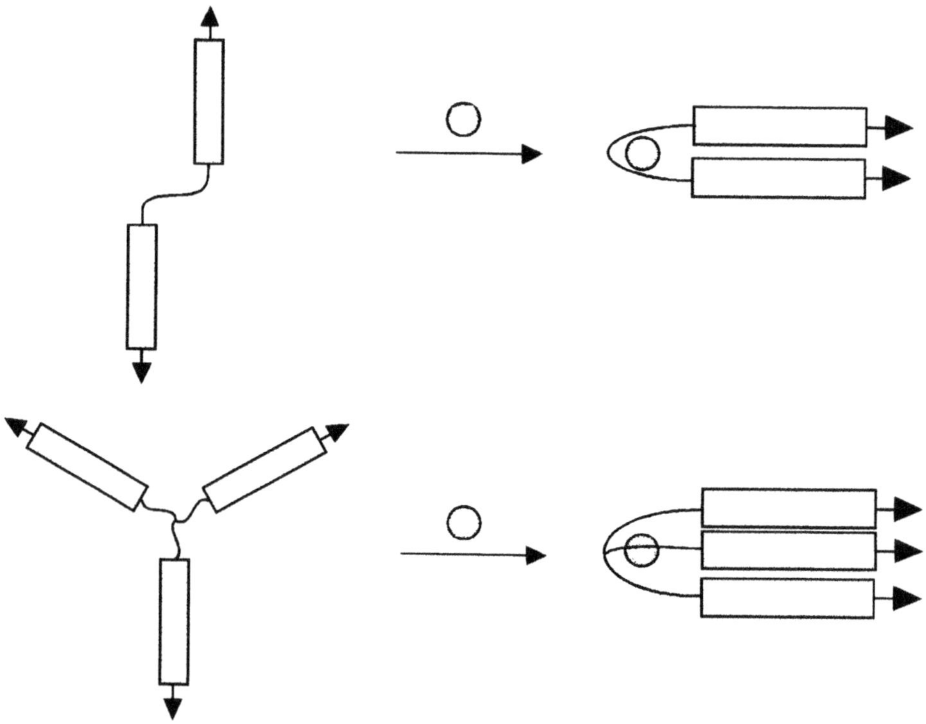

Figure 2. Transition metal ions may also organize a molecular receptor stabilizing a particular conformation.

We have investigated systems able to interact with transition metal ions following both approaches, studied the modification of their properties because of the interaction with transition metals and used several of the resulting complexes as catalysts of the cleavage of carboxylate esters. Some of the systems we have studied so far will be briefly illustrated.

2. Macrocycles

Our efforts concentrated on the synthesis of a molecule able to bind two transition metal ions and show co-operativity between the two of them. In the early eighties Lehn [2] anticipated that a properly designed macrocycle could be capable of binding two metal ions (not necessarily identical) with such a geometry that they co-operate in the process of recognition of a substrate. Several examples of molecules operating on these basis have been reported so far.[3] A step forward is realized on going from recognition to modification of a substrate. For instance, it has been recently shown that two metal centers providing double electrophilic assistance act as powerful catalysts of the cleavage of phosphate diesters and RNA or DNA models as well.[4] Making a receptor not only able to recognize substrates but also to perform a (possibly) catalytic modification was our goal. We started with a conceptually very simple approach: design a simple macrocycle able to bind two Cu(II) ions and rigid enough to keep them properly spaced so that an ω-amino acid ester could interact with both ions *via* the amino and C==O groups.

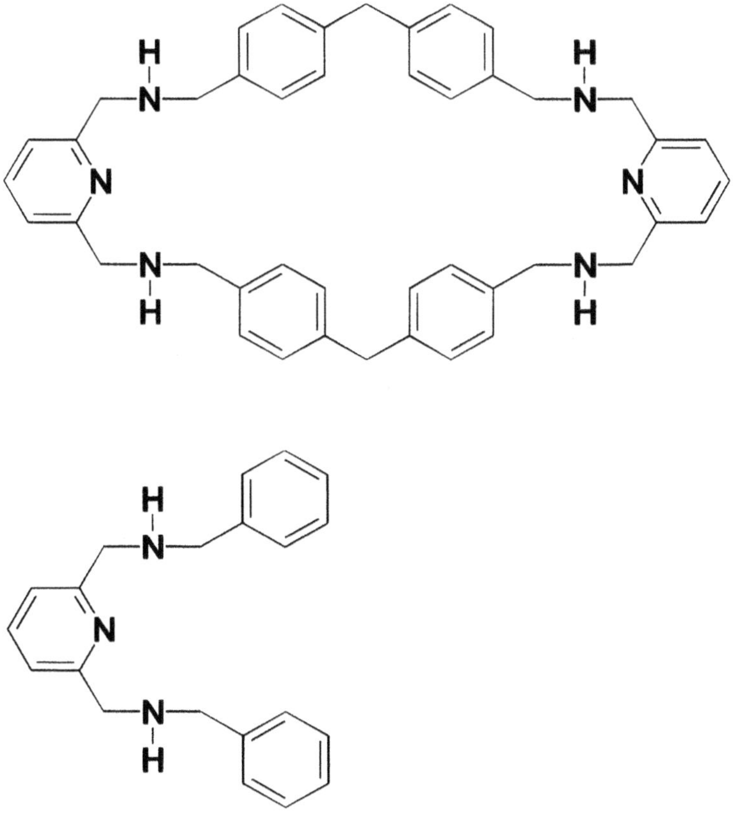

a

b

Figure 3. Our simple macrocycle capable of binding two Cu(II) ions (a) and the acyclic "half" model (b).

We ended up with the synthesis of the macrocycle shown in Figure 3.[5] Our choice was based on the following points: i) the 2,6-diaminomethylpyridine units are known[6] to have binding constants for Cu(II) (1:1 complex) as high as 10^{15} M^{-1}; ii) the two diphenylmethane units are rigid enough to avoid the collapse of the structure; iii) the distance between the two Cu(II) ions is, according to molecular models 10 Å, suitable for binding a β-amino acid ester, in its all s-trans conformation, between the two metal ions. Following this idea our system was tested as a catalyst of the cleavage of the *p*-nitrophenyl ester of two amino acids: α-leucine (α-LeuPNP) and β-alanine (β-AlaPNP). Figure 4, reporting the kinetic results, clearly shows that the dicopper complex is a better catalyst than free Cu(II) only with the β-amino ester. In the case of the α-amino ester not only is free Cu(II) more active than the binuclear complex but there is no evidence of co-operativity between the two metals in catalyzing the cleavage process.

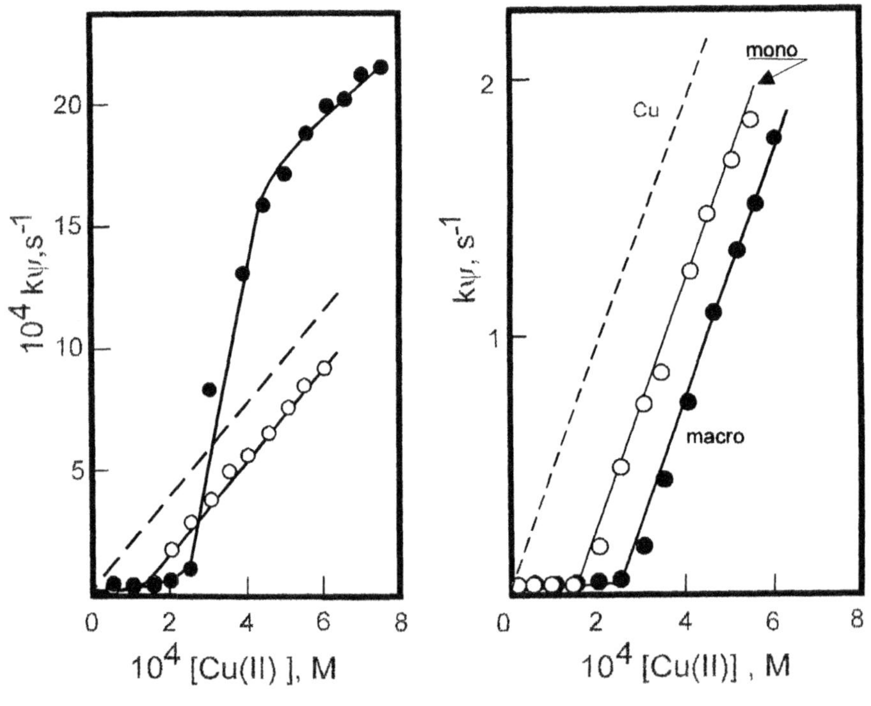

Figure 4. Kinetic effects of the addition of Cu(II) ions to solutions of β-AlaPNP (left) and α-LeuPNP (right) containing the macrocycle ($2x10^{-4}$ M, ●), the acyclic ligand ($2x10^{-4}$ M, ○) or no additive (dotted lines).

The rate accelerations provided by the system in the case of β-AlaPNP is no more than ten fold when compared with the mononuclear complex. A possible explanation comes from the analysis of the X-ray structure of the dinuclear complex. As shown in Figure 5 the two diaminopyridine subunits do not point one against the other, but rather lay in planes almost perpendicular to the macrocycle plane as defined by the two diphenylmethane units. If this is the most stable conformation also in solution the β-amino ester can take advantage of the interaction with the two Cu(II) ions only *via* weak co-ordination positions of copper[7] or forcing the two pyridines to rotate so as to lay in the macrocycle plane.

We thought that a less flexible macrocyclic structure could allow us to overcome this problem. For this reason we tested the ligand shown in Figure 6, reported several years ago by Lehn and his associates,[8] both as a Cu(II) and a Zn(II) complex. To our surprise and disappointment the system was even less reactive than our simple macrocycle. The work is still in progress and conclusions can not be drawn; however it is possible that with a molecule like that shown in Figure 6, steric problems concerning either the binding of substrate or the co-ordination of the nucleophile may become very stringent.

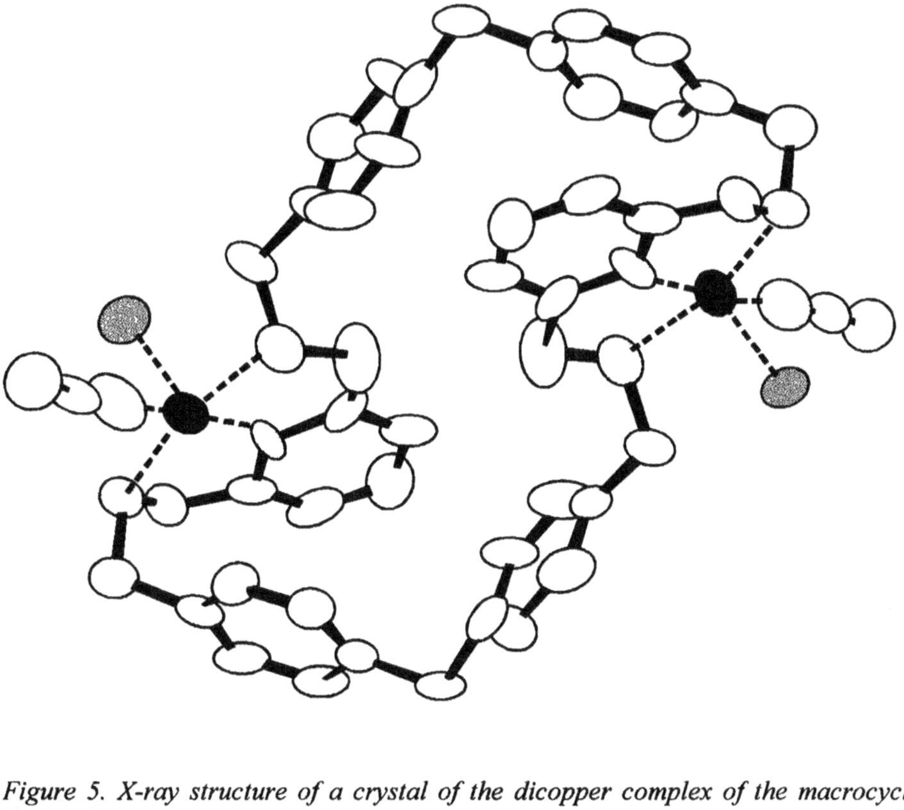

Figure 5. X-ray structure of a crystal of the dicopper complex of the macrocycle of Figure 3.

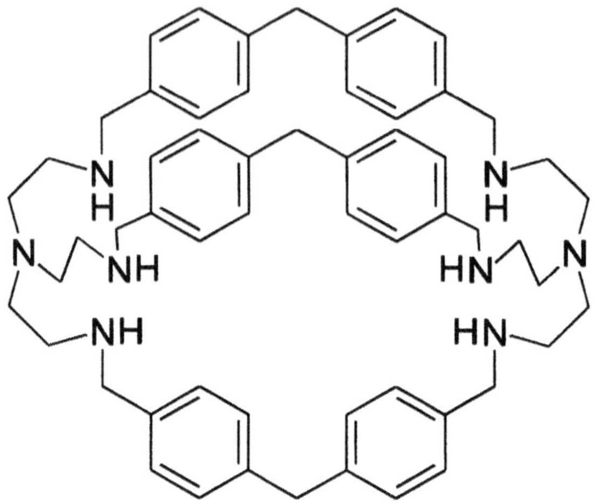

Figure 6. The macrobicyclic ligand described by Lehn[8].

There is a quite simple lesson that we learned from these experiments: the design of rigid catalysts with very limited flexibility must be very accurate and, at any rate, there is an high chance that the results are disappointing and frustrate all the efforts to obtain a very active system because of a failure to take into account even minor details.

3. Metal-organized systems

In search of more flexible receptors we turned our attention to acyclic molecules where the metal ion could be used to fix a particular conformation so that specific functions could interact with a substrate.

3.1 TWEEZERS-LIKE RECEPTOR

The first molecule we synthesized[9] is depicted in Figure 7a: it is the acyclic version of the macrocycle of Figure 3 with one of the two pyridine units replaced by two amino groups. Ammonium groups were also added to ensure water solubility at all concentrations without the need of using mixed solvents. Because of the similarity with the bispyridine macrocycle we decided to test our compounds as catalyst of the cleavage of β-aminoacid ester β-AlaPNP on the assumption that upon binding a Cu(II) ion to the pyridine subunit, the molecule would bind a second one through the co-ordination of the two amine of the diphenylmethane arms.

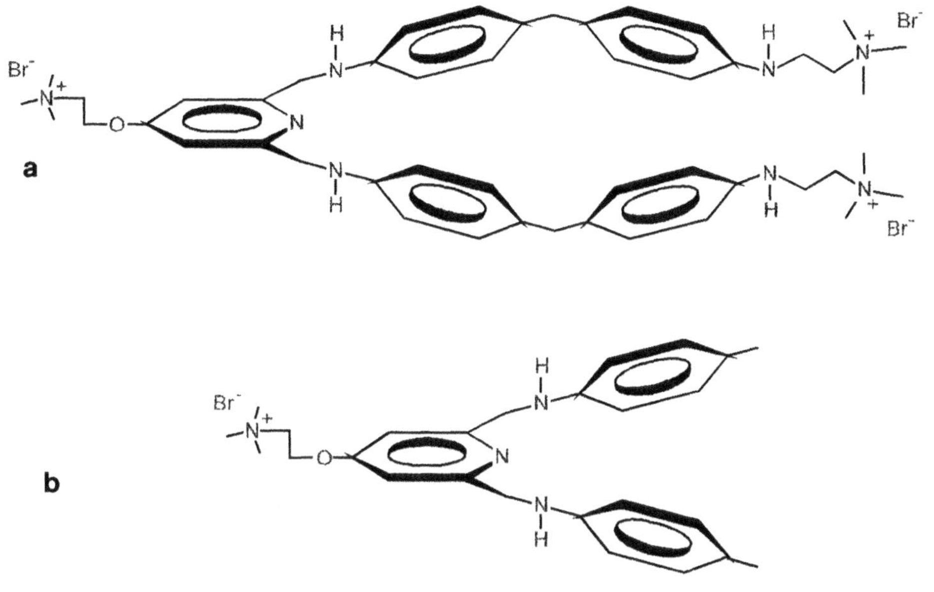

Figure 7. Our twizers like ligand (a) and its reduced model (b).

Kinetics were performed adding increasing concentrations of ligand to a pH=6.3 buffered solution containing a fixed amount of Cu(II). The results are reported in Figure 8 together with the rate constants measured using the ligand shown in Figure 7b which is able to bind only a single Cu(II) ion. It was particularly gratifying to see that only the binuclear complex was a better catalyst than free Cu(II) while the mononuclear one slows down the reaction rate. If we compare the observed rate constants for the two systems at the highest ligand concentration explored we observe that the dinuclear complex is ten times more active than the mononuclear one. The effective involvement of the second Cu(II) ion in the process is clearly highlighted by the kinetic version of the well known Job plot[10] shown in the inset where the maximum rate acceleration is obtained for a ligand to copper ratio of 1:2. So, without the hassle of the macrocyclization we were able to obtain a supramolecular receptor where the metal ions play the role of: a) organizing the molecule with a specific geometry and b) catalyzing the hydrolytic process. Again, in analogy with our speculation on the mechanism operating with the macrocyclic system, we propose that within the active species the amino and C==O groups of the substrate interact with the two Cu(II) ions. The same kinetic experiments carried out with α-LeuPNP showed that both complexes are less active than aquo Cu(II) in accord with the fact that α-amino acids prefer the formation of chelates with Cu(II)[11] and accordingly, do not form complexes involving the two ions like the one proposed for the β-amino acid derivative.

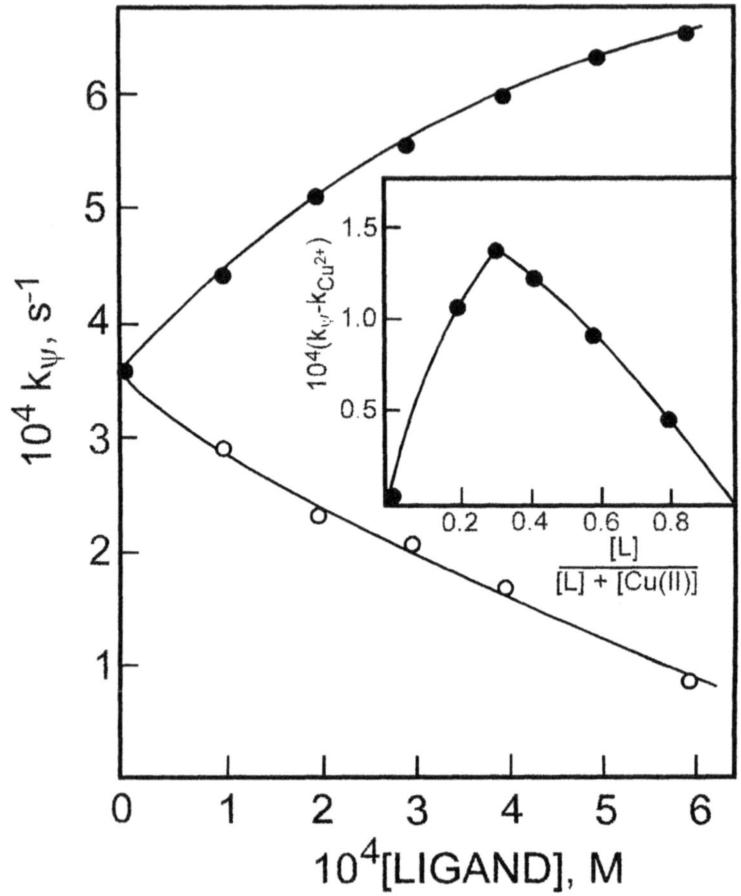

Figure 8. Kinetic effects on the hydrolysis of β-AlaPNP by the tweezers-like receptor (●) and its reduced model (O) in the presence of 5x10-04- M Cu(II).

3.2 TRIPODAL RECEPTORS

The conformational modification triggered by the complexation of the metal ion may, obviously, bring in close proximity to the bound substrate any functional group and not only a second metal ion as in the case illustrated above. For instance we recently synthesized the TREN-based ligand depicted in Figure 9 (TREN, trisaminoethylamine).[12] In such a molecule it is expected that upon binding of a metal ion to the TREN platform the three phenol units get close together roughly defining a cavity.

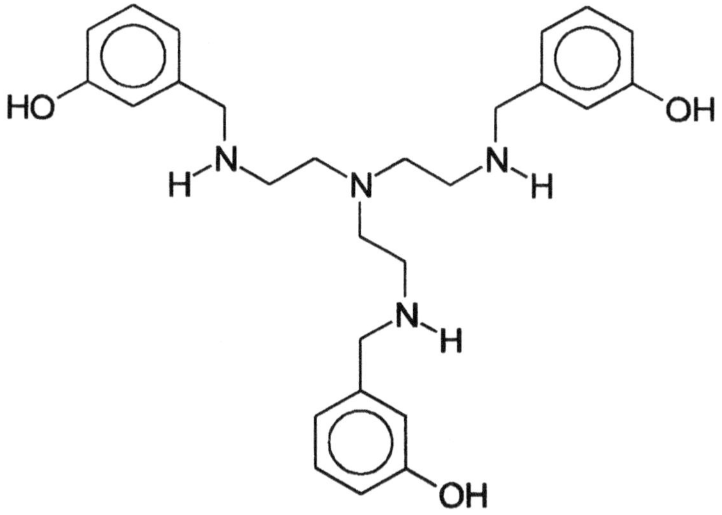

Figure 9. Functionalization of TREN with three molecules of 3-hydroxybenzaldehyde followed by reduction of the tris-imine derivative led to the synthesis of this ligand.

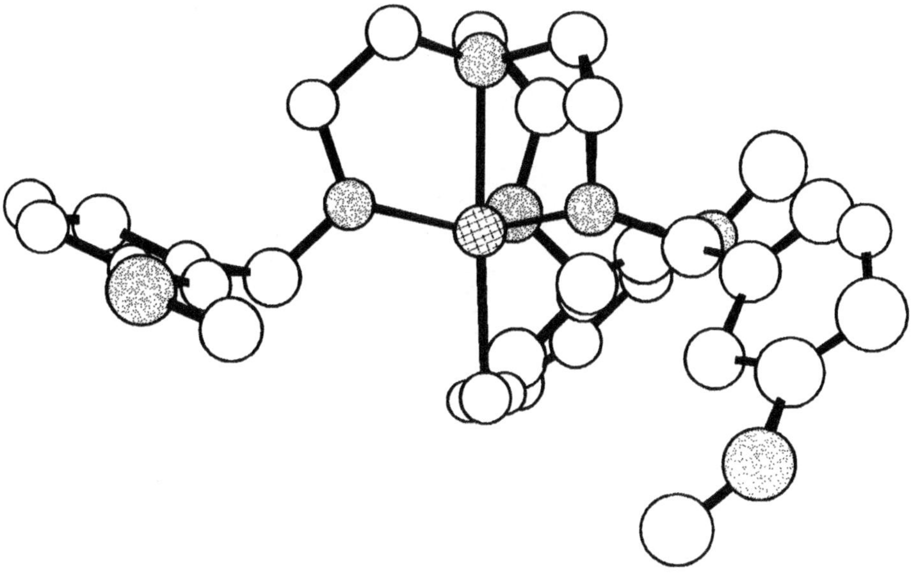

Figure 10. X-ray crystal structure of the Zn(II) complex of the methoxy derivative of the ligand of Figure 9.

This could be demonstrated for the Zn(II) complex of the methoxy derivative for which the X-ray structure of a single crystal could be solved (Figure 10). From the analysis of Figure 10 we note that the co-ordination sphere of Zn(II) is not completely saturated by the four amino groups and one apical position is still available (in the crystal it is occupied by a Cl^- ion). We anticipated that such a Cl^- could be easily exchanged in aqueous solution by water itself or a substrate having a properly positioned donor atom. That water can be co-ordinated to Zn(II) in the complex could be demonstrated by titration experiments (we estimated a pK_a of *ca.* 8.2 for this water molecule). However, these experiments clearly showed that one of the three phenolic groups is more acidic than the others (pKa ≈8.5) likely because of some interaction with the metal ion. Addition of the Zn(II) complex to a aqueous solution of pH=8.3 of the *p*-nitrophenylester of 4-pyridinecarboxylic acid (PNPIN) resulted in a large rate acceleration in the cleavage of this substrate. Similar effects could be observed with the *p*-nitrophenyl ester of 3-pyridine carboxylic acid. The rate acceleration disappears when: i) the phenolic hydroxyls are methylated; ii) a mixture of native TREN and phenol is used. Furthermore 4-methylpyridine is an inhibitor of the process likely because it competes with the substrate for the occupation of the co-ordination site to the Zn(II) ion. All these observations led us to formulate the mechanism depicted in Figure 11 as the most likely one in order to explain the accelerated cleavage of PNPIN by the Zn(II) complex of our trisphenolic derivative of TREN.

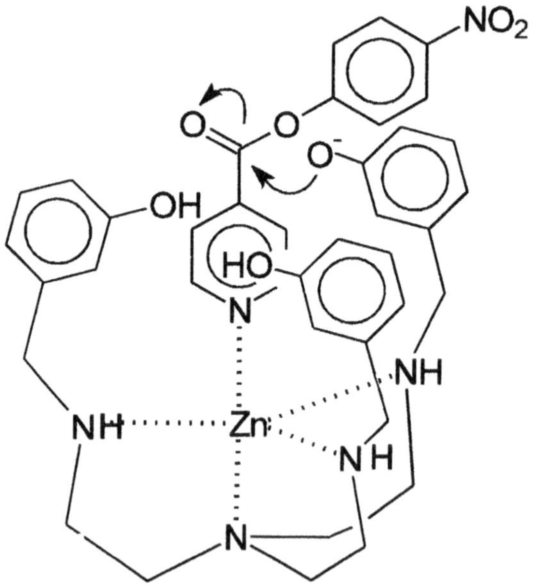

Figure 11. Proposed mechanism for the cleavage of PNPIN by the Zn(II) complex of the TREN derivative of Figure 9.

First the substrate is bound to the complex to form the supramolecular complex. Next a pseudo-intramolecular attack of one of the phenols to the C==O of the substrate leads to the formation of a transacylation intermediate. This latter is eventually hydrolyzed turning over the catalyst. The use of the term catalyst is here appropriate as preliminary experiments with excess substrate showed that the turn-over step is at least as fast as the transacylation process so that there is not accumulation of the intermediate during the overall process.

3.3 ARTIFICIAL PROTEINS

By further implementing the conformational changes brought about to the TREN architecture by metal binding we thought that new proteins could be realized by simply functionalizing the primary amines of TREN with relatively small peptides. In collaboration with the group of C. Toniolo we synthesized a derivative of TREN bearing three decapeptide units connected *via* a benzammide spacer (Figure 12).[13]

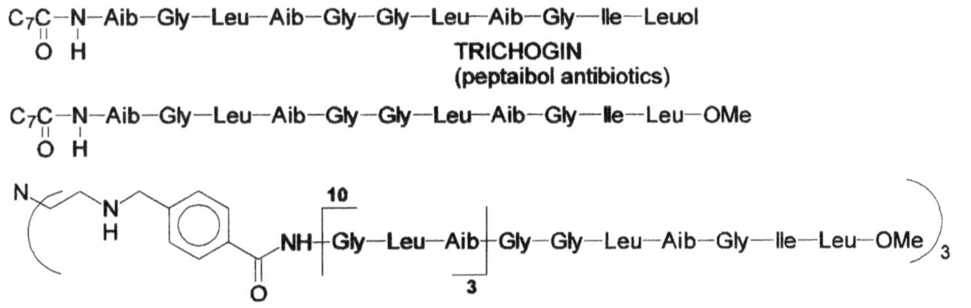

Figure 12. Structure of Trichogin (top), the synthetic peptide we have used for our experiments bearing the terminal Leucine esterified and not reduced (middle) and the two TREN tris-peptides we have synthesized (bottom).

The decapeptide units were synthesized with an amino acid sequence similar to that found in the natural antibiotics belonging to the class of peptaibols.[14] These molecules are known to influence the permeability of biological membrane and because of the relatively high content of α-disubstituted amino acid α-aminoisobutirric acid (Aib) they adopt an helical conformation.[15] The length of the natural peptides ranges from 19 amino acids (alamethicin) to 10 amino acids (trichogins). These latter bear a *n*-octanoyl hydrocarbon chain at the N-terminus and, because of this, are often referred to as lipopetides. It is believed that the longest peptides, spanning in length the phospholipid bilayer, form clusters so creating pores in the membrane with the interior more hydrophilic than the exterior. This is supported by the amino acid sequence that allows the glycine residues to place themself preferentially on one side of the helix. Shorter peptides, like the trichogins, form pores, too; however, they span only a single leaflet of the bilayer and in order to allow permeation of ionic species two of such pores

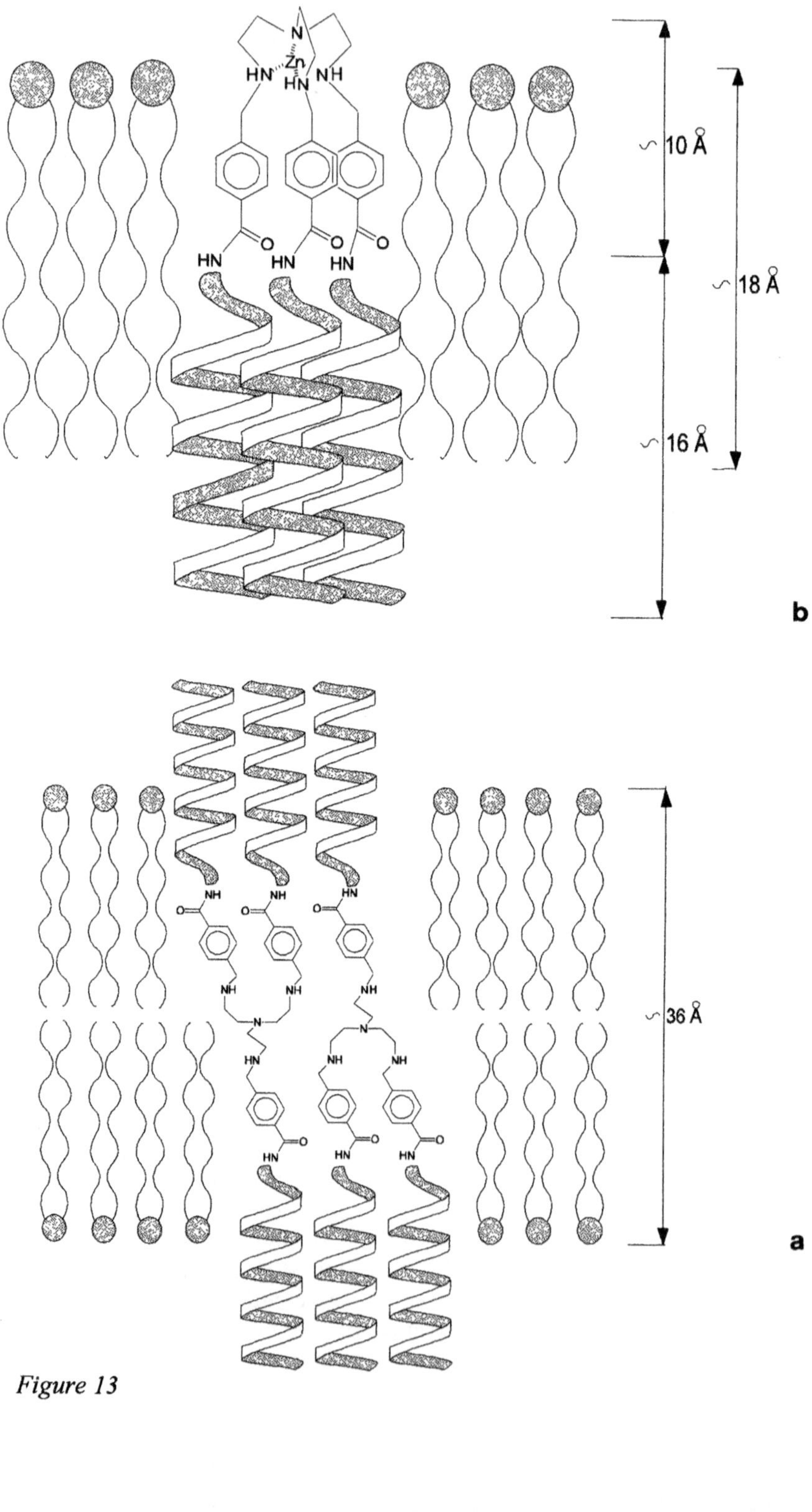

Figure 13

must align. Likely interdigitation of the hydrocarbon portion is a key feature for the alignment of the two half pores.[16] As a matter of fact a decapeptide devoid of the lipid tail is not active at all. We argued that connecting three decapeptides to TREN would result in a very effective system because the length of the molecules would allow, in an extended conformation, the complete spanning of the membrane (Figure 13a); upon binding of a metal ion however such an extended conformation is no longer possible (Figure 13b) and the shorter pores would affect permeation at a lesser extent. In short we expected that permeation could be controlled by the formation of the metal complex. Experiments were carried out monitoring the permeation of carboxyfluorescein entrapped in sonicated liposomes made of a 70:30 mixture of phosphatidylcholine and cholesterol.[17] After addition of the peptide the amount of released dye was spectrofluorimetrically determined after 20 minutes. Figure 14 reports the data obtained as a function of the concentration of the TRIS-tris-decapeptide alone and as the Zn(II) complex, the octanoyl-decapetide and TRIS functionalized with a shorter tripeptide.

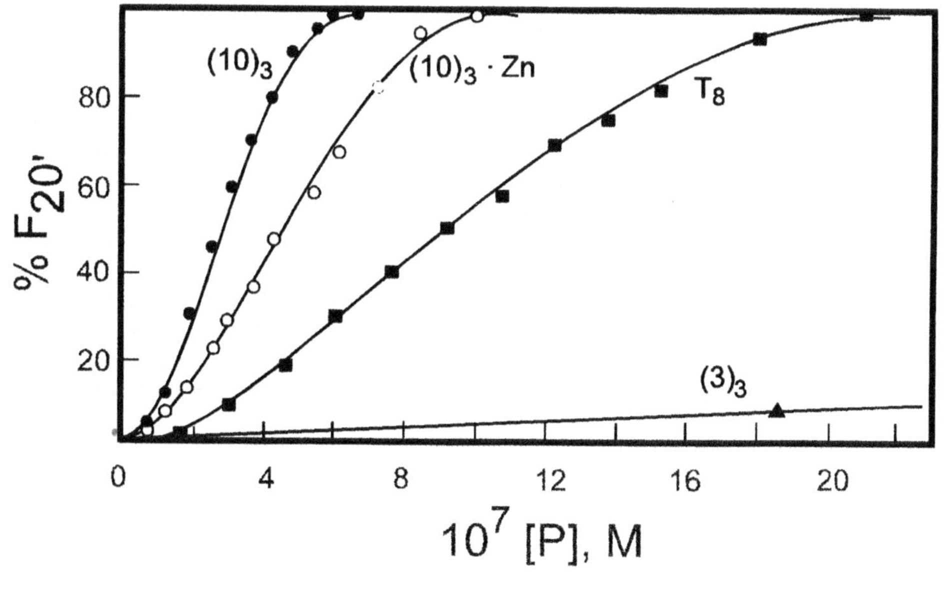

Figure 14. Dye release experiments performed with different peptides: (●)TREN functionalized with the tris-decapeptide; (O) TREN functionalized with the tris-decapeptide as Zn(II) complex; (■) Trichogin model represented in the middle of Figure 12; (▲) TREN functionalized with the tris-tripeptide.

TRIS functionalized with the decapeptide is clearly the most active system while its complex with Zn(II), though still active, is almost seven times less effective (at 4×10^{-7} M concentration). The shorter peptide is not effective at all. Experiments carried out following the time course of the release of carboxyfluorescein show that the process is biphasic with an initial burst of fluorescence followed by a much slower process. Likely the slower process indicates a control of the release of the dye by the rate of intake of Cl^-, a much more hydrophilic ion than carboxyfluorescein. The inward flow of anions is required for compensating the electric potential formed after the first fast flux of anionic dye has occurred. When the experiment is started by adding the Zn(II) complex of the decapeptide derivative the relatively slower release of dye can be speeded up by addition of excess EDTA which complexes Zn(II) transforming the tris-decapeptide in a more active species. This demonstrated that our goal of being able to control the permeability of the membrane through the addition (or removal) of a metal ion could be experimentally achieved.

4. Conclusion

Very simple systems where transition metal ions play quite different roles from the organization of molecular structures to recognition of substrates and modification of reaction rates have been realized in our laboratories. The design of open structures properly assembled by the metal ions appears a more flexible strategy for the realization of new supramolecules bearing different functional groups able to co-operate together. It is conceivable that in the next future these basic concepts will be further exploited in ours and other laboratories as well. As recently stated by Stoddard in an opening address at a Conference on Supramolecular Chemistry: "Chemistry can be linked to 'language'. The atoms are the 'letters'. The molecules are the 'words'. Assemblies of molecules make up the 'sentences'. The sets of assembled molecules or supramolecules are the 'paragraphs'. The ways in which the molecular assemblies and supramolecular arrays contain and express information are the 'chapters'. ... Ultimately, chemistry has to tell a 'story'". Clearly we are only at the very beginning.

5. Acknowledgments

Technical assistance by Mr. E. Castiglione and financial support by the Ministry of University, Scientific and Technological Research (MURST) and the National Research Council (CNR, Rome) in the frame of the Special Project "Chimica Fine II", are gratefully acknowledged.

6. References

1. Hay, R. W. (1984) *Bio-inorganic Chemistry*, Ellis Horwood, Chichester;

2. Lehn, J. M. (1985) *Science*, **227**, 849.

3. (a) Drew, M. G. B. (1980) *J. Chem. Soc., Chem. Commun.* 1122; (b) Motekaitis, R. J., Martell, A.E., Lehn, J.-M., and Watanabe, E.-I (1982), *Inorg. Chem.* **21**, 4253; (c) Agnus, Y., Louis, R., Gisselbrecht, J.-P., and Weis, R. (1984) *J. Am. Chem. Soc.* **106**, 93; (d) Motekaitis, J. R., Martell, A. E., Dietrich, B., and Lehn, J.-M. (1984) *Inorg. Chem.* **23**, 1588; (e) Sutherland, I. O. (1989) *Pure Appl. Chem* **61**, 1547; (f) Vigato, P. A., Tamburini, S., and Fenton, D. E. (1990) *Coord. Chem. Rev.* **106**, 25; (g) Verbroom, W., Rudkevich, D. M., and Reinhoudt, D. N. (1994) *Pure Appl. Chem* **66**, 679; (h) Martell, A. E.; Motekaitis, R. J., Rockcliffe, D., Menif, R. and Ngwenya, P. M. (1994) *Pure Appl. Chem.* **66**, 859; (i) van Veggel, F. C. J. M., Verboom, W., and Reinhoudt, D. N. (1994) *Chem. Rev.* **94**, 279; (j) Anderson, S., Anderson, H. L. and J. K. M. Sanders (1993) *Acc. Chem. Res.* **26**, 469.

4. Linkletter, B. and Chin, J. (1995) *Angew. Chem. Int. Ed. Engl.* **34**, 472 and references therein.

5. Scrimin, P., Tecilla, P., Tonellato, U., Valle, G., and A. Veronese (1995) *Tetrahedron* **51**, 527.

6. Couturier, Y. and Petitfaux, C. (1978) *Bull. Soc. Chim. France* I-435.

7. Hathaway, B. J. (1987) in Wilkinson, G., Gillard, R. D., and McCleverty, J. (eds), *Comprehensive Coordination Chemistry*, Pergamon Press, Oxford, p. 596.

8. Jazwinski, J., Lehn, J.-M., Lilienbaum, D., Ziessel, R., Guilhelm, J., and Pascard, C. (1987) *J. Chem. Soc., Chem. Commun.* 1691.

9. Scrimin, P., Tecilla, P., Tonellato, U., and Vignaga, N. (1991) *J. Chem. Soc., Chem. Commun.* 449.

10. Connors, K. A. (1987) *Binding Constants*, John Wiley, New York, pp. 24-28.

11. Demaret, A., Abello, L., Fourati, M. and Lapluye, G. (1978) *J. Chem. Res.*, **9**, 354.

12. Scrimin, P., Tecilla, P., Tonellato, U., Valle, G. and A. Veronese (1995) *J. Chem. Soc., Chem. Commun.* 1163.

13. Formaggio, F., Scrimin, P., Tecilla, P., Tonellato, U., Toniolo, C. and Veronese, A. (1995) submitted for publication.

14. Benedetti, E. et al. (1982) *Proc. Natn. Acad. Sci. USA* **79**, 7951.

15. Toniolo, C., Peggion, C., Crisma, M., Formaggio, F., Shui, X., and Eggleston, D. S. (1994) *Struct. Biol.* **1**, 908.

16. Bodo, B., Rebuffat, S., Goulard, C., Peggion, C., Monaco, V., Formaggio, F., Crisma, M., and Toniolo, C. (1995) in Maia, H. L. S. (ed) *Peptides 1994*, ESCOM Science Publisher, Leiden, The Netherlands.

17. Auvin-Guette, C., Rebuffat, S., Prigent, Y. and Bodo, B. (1992) *J. Am. Chem. Soc.* **114**, 2170.

NEUTRAL ORGANOBORON RECEPTORS WITH ORIENTED DIPOLES

BRADLEY D. SMITH,* JEFFREY T. BIEN, and MARTIN PATRICK HUGHES
Department of Chemistry and Biochemistry, University of Notre Dame, Notre Dame, IN 46556

1. Catalysts with Oriented Dipoles

Electrostatic effects within an enzyme active-site play an important role in stabilizing reactive intermediates.[1] This article concerns the stabilization that is gained by having a binding surface with a preorganized arrangement of strong dipoles. Perhaps the best known enzymological example is the oxyanion hole which operates in serine proteases.[2] A protease enzyme expends folding energy to align the N-H dipoles associated with its oxyanion hole towards the carbonyl group of a reacting substrate. This prealignment of dipoles minimizes the energetic cost associated with reorganizing the solvent molecules around the developing negative charge. Thus, enzymatic stabilization of the anionic tetrahedral intermediate is achieved.

Another intriguing case is the macrodipole generated by a peptidyl α-helix which is often invoked as an electrostatic influence.[3] A recent example of this comes from the work of Knox who solved the X-ray crystal structure of the enzyme D-alanine:D-alanine ligase with adenosine diphosphate (ADP) and a phosphinate transition-state-analogue inhibitor bound in the active site.[4] Using this enzyme-inhibitor crystal structure, the workers were able to propose a mechanism for binding and catalysis. As shown in Figure 1, the first alanine is phosphorylated by ATP to give an activated acylphosphate which is

C. S. Wilcox and A. D. Hamilton (eds.), Molecular Design and Bioorganic Catalysis, 227–242.

then attacked by the second alanine. For this attack to occur the alanine ammonium has to be converted to its more nucleophilic amine form. The crystal structure shows no apparent basic residue within proximity to facilitate deprotonation. Instead it appears that the ammonium is hydrogen bonded to the triad Tyr^{216}-Ser^{150}-Glu^{15}. In addition, the N-terminal end of α-helix 11, represented in Figure 1 as a rectangle, is pointed directly at the attacking alanine. The authors postulate that the macrodipole associated with helix 11 is able to bind and position this second alanine for reaction, and that the combination of the helix dipole and the H-bonded triad is able to alter pKa's enough to induce ammonium deprotonation and subsequent N-acylation.

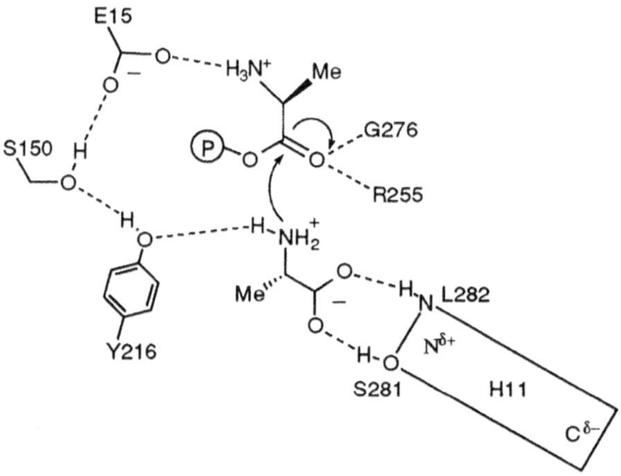

Figure 1. Proposed mechanism for D-ala:D-ala ligase.[3]

Whether the above explanation is correct remains to be seen; however, it does provide inspiration for biomimetic chemists trying to invent artificial enzymes. To date, most of the work on artificial enzymes has attempted to facilitate reaction by using proximity effects, proton transfer, and/or nucleophilic catalysis.[5] The development of catalytic receptors that utilize electrostatic effects to stabilize reactive intermediates is a surprisingly recent approach (the use of metal cations to polarize and activate reactants is considered a separate case). The research program initiated by the Wilcox group is a specific example. These workers are attempting to design artificial catalysts that incorporate intramolecular ion-pairs within their structures.[6,7] As summarized in Figure 2, the dipoles associated with these ion-pairs are able to stabilize polar intermediates, and thus facilitate reaction. This is an appealing strategy for catalysts designed to operate in organic solvents, since electrostatic effects are generally stronger in non-polar environments. In addition, supramolecular research in organic solution has produced, over the last fifteen years, a large portfolio of well-characterized reversible binding systems. Therefore, the challenge is to design catalytic systems with the appropriate combination of binding and stabilization properties. We are particularly intrigued by the problem of introducing the electrostatic components. As demonstrated by Wilcox, formally charge-neutral receptors that incorporate ion-pairs can exhibit strong catalytic effects using the binding cycle described in Figure 2.[7] However, the poor dipole directionality generally associated with an ion-pair means the catalyst will be inherently inefficient and lacking in substrate structural selectivity. In many cases, non-selective catalysts are the preferred choice; however, if reaction efficiency and substrate selectivity are important goals then the dipoles need to be correctly prealigned to provide maximum catalytic effect.

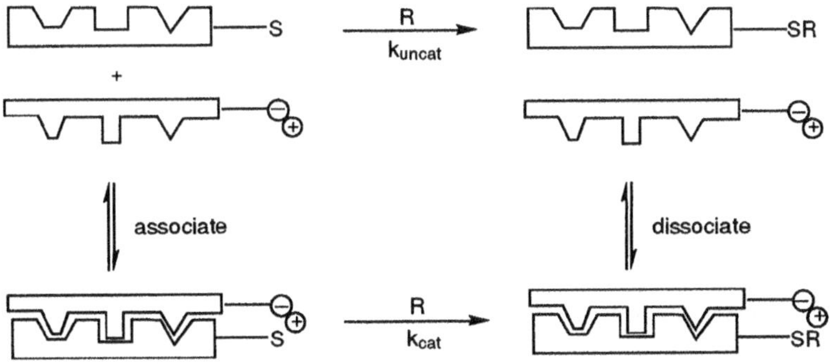

Figure 2. Reaction of a substrate S with a reactant R is catalyzed when S associates with an ion-pair-containing receptor.[6]

In this article we describe our initial studies of two new classes of organoboron receptors that incorporate highly oriented dipoles. We show that the electrostatic fields generated by these receptors produce not only enhanced binding abilities, but also substrate binding selectivities that are complementary to the spatial arrangement of the dipoles. These exercises in small molecule design are our first steps towards the eventual construction of artificial catalysts that utilize the full repertoire of stabilizing factors to facilitate chemical reaction.

2. Neutral Acetate Receptor

The first example is a neutral receptor designed for carboxylate and phosphate anions. Neutral receptors for anions have been recently reported by the research groups of Reinhoudt,[8] Gellman,[9] and Schmidtchen.[10] Macrocyclic receptor **1**, described by Schmidtchen and coworkers is particularly relevant, as its binding mechanism is related to the systems described below. The formally uncharged receptor uses the convergent dipoles formed by four preorganized borane-amine adducts to selectively bind small spherical anions, such as halides, at its center.[10]

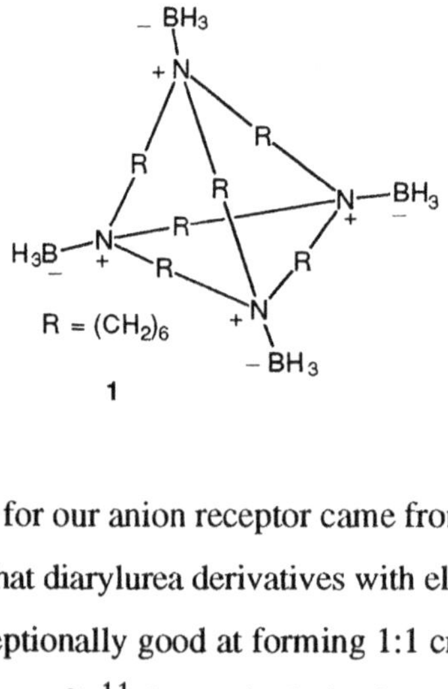

1

The initial inspiration for our anion receptor came from a study by Etter and coworkers, who discovered that diarylurea derivatives with electron withdrawing groups in the *meta* positions were exceptionally good at forming 1:1 crystalline complexes with hydrogen bond acceptors (Figure 3).[11] Isomeric derivatives with *ortho* or *para* substituents did not form such co-crystals. This "meta effect" was attributed to a substituent induced intramolecular C-H···O hydrogen bonding interaction between the carbonyl oxygen and the acidified *ortho* hydrogens (Figure 3). Such an intramolecular interaction was postulated to weaken the hydrogen bond accepting ability of the urea carbonyl. A subsequent study of the complexes in solution by Wilcox and coworkers, however, lead to an alternative explanation.[12] They found that the "meta effect" vanished in the solution-state, and that observed binding affinities for the various substituted derivatives could be correlated with calculated surface electric potentials.

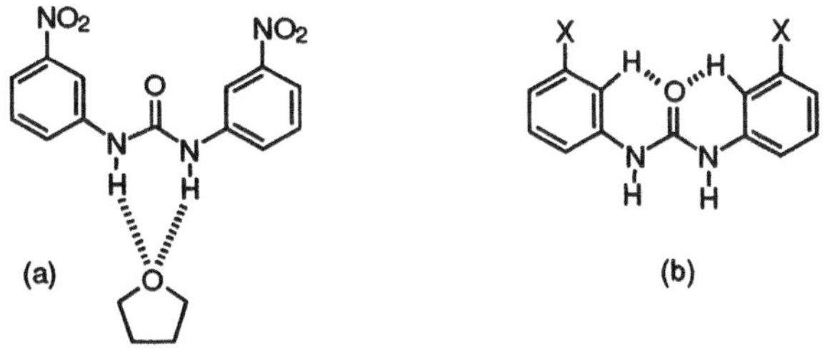

Figure 3. (a) Typical 1:1 crystalline complex observed for 1,3-bis(*m*-nitrophenyl)urea. (b) Electron withdrawing groups X strengthen intramolecular C-H···O hydrogen bonding.

Although the Etter explanation now appears to be incomplete, it provided us with an idea. We reasoned that a C-H group was a very poor Lewis acid, and that one could dramatically increase the putative effect by using a stronger Lewis acid. Thus, we decided to investigate receptors whose structures were related to the generic compound **2** (Figure 4). In line with our current interest in organoboron receptors it was an obvious step to propose a study of the structural class **3**. The structure for receptor **3** can be represented by two limiting forms, **3a** or **3b** (Figure 4). Recent work by Groziak and coworkers suggested that **3b** would be the more likely structure, as they found that when compound **4** was dissolved in methanol the resulting product was the zwitterionic **5b** as proved by NMR and X-ray studies (Figure 5).[13] Thus, we initiated a synthesis of receptor **6** (Figure 6). This entailed treating 2-aminophenylboronic acid, **7**, with methylisocyanate to give compound **8** (structure proved by X-ray crystallography). Condensation of **8** with pinacol gave an organic soluble product whose spectral characteristics were consistent with either **6a** or **6b**. The subsequent NMR titration behavior with acetate anion indicated the structure to be **6b**. In any case, compound **6** was highly water-sensitive and readily hydrolyzed to regenerate compound **8**.

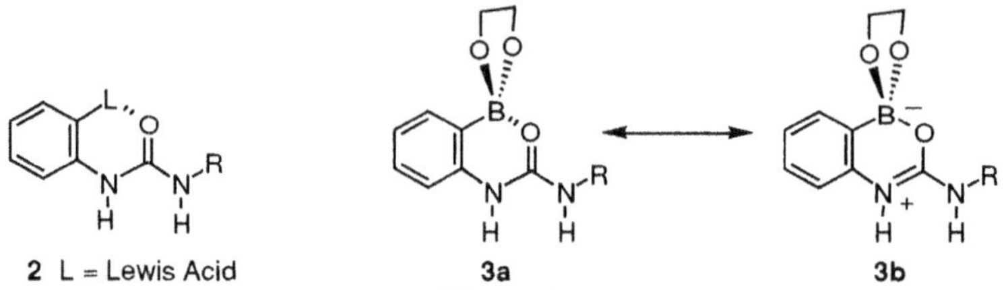

Figure 4.

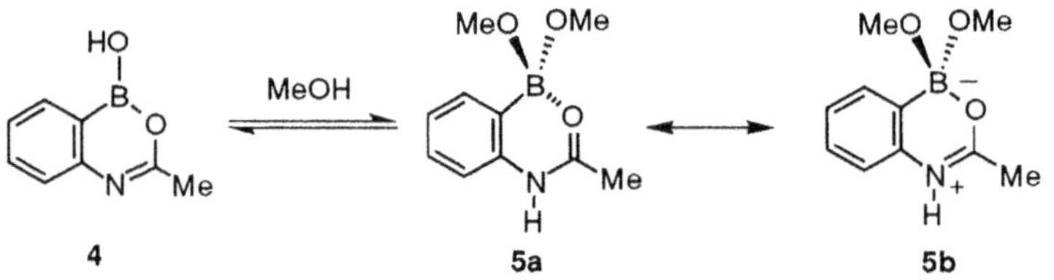

Figure 5.

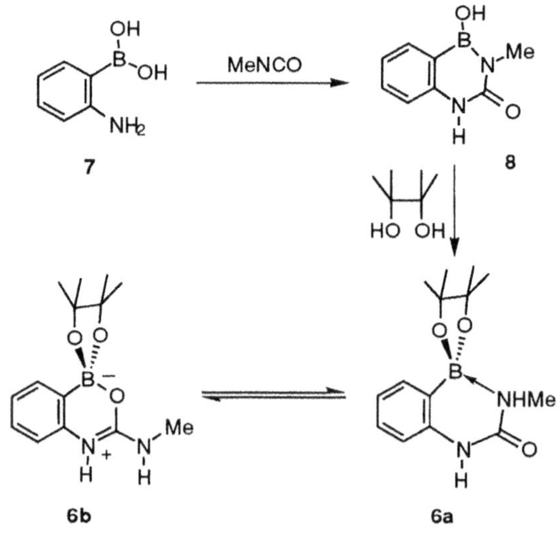

Figure 6.

Binding studies with compound **6** are still underway, and only qualitative trends will be discussed here. As summarized in Table 1, the acetate binding ability of **6** has been determined in DMSO, and compared to the binding ability of control compounds **9** and **10**. Association constants were measured by ^{1}H NMR titration studies.[14,15]

Specifically, the downfield changes in host N-H resonances were monitored upon addition of aliquots of tetrabutylammonium acetate. In each case 1:1 binding was verified by a Job plot. The association constant for **6** has been corrected to allow for the loss of functional host, due to hydrolysis, as the titration experiment progressed. This is a reasonable correction since the products from the hydrolysis, namely pinacol and **8**, have negligible acetate binding ability in DMSO. In addition, control experiments showed that acetate anion has no affinity for the trigonal boron in **9**.

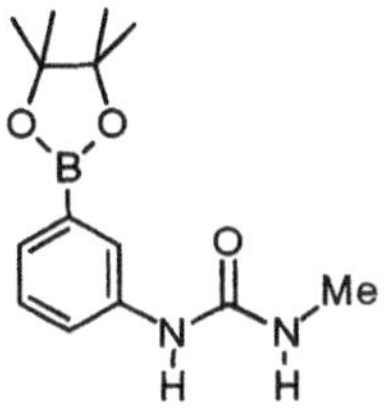

9

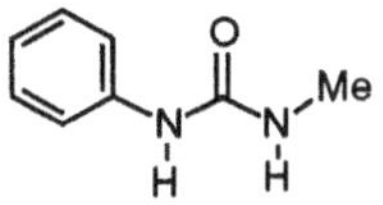

10

TABLE 1. Acetate association constants, and $\Delta\delta_{max}$ values from ^{1}H NMR titrations in DMSO at 296 K.

Host	K_{assn} (M^{-1})[a]	ΔG_{296} (kcal/mol)	$\Delta\delta_{max}$ (ppm)[b]
6	$(2.0 \pm 0.5) \times 10^3$	-4.4	3.63
9	$(6.0 \pm 0.6) \times 10^2$	-3.7	2.02
10	$(4.0 \pm 0.4) \times 10^2$	-3.5	2.12

[a]Average of three independent trials. [b]Average of $\Delta\delta_{max}$ for both N-H residues, derived for fully saturated host.

In qualitative agreement with our initial hypothesis, host **6** binds acetate better than controls **9** and **10**. Thus, it appears the strategically placed boron in **6** is able to polarize the urea carbonyl, and increase acetate binding by acidifying the N-H residues and enhancing the ion-dipole interaction between host and acetate. The most likely structure for the **6**:actetate complex is shown in Figure 7. The fact that **6** is a poorer acetate binder than guanidinium cation ($K_{assn} \sim 10^4$ M^{-1} in DMSO)[16] is understandable because ion-dipole interactions are inherently weaker, and fall off more quickly than ion-ion interactions.[17]

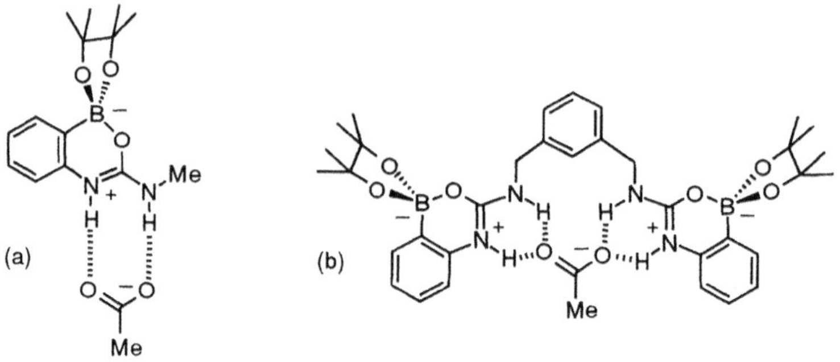

Figure 7. (a) Likely structure for **6**:acetate complex. (b) Proposed structure for **12**:acetate complex.

To better appreciate the changes in host electrostatics induced by the intramolecular Lewis acid, surface electric potentials were calculated for hosts **6**, **9**, and **10**, using Hartree-Frock methods (AM1 geometry optimization / 6-31G* HF potentials).[12,18] The results of these calculations are shown in Table 2. As predicted, the ortho boron residue is able to dramatically increase the magnitude of the molecular dipole, as well as the positive electric potential surrounding the urea N-H residues. Both factors contribute to the enhanced binding of the anionic guest, acetate.

TABLE 2. Molecular dipole, and maximum electrostatic surface potential for N-H residues.[a]

Compound	Electrostatic Potential (kcal/mol)	Molecular Dipole (D)
6	+ 81.0	6.1
9	+ 67.1	2.4
10	+ 67.6	3.8

[a]AM1 geometry optimization / 6-31G* HF potentials.[12,18]

As already stated, the project is ongoing and many questions still need to be addressed. For example, How much does the intramolecular Lewis interaction in **6** increase the acidity of the urea N-H groups? Is there any evidence for proton transfer in the host:guest complex? Unfortunately, the hydrolytic instability of **6** prohibits pK_a determinations. For this measurement, a more stable compound is needed. Current efforts are focused on the guanidine-based receptor **11**, which has been synthesized according to Figure 8. Preliminary results indicate this compound has increased hydrolytic stability.

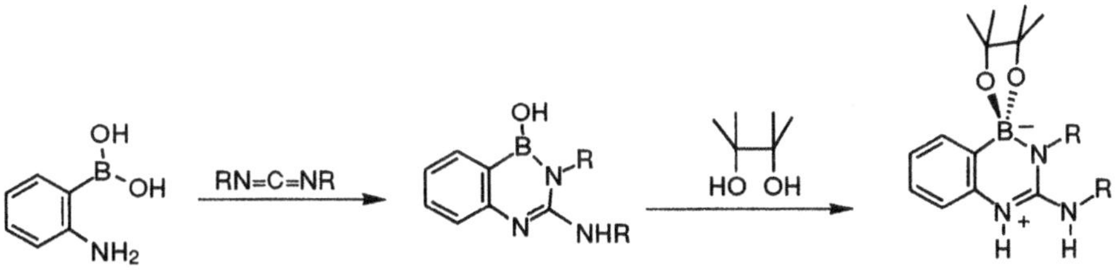

Figure 8. Synthesis of guanidinium derivative **11**.

We believe this strategy of using an intramolecular Lewis acid to polarize and magnify the binding abilities of well-known recognition motifs such as ureas, guanidiniums, amides, imides, etc., should be reasonably general. In addition, it provides a way of introducing strong dipoles into neutral receptors, whose directions can be controlled by structural design. Our goal is devise more sophisticated receptors (*e.g.*, we have synthesized the dimeric receptor **12** in Figure 7), and examine their abilities to catalyze reactions that develop negatively charged transition-states (*e.g.*, phosphate ester hydrolysis).

3. Neutral Paraquat Receptor

Whereas uncharged hosts for anions are quite rare, neutral hosts for cations are the largest class of artificial receptors known. Consider for example the crown ethers.[19] As shown in Figure 9, when 18-crown-6 adopts its symmetrical crown conformation, all the dipoles associated with the polarized C-O bonds point directly towards a central position, enhancing the compound's ability to bind metal cations. This section focuses on an organoboron receptor that uses the dipole of a dative B-N bond in a similar manner, that is, the negative end of the dipole is oriented towards a cationic center. Specifically, we have designed and synthesized the ditopic receptor **13** whose structure incorporates two dative B-N dipoles fixed in convergent orientations that are complementary with the two cationic centers of the aromatic dication paraquat, **14** (methylviologen). Since the details of this work have been published elsewhere, only the conclusions will be emphasized here.[20]

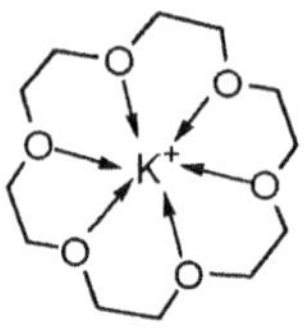

Figure 9.

Paraquat forms charge-transfer complexes with a range of organic and inorganic anions, as well as neutral π-donor molecules.[21] Of the more sophisticated neutral paraquat hosts, the best characterized are those designed by Stoddart and coworkers.[22] For example, the macrocycle **15** forms a 1:1, face-to-face complex with paraquat in polar, aprotic solutions (K_{assn} = 730 M^{-1} in acetone) and in the solid state. The complex is stabilized by ion-dipole interactions between the cationic nitrogens and the polyethylene ethers, π -π interactions, and CH···O hydrogen bonding between the polyethylene oxygens and the relatively acidic C-H groups adjacent to the nitrogens.

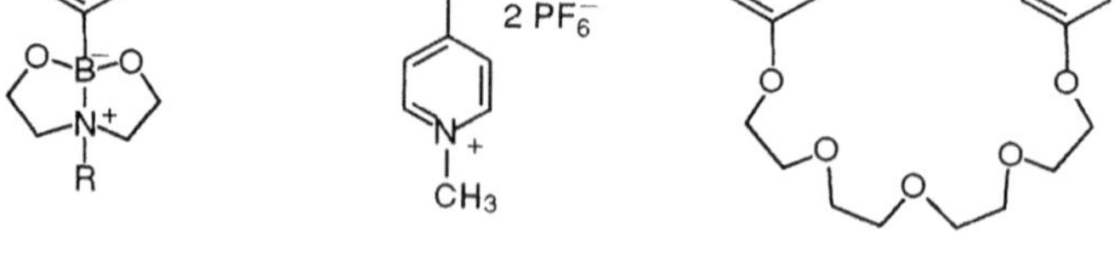

13 R-R = $(CH_2)_{12}$ **14** **15**
16 R = $(CH_2)_4CH_3$

Host **13** contains two boronic acid diethanolamine esters. There is compelling spectroscopic and X-ray crystallographic evidence that the diethanolamine esters of boronic acids form dative B-N bonds.[23,24] This Lewis acid-base interaction produces a strong dipole directed along the B-N bond. In the case of arylboronic esters, steric hindrance fixes the dipole in an orientation that points away from the plane of the aryl ring (Figure 10).

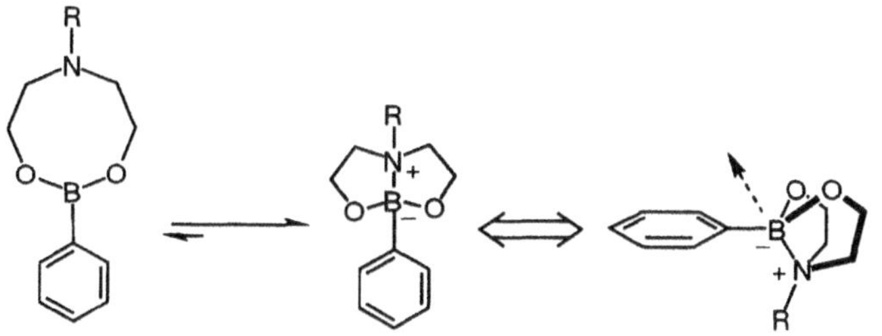

Figure 10. Structure of arylboronic acid diethanolamine ester.

With this knowledge in mind, receptor **13** was envisioned as having a binding shape that is complimentary to a dicationic paraquat guest. The postulated **13**:paraquat binding complex is shown schematically in Figure 11, and a space-filling representation is provided in Figure 12. The complex is held together by a combination of intermolecular forces: (i) The negative ends of the two dative B-N dipoles in **13** are pointed directly towards the two cationic centers of paraquat, (ii) The electron-rich and electron-poor biphenyls produce an attractive π-π stacking interaction, (iii) C-H···O hydrogen bonding can occur between the relatively acidic paraquat methyl groups and the basic boronate oxygens.

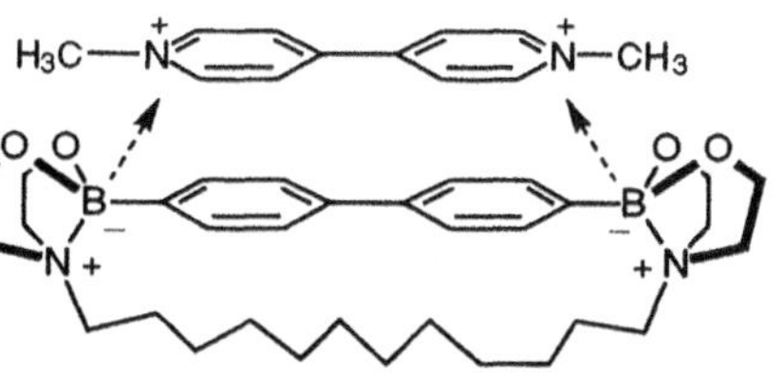

Figure 11. Postulated **13**:paraquat complex.

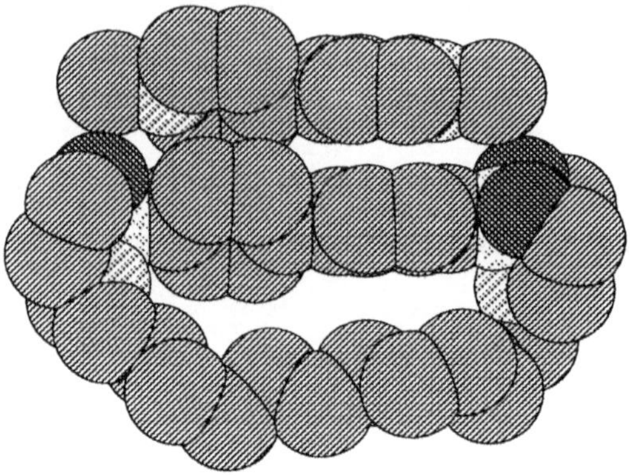

Figure 12. Computer generated space-filling representation of the **13**:paraquat complex with hydrogens omitted for clarity.

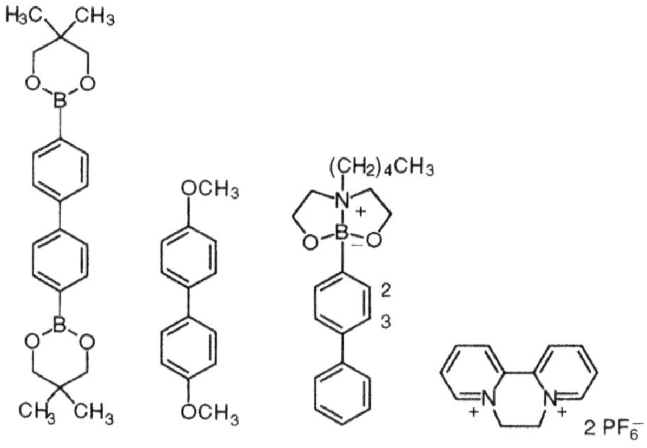

17 **18** **19** **20**

Receptor **13** and control hosts **16** - **19** were synthesized by standard methods. All analytical and spectral data were in agreement with the structures as drawn.[20] Paraquat binding abilities were determined in acetone. At sufficiently high concentrations, all host-guest solutions were yellow-orange, indicative of charge-transfer interactions. Association constants were determined by ^{1}H NMR titrations and subsequent iterative curve fitting procedures.[14,15] As described in Table 3, receptor **13** bound paraquat bishexafluorophosphate with an association constant of 320 ± 30 M^{-1}. A Job plot indicated 1:1 binding as the dominant complex stoichiometry. The upfield change in chemical shifts for the aromatic signals of both host and guest is evidence for an aromatic face-to-face orientation as the major binding structure. Face-to-edge binding can be ruled out as the major complex structure due to the absence of downfield shifts in one of the binding partners.[26] The magnitude of the complexation-induced shifts (-0.18 ppm) are significantly smaller than those observed for other paraquat binding systems such as the macrocyclic host **15** (~ -0.4 ppm).[22] The difference can be rationalized in terms of a "loosely bound" complex between paraquat and **13** with weak NMR shielding effects, as compared to a tighter, more oriented complex with **15** that results in stronger shielding.[22] Unfortunately, numerous attempts at host:guest co-crystallization were unsuccessful in producing material suitable for X-ray analysis.

TABLE 3. Association constants, and $\Delta\delta_{max}$ values from ^{1}H NMR titrations in acetone at 296 K.

Host	Guest	K_{assn} (M^{-1})[a]	ΔG_{296} (kcal/mol)	$\Delta\delta_{max}$ (ppm)[b]
13	14	320	-3.4	-0.16,[c] -0.18[d]
16	14	230	-3.2	-0.15,[c] -0.14[d]
17	14	< 5	> -1	-
18	14	< 5	> -1	-
19	14	25	-1.9	-0.13,[c] -0.13[d]
13	20	120	-2.8	-0.11[c]
16	20	85	-2.6	-0.10[c]

[a]In most cases K_{assn} is the average of values obtained from host H-2 and H-3 titration curves. All duplicate runs were within ±10 %. [b]Derived for fully saturated host. [c]Host resonance H-2. [d]Host resonance H-3.

The less preorganized control receptor **16** was found to be a poorer paraquat binder with an association constant of 230 ± 20 M^{-1}. Receptor **16** suffers, of course, because it can adopt either of two low-energy conformations; a high affinity *syn* conformation with convergent dipoles or a low affinity *anti* conformation with divergent dipoles. This problem is eliminated in receptor **13** because the macrocycle enforces the convergent *syn* dipole relationship (Figure 11). The increase in binding energy gained by moving to the more preorganized receptor is smaller than expected, and is suggestive of an enthalpy-entropy compensation effect, however, this has not yet been proved.[27]

To probe the relative importance of the various contributors to host/guest binding, association constants were determined for a number of control hosts. As summarized in Table 3, the trigonal diboronate ester **17** exhibited a very weak affinity for paraquat ($K_{assn} < 5\ M^{-1}$). A similar association was obtained with the electron-rich 4,4'-dimethoxybiphenyl, **18**. Thus the electrostatic components associated with the B-N dative bonds in **13** appear to have a major effect on host:guest binding. The importance of the ditopic nature of receptor **13** is highlighted by the much lower association ($K_{assn} = 25\ M^{-1}$) obtained with the monotopic receptor **19**. The guest selectivities of ditopic receptors **13** and **16** were probed by comparing their paraquat binding affinities with those for the structurally related diquat, **20** (Table 3). In both cases, binding selectivities of 0.6 kcal/mol in favor of paraquat were obtained, emphasizing the directional nature of the oriented B-N dipoles, and the associated shapes of the host electrostatic surfaces.

4. Conclusion

Two new classes of organoboron receptors have been described which share a common structural feature; they both incorporate highly oriented dipoles which are generated by intramolecular Lewis acid-base interactions with the boron atoms. The receptors exhibit enhanced binding abilities, and substrate binding selectivities that are complementary to the spatial arrangement of the dipoles. The long-term goal of this work is to use these dipole-producing motifs as stabilizing modules that can be combined, either rationally or by library methods, with other modular units such as binding motifs (*e.g.*, H-bonding, aryl stacking, etc.), and catalytic motifs (*e.g.*, proton transfer, nucleophilic residues) to produce sophisticated later-generation enzyme mimics.

5. Acknowledgments

This work was supported by a grant from the US National Science Foundation.

6. References

1. Warshel, A. (1981) Electrostatic basis of structure-function correlation in proteins, *Acc. Chem. Res.* **14**, 284-290.
2. Warshel, A. (1989) How do serine proteases really work, *Biochemistry* **28**, 3629-3637.
3. Lodi, P. and Knowles, J. R. (1993) Direct evidence for the exploitation of an α-helix in the catalytic mechanism of triosephosphate isomerase, *Biochemistry* **32**, 4338-4343 and references cited therein.
4. Chang, F., Moews, P. C., Walsh, C. T. and Knox, J. R. (1994) Vancomycin resistance: structure of D-alanine:D-alanine ligase at 2.3 Å resolution, *Science* **266**, 439-443.
5. Page, M. I. and Williams, M. (eds) (1987), *Enzyme Mechanisms*, Royal Society of Chemistry, London.
6. Smith, P. J. and Wilcox, C. S. (1991) The chemistry of functional group arrays. Electrostatic catalysis and the "intramolecular salt effect", *Tetrahedron* **47**, 2617-2628.
7. Smith, P. J., Kim, E. and Wilcox, C. S. (1993) Substrate-specific catalysis by ion-pairs, *Angew. Chem. Int. Ed. Engl.* **32**, 1648-1650.
8. Scheerder, J., Fochi, M., Engbersoen, J. F. J. and Reinhoudt, D. N. (1994) Urea-derivatized p-tert-butylcalix[4]arenes: neutral ligands for selective anion complexation, *J. Org. Chem.* **59**, 7815-7820 and references cited therein.
9. Savage, P. B., Holmgren, S. K. and Gellman, S. H. (1994) Anion and ion pair complexation by a macrocyclic phosphine oxide disulfoxide, *J. Am. Chem. Soc.* **116**, 4069-4070.
10. Worm, K., Schmidtchen, F. P., Schier, A., Schäfer, A. and Hesse, M. (1994) Macrotricyclic borane-amine adducts: the first uncharged synthetic host compounds without lewis acid character, for anionic guests, *Angew. Chem. Int. Ed. Engl.* **33**, 327-329.
11. Etter, M. C., Urbanczyk-Lipkowska, Z., Zia-Ebrahimi, M., Panunto, T. W. (1990) Hydrogen bond directed cocrystallization and molecular recognition properties of diarylureas, *J. Am. Chem. Soc.* **112**, 8415-8426.
12. Wilcox, C. S., Kim, E. K., Romano, D., Kuo, L. H., Burt, A. L. and Curran, D. P. (1995) Experimental and theoretical studies of substituent effects in hydrogen bond based molecular recognition of a zwitterion by substituted arylureas, *Tetrahedron* **51**, 621-634.
13. Groziak, M. P., Ganguly, A. D. and Robinson, P. D. (1994) Boron heterocycles bearing a peripheral resemblance to naturally-occurring purines: design, syntheses, structures, and properties, *J. Am. Chem. Soc.* **116**, 7597-7605.
14. Conners, K. A. (1987) *Binding Constants, The Measurement of Molecular Complex Stability*, Wiley, New York.
15. Wilcox, C. S. (1991) Design, evaluation of an efficacious functional group dyad. Methods and limitations in the use of NMR for measuring host-guest interactions, in Schneider, H. -J., Dürr, H., (eds) *Frontiers in Supramolecular Organic Chemistry and Photochemistry*, VCH, Weinham, pp. 123-143.
16. Fan, E., Arman, S. A., Kincaid, S. and Hamilton, A. D. (1993) Molecular recognition: hydrogen-bonding receptors that function in highly competitive solvents, *J. Am. Chem. Soc.* **115**, 369-370.
17. Schneider, H. -J. (1991) Mechanisms of molecular recognition: investigations of organic host-guest complexes, *Angew. Chem. Int. Ed. Engl.* **30**, 1417-1436.
18. Spartan, Version 3.1, Wavefunction Inc.: Irvine, CA, 1994.

19. Izatt, R. M., Pawlak, K., Bradshaw, J. S. and Breuning, R. L. (1991) Thermodynamics and kinetic data for macrocycle interaction with cations and anions, *Chem. Rev.* **91**, 1721-2085.
20. Bien, J. T., Eschner, M. J. and Smith, B. D. (1995) A neutral parquat receptor that uses oriented dipoles produced by dative B-N bonds, *J. Org. Chem.*, in press.
21. White, B. G. (1969) Bipyridylium quaternary salts and related compounds, *Trans. Faraday Soc.* **65**, 2000-2015.
22. Allwood, B. L., Spencer, N., Zavareh, H. S., Stoddart, J. F. and Williams, D. J. (1987) Complexation of paraquat by a bisparaphenylene-34-crown-10 derivative, *J. Chem. Soc. Chem. Commun.* 1064-1066.
23. Rettig, S. J. and Trotter, J. (1975) Crystal and molecular structure of B-phenyl-diptychboroxolidine, *Can. J. Chem.*, 53, 1392-1401.
24. Csuk, R., Müler, N. and Sterk, H. (1985) Stabilitätsuntersuchungen an bor-stickstoff-sauerstoff-heterocyclen mittels 11-B NMR spektroscopie, *Z. Naturforsch.* **40b**, 987-989.
25. Mancilla, T.; Rosalinda, C. and Wrackmeyer, B. (1986) The N-B coordination in hindered cyclic thexylboronic esters derived from diethanolamines, *J. Organomet. Chem.* **307**, 1-6 and references cited therein.
26. Schneider, H. -J. and Schiestel, T., Zimmerman, P. (1992) The incremental approach to noncovalent interactions: coulomb and van der waals effects in organic ion pairs, *J. Am. Chem. Soc. 114*, 7698-7703.
27. Zimmerman, S. C., Mrksich, M. and Baloga, M. (1989) Highly efficient complexation of a π-acceptor by a molecular tweezer containing two π-donors: the role of preorganization, *J. Am. Chem. Soc.* **111**, 8528-8530.

REGULATION OF PROTEIN SYNTHESIS BY ANTITUMOR PEPTIDES

P. L. TOOGOOD
Department of Chemistry
University of Michigan
Ann Arbor, MI 48109-1055

Introduction

The biosynthesis of proteins *in vivo* is a complex and tightly regulated process.[1-4] The genetic code, in the form of mRNA, is translated into polypeptides on large subcellular organelles known as ribosomes, which comprise a complex of approximately 75 proteins and 4 RNA molecules.[5] The ribosomes are assisted by more than 20 additional polypeptides, called initiation factors and elongation factors, which variously catalyze (i) assembly of the ribosome on an RNA message, (ii) delivery of the first amino acid, methionine, as its aminoacyl-tRNA ester, and (iii) polypeptide elongation.[1] Several of these proteins are post-translationally modified e.g., by phosphorylation, and evidence is accumulating for the importance of some of these modifications in translational control.[3,4] The remarkable ability of cells to rapidly and reversibly modulate the rate of protein synthesis, either generally or for specific proteins, contributes to the importance of translational regulation in the control of critical cellular proceses, such as progression through the cell cycle, cell differentiation, and response to stress.

The central role of translation also makes it a natural target for cytotoxins.[6] Many microorganisms produce inhibitors of protein synthesis, presumably as weapons against competing microorganisms. Naturally occurring inhibitors of protein synthesis include catalytic proteins such as ricin and colicin E3,[7] which cleave ribosomal RNA,

C. S. Wilcox and A. D. Hamilton (eds.), Molecular Design and Bioorganic Catalysis, 243–262.

and small molecules that bind to ribosomes or translation factors, preventing them from performing their normal functions.[6] The discovery of inhibitors that act selectively upon prokaryotic cells has led to their widespread use as antibiotics.[8]

Many compounds that inhibit protein synthesis in eukaryotic cells have been identified,[6] and several of these compounds have been investigated as potential antitumor agents. However, achieving selectivity for tumor cells *versus* healthy proliferating cells is difficult, and non-selective inhibitors can cause severe toxicity and display poor clinical efficacy. Despite these problems, inhibitors of protein synthesis continue to be studied in human clinical trials because the possibility exists that one or more of these agents may display sufficient selectivity to be therapeutically useful. Since detailed understanding of a compound's mechanism of action can frequently indicate approaches for achieving improved potency and/or selectivity, we have embarked upon a program to elucidate the mechanisms of action for two classes of protein synthesis inhibitors, each of which contains at least one candidate currently undergoing clinical trials. These two classes of compounds are the didemnins and the RA-peptides.

The didemnins

The didemnins (Fig. **1**) are cyclic depsipeptides, isolated from marine tunicates, that display potent cytotoxic and antiproliferative effects *in vitro*.[9] These compounds were first reported by Rinehart *et al.* in 1981,[9] and have been the subject of considerable research.[10] Total syntheses of didemnins A, B, and C and nordidemnin have been achieved,[11-15] and the three-dimensional structure of didemnin B, the most potent member of this class of compounds, has been studied both in solution and in the solid state.[16-18]

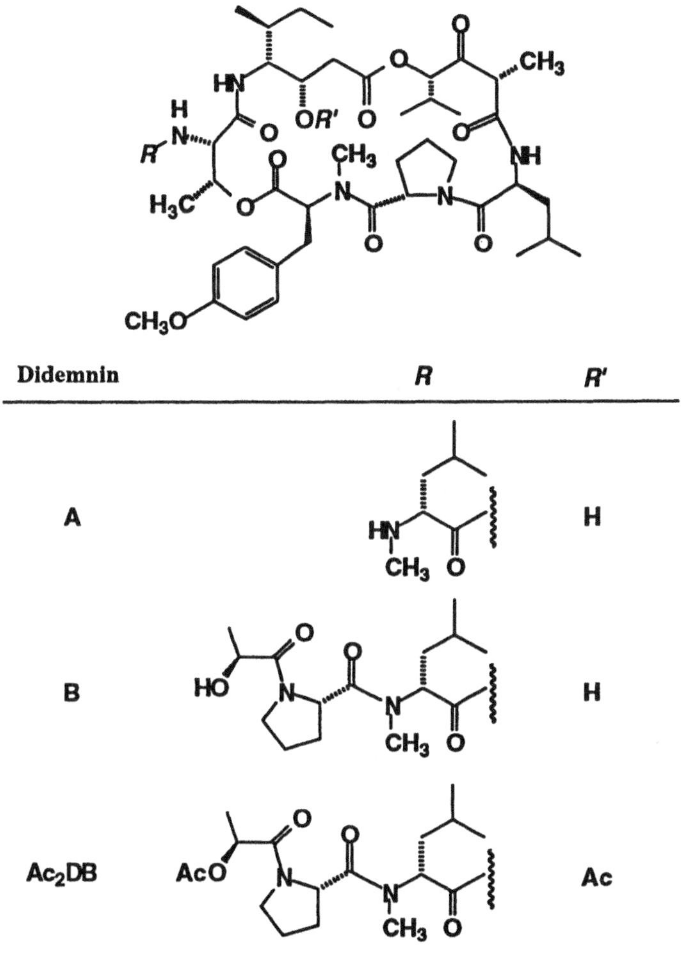

Figure 1. Some representative didemnins

Mechanistic studies are in progress in several laboratories. Didemnin B inhibits DNA and protein synthesis in intact cells[19-22] and preliminary work indicated that inhibition of protein synthesis represents the primary mode of cytotoxicity for this compound.[19, 23] However, the antiproliferative effects of didemnin B appear to arise independently of any influence upon protein synthesis.[22] Didemnin B inhibits blastogenesis in Nb2 cells,[22] and *myo*-inositide metabolism in WRK_1 cells (a rat mammary tumor cell line),[24] at concentrations that are significantly lower than required to inhibit protein synthesis in the same cells. The relationship between these various biological effects is unclear.

Several pieces of evidence favor the hypothesis that the cytotoxic effects of didemnin B arise due to direct interactions between this molecule and the translational machinery. First, the time-course for inhibition of protein biosynthesis in L1210 murine leukemia cells correlates closely with inhibition of cell growth.[19] Second, in a study using a didemnin affinity column to discover didemnin targets in bovine brain cell lysates, Schreiber and coworkers showed that the major protein isolated is translation factor EF-1α.[25] It was further demonstrated that N-acetyl-didemnin A binds to EF-1α in a GTP dependent manner with $K_d = 15\ \mu M$. This protein is very likely to be the same abundant cytosolic protein previously studied by Shen and coworkers and proposed to mediate the cytotoxic effects of didemnin B.[22]

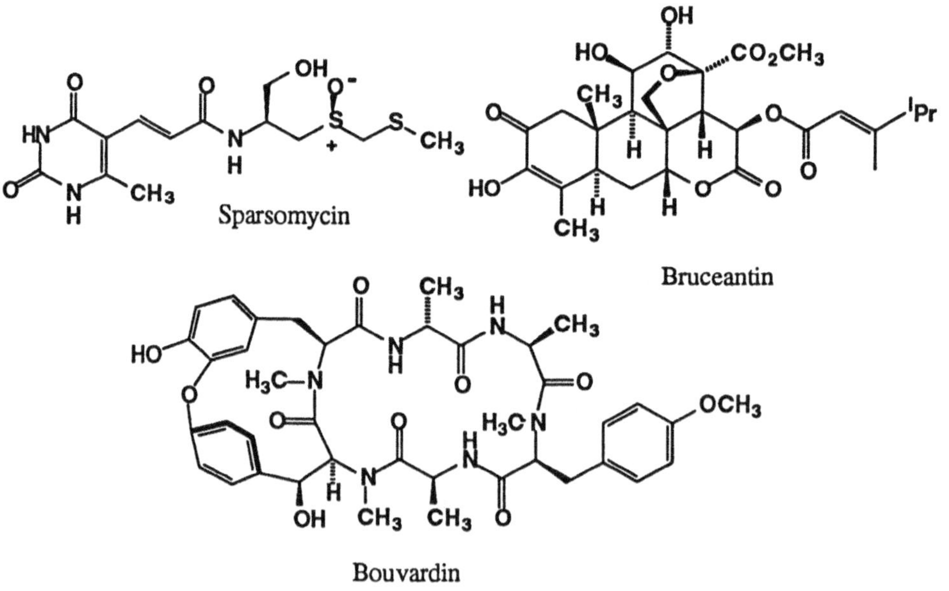

Figure 2. Compounds identified by mean graph analysis to show a similar spectrum of cytotoxicity to didemnin B.

Finally, comparison of didemnin B with other cytotoxins in the NIH database (mean graph analysis) indicates that it displays a similar spectrum of activity against 60 human tumor cell lines to the known protein synthesis inhibitors bouvardin, sparsomycin and bruceantin, all of which are structurally quite dissimilar to didemnin B (Figure 2).[26] Such correlations in activity may indicate similar mechanisms of action. Considered together, these results provide compelling support for a study of the mechanism of protein synthesis inhibition by didemnin B.

Didemnin B as an inhibitor of translation

We have employed *in vitro* biochemical assays to investigate the effect of didemnin B upon the translational machinery. Didemnin B inhibits protein synthesis in cell-free systems from rabbit reticulocytes, wheat germ, yeast and L1210 cells with IC_{50} values between 1-10 μM.[27] No inhibition is observed in an *E. coli* cell-free system at concentrations of didemnin B up to 50 μM, suggesting that inhibition is specific for the eukaryotic translational machinery. A time-course analysis of the effect of didemnin B in rabbit reticulocyte lysates indicates that inhibition occurs at the stage of polypeptide elongation since inhibition of an active translation system occurs immediately upon addition of the inhibitor. In contrast, inhibitors of polypeptide initiation display a lag of approximately 8-10 minutes following addition of the drug and prior to the onset of inhibition, as previously initiated ribosomes continue to the end of their RNA message.

Polypeptide elongation involves three discrete steps per cycle (Figure 3): delivery of the aminoacyl-tRNA, catalyzed by the GTP dependent protein elongation factor-1α, (EF-1α); peptide bond formation, catalyzed by the large ribosomal subunit (peptidyltransferase); translocation of the peptidyl-tRNA from the A- to the P- site on the ribosome.[1] Each of these steps may be assayed independently *in vitro.*[28]

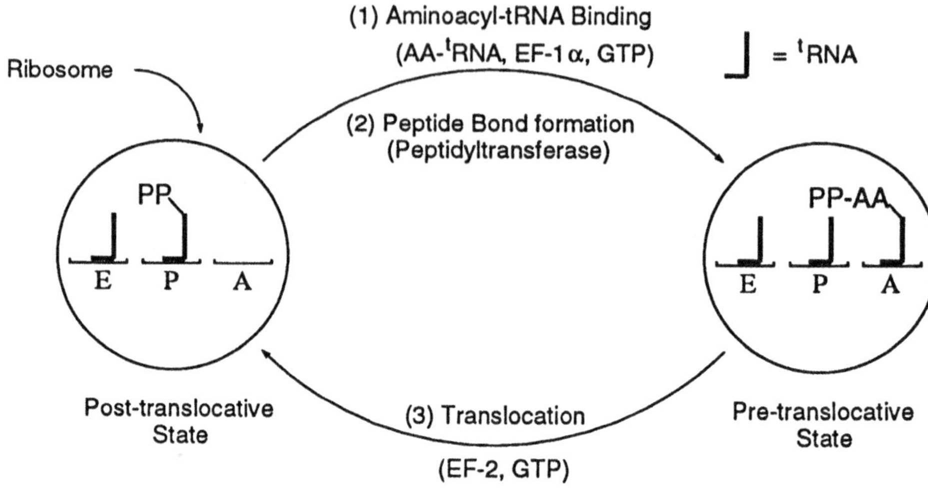

Figure 3. Steps in polypeptide elongation. Following binding of a new aminoacyl-tRNA (AA-tRNA) to the ribosomal A-site and peptide bond formation to extend the polypeptide chain, the ribosome and mRNA translocate to move the new peptidyl-tRNA (PP-tRNA) from the A- to the P-site. The uncharged tRNA moves from the P- to the E-site. Didemnin B stimulates aminoacyl-tRNA binding (step 1) and inhibits translocation (step 3).

Didemnin B has no effect upon the intrinsic ribosomal peptidyltransferase activity as determined using the puromycin reaction. In this experiment, a P-site bound acetyl-phenylalanyl-tRNA is reacted with the antibiotic puromycin, an analog of the 3'-terminal of an aminoacyl-tRNA, to form acetyl-phenylalanyl-puromycin (Figure 4). This reaction is catalyzed by the ribosomal peptidyltransferase and does not occur when the reagents are free in solution or when the aminoacyl-tRNA is bound in the ribosomal A-site. Didemnin B also does not inhibit EF-1α catalyzed aminoacyl-tRNA binding to ribosomes, but under the conditions of the assay (ribosomes in excess over [^{14}C]-Phe-tRNA), the amount of aminoacyl-tRNA bound to ribosomes is higher in the presence of didemnin B than in its absence. This effect is particularly apparent when GTP, which is required for formation of the EF-1α•aminoacyl-tRNA complex, is substituted with a non-hydrolyzable analog (5'-guanalylimidodiphosphate; GDPNP).

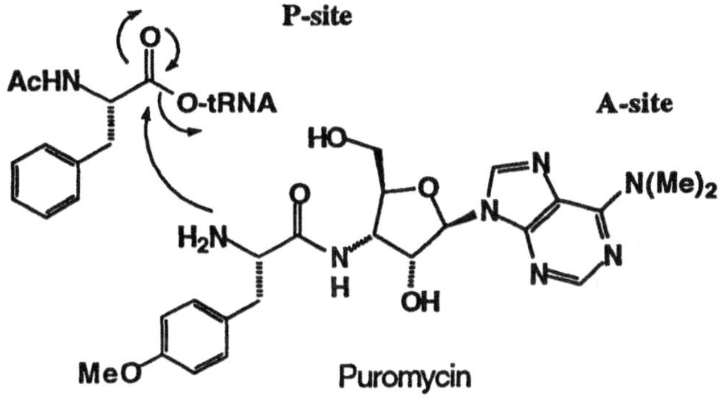

Figure 4. The puromycin reaction. The puromycin primary amine reacts with aminoacyl-tRNA to form a new amide bond with displacement of the tRNA.

The effect of didemnin B upon ribosomal translocation was investigated using poly(U) programmed ribosomes and phenylalanyl-tRNA. Translocation of phenylalanyl-tRNA from the A- to the P-site was measured by determining the increase in puromycin reactivity following addition of EF-2 to ribosomes containing phenylalanyl-tRNA bound in the A-site. Didemnin B (50 µM) completely inhibits translocation and this effect can be mitigated by increasing the concentration of EF-2. However, inhibition of translocation by didemnin B appears also to be a function of EF-1α. Thus, when phenylalanyl-tRNA is delivered to the ribosomal A-site independently of EF-1α by raising the magnesium ion concentration, subsequent EF-2 dependent translocation is unaffected by didemnin B (Figure 5). This observation, together with the stimulation of EF-1α dependent aminoacyl-tRNA binding, suggests that the target of didemnin B is EF-1α. This conclusion is consistent with the results of concurrent work published by Crews and coworkers who identified EF-1α as a didemnin target using affinity chromatography (*vide supra*).[25]

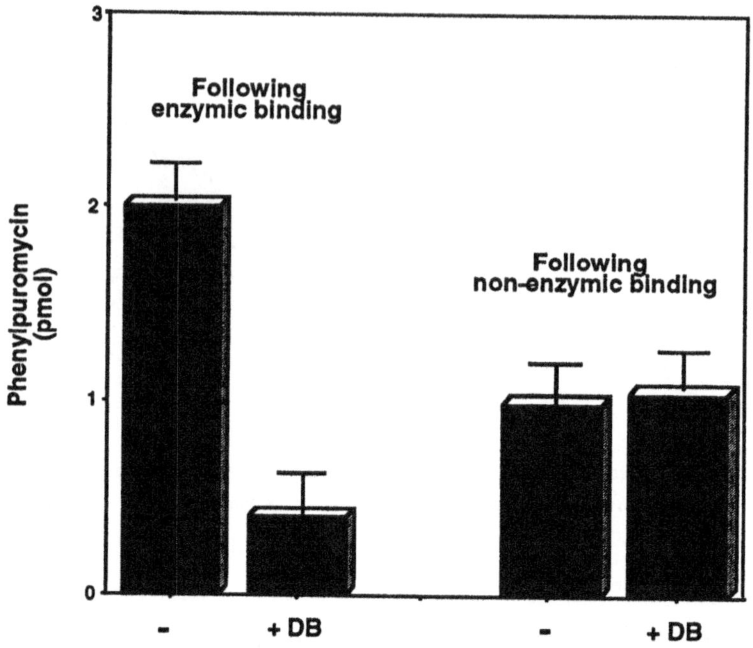

Figure 5. Effect of didemnin B (50 μM) upon translocation following enzymic and non-enzymic binding to ribosomal A-sites, shown as extent of phenylalanyl-puromycin formation.

Comparison of didemnin B to known inhibitors of protein synthesis suggests that it is most similar mechanistically to the antibiotic kirromycin (Figure 6).[29] Kirromycin binds to EF-Tu, the prokaryotic homolog of EF-1α, sequestering the aminoacyl-tRNA and EF-Tu at the ribosomal A-site and inhibiting all subsequent events, including peptide bond formation. Since didemnin B does not inhibit the intrinsic peptidyltransferase activity in the absence of any elongation factors, it was of interest to re-examine the effect of this inhibitor upon peptide bond formation in the presence of elongation factor EF-1α. This experiment was performed in two ways, analyzing either for phenylalanyl-phenylalanine formation or acetylphenylalanyl-phenylalanine formation, in order to make a direct comparison with kirromycin.

In the first experiment, phenylalanyl-tRNA was delivered to the ribosomal P-site using both EF-1α and EF-2, and a second phenylalanyl-tRNA was bound to the A-site in the presence or absence of 50 μM didemnin B. The products were analyzed by hydrolysis of the aminoacyl- and peptidyl- tRNA esters followed by paper chromatography of the amino acid products. As much phenylalanyl-phenylalanine was obtained in the presence of didemnin B as in control reactions containing no inhibitor. Reactions performed in the absence of ribosomes produced no dipeptide product. Similarly, following binding of an acetylphenylalanyl-tRNA to the ribosomal P-site, phenylalanyl-tRNA was delivered to the A-site using EF-1α, and peptide bond formation was allowed to occur. Again, the amount of dipeptide formed in the presence of didemnin B matched the amount observed in control reactions. In control experiments, peptide bond formation is inhibited completely by 150 μM sparsomycin, as expected from previous work.[30]

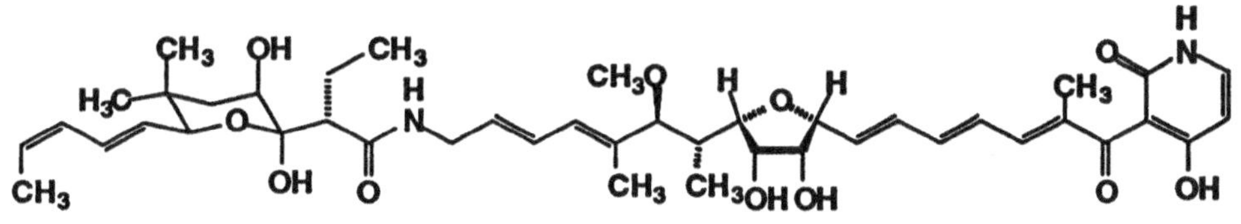

Figure 6. Kirromycin

We conclude that, unlike kirromycin, didemnin B does not inhibit peptide bond formation. Thus, didemnin B appears to inhibit protein synthesis *via* a previously uncharacterized mechanism and therefore may be a valuable tool for studying further details of the translational process.

Based on the current data, we propose that didemnin B binds to a complex containing the ribosome, EF-1α, aminoacyl-tRNA and either GTP or GDP, increasing the affinity of the aminoacyl-tRNA for the ribosomal A-site (Figure 7). Peptide bond formation still occurs, but translocation of the newly formed peptidyl-tRNA is prevented. Further experiments are required to establish the precise composition of the inhibited complex, and the timing of the various binding events. For example, does didemnin bind to EF-1α•GTP, as shown by Crews and coworkers,[25] followed by complexation of aminoacyl-tRNA and binding to the ribosome, or does didemnin B bind to a pre-organized recognition site containing the elongation factor and aminoacyl-tRNA already bound to the ribosome? These issues are currently under investigation.

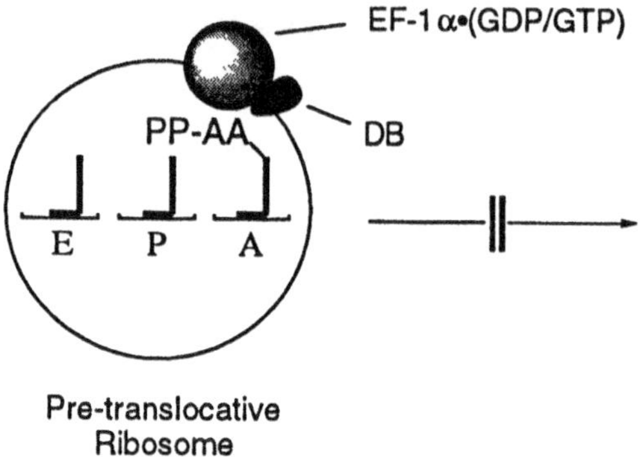

Figure 7. Mechanistic model for inhibition of protein synthesis by didemnin B.

Implications for the effect of didemnin B *in vivo*

While it is clear that the didemnins can inhibit protein synthesis by binding directly to the translational machinery, it is not obvious that this mechanism is followed *in vivo*. In fact, several pieces of evidence indicate that the primary mechanism of action for didemnin B in cells has yet to be defined. First, the potency and trends in potency of didemnins as inhibitors of L1210 cell growth are not reproduced in cell-free translation experiments.[27] Didemnins A and B are approximately three orders of magnitude weaker inhibitors of protein synthesis *in vitro* than in intact cells. Furthermore, as inhibitors of cell growth, Ac_2DB > DB > DA but for inhibition of protein synthesis *in vitro* DB~DA > Ac_2DB (Fig. 1). Even more compelling, nordidemnin (Figure 8) is approximately 10^3-fold better as an inhibitor of cell growth than epi-nordidemnin which differs only in the stereochemistry at one asymmetric carbon atom.[31] However, these molecules are essentially equipotent as inhibitors of *in vitro* protein synthesis in a rabbit reticulocyte cell-free system.[32]

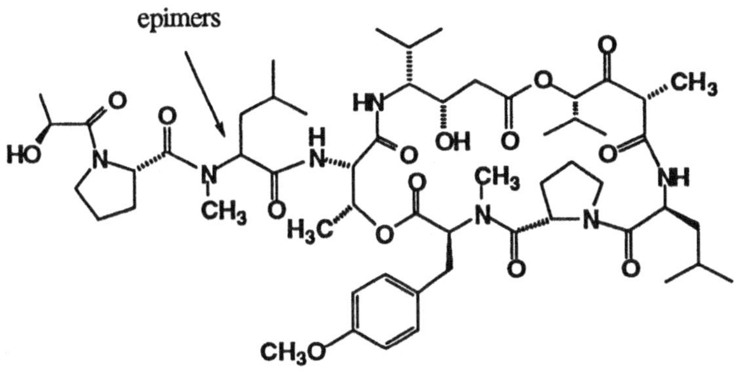

Figure 8. Nordidemnin (N-Methyl-D-leucine) and epi-nordidemnin (N-Methyl-L-leucine).

Some of these results can be rationalized without invoking different mechanisms for didemnin B *in vitro* and *in vivo*. For example, the relative potency of didemnin B and its diacetyl derivative could reflect the relative hydrophobicities of these molecules and consequently their relative ability to cross the cell membrane. Similarly, the approximately 10^3-fold difference in absolute potency of didemnins for inhibition of protein synthesis in intact cells and inhibition of protein synthesis *in vitro*, might be explained by the accumulation of didemnins inside cells, leading to a higher intracellular concentration of didemnin than exists externally; evidence to support this hypothesis has been presented by Orlowski *et al.*[33]

It is possible also that the complex organizational state of the translational machinery inside a cell renders it more susceptible to inhibition than in a cell lysate, perhaps due to pre-organization of a specific binding site.[34] Didemnin B is approximately 10-fold more effective at inhibiting protein synthesis in permeabilized CHO cells than in rabbit reticulocyte lysates.[35] Unfortunately, no data are available regarding the potency of didemnin B as an inhibitor of protein synthesis in CHO cell lysates so these results are difficult to interpret. Furthermore, it is not clear how the organizational state of the translational machinery affects the rate of protein synthesis in rabbit reticulocytes since in this case the cell-free system is almost as efficient as the intact cells.[36] It should be noted that while the reliability and high activity of rabbit reticulocyte lysate favor its use in studies of protein synthesis inhibitors *in vitro*, it is not known whether this system provides a good model for intact tumor cells.

Despite these caveats regarding the interpretation of data obtained for didemnins, the discrepancy between results obtained for nordidemnin and its epimer as inhibitors of protein synthesis *in vitro* and in L1210 cells remains unexplained. Substitution of N-methyl-D-leucine by its epimer in didemnin B causes only minor structural perturbations, but similarly results in dramatically reduced biological potency.[17] The close structural similarity between these compounds strongly suggests that any difference in their ability to inhibit cell growth derives from differential binding to a specific receptor molecule. Since marked differences in activity are not observed for nordidemnin and epi-nordidemnin in our *in vitro* assays, it seems likely that inhibition of cell growth and inhibition of protein synthesis *in vitro* occur *via* different mechanisms.

Other workers have argued that inhibition of protein synthesis does not represent the primary mechanism of action of didemnin B in cells. Montgomery and coworkers showed that lymphocyte blastogenesis is inhibited by picomolar concentrations of didemnin B, i.e., approximately 10^3-fold lower concentrations than required to inhibit protein synthesis in the same cell line.[37] These authors also provided evidence for a cytosolic receptor for didemnin B that mediates the effect of this molecule upon protein and DNA synthesis (*vide supra*).[22] The affinity of a tritiated didemnin for this receptor was determined to be K_d = 180 nM. Thus, in lymphoid cells, didemnin B appears to inhibit cell growth and cell proliferation by different mechanisms, mediated by different macromolecular targets. In related work, Dominice and coworkers recently reported data indicating that nordidemnin interferes with *myo*-inositol metabolism and inhibits *myo*-inositol uptake into WRK_1 cells, independently of any effect upon general protein synthesis.[24] To further complicate matters, didemnin B inhibits phorbol ester stimulated phosphorylation of elongation factor EF-2.[38, 39] Since phosphorylated EF-2 is inactive,[3] this effect of didemnin B appears to be inconsistent with inhibition of protein synthesis.

Structure-function studies

We propose that the didemnins can bind to two separate (but not necessarily independent) receptors using different loops of the depsipeptide ring. The loop that contains the threonine-tyrosine-proline residues, and to which the lactyl-prolyl-leucine side-chain is attached, recognizes an unidentified receptor that mediates the antiproliferative effects of didemnins including inhibition of T-cell proliferation and inhibition of blastogenesis. In contrast, the "backside" of the molecule, which contains the isotatine and hydroxyisovalerylpropionate residues, recognizes a binding site on elongation factor EF-1α that mediates the inhibition of protein synthesis. This hypothesis will be tested using didemnin analogs and affinity labeling reagents.

Several laboratories are interested in delineating a structure-function profile for the didemnins. Most of the didemnin analogs that have been prepared to date incorporate differences in the side chain function that distinguishes the various naturally occurring didemnins.[31, 40] Consistent variation in biological potency has been found commensurate with changes in the side chain structure, including some molecules that are more potent than the most potent natural product in this series, didemnin B.[40] Interestingly, inhibition of cell growth and inhibition of T-cell proliferation appear to follow the same trend with alteration in the side chain structure, possibly suggesting that these biological properties *are* mediated through the same mechanism. Alterations in the macrocyclic depsipeptide ring of didemnins have been described by Shioiri and coworkers, however no biological data for these compounds have been reported.[41] Much work remains to be done in this area.

Comparison of didemnin B to RA-peptides

As mentioned above, a mean graph analysis for didemnin B identified bouvardin, bruceatin, sparsomycin and deoxybouvardin, as the four compounds that best correlate with didemnin B in their spectrum of cytotoxicity. The first three of these compounds have been demonstrated to inhibit protein synthesis specifically at the elongation stage through direct interactions with the translational machinery.[30, 42, 43] Deoxybouvardin is structurally homologous to bouvardin and probably displays a similar mechanism of action. Several interpretations may be placed upon this mean graph analysis: (i) the correlation between didemnin B and the RA-peptides is fortuitous and unrelated to their mechanisms of action; (ii) didemnins and RA-peptides all interfere directly with the translational machinery to inhibit protein biosynthesis; (iii) didemnins and RA-peptides possess similar mechanisms of action but these mechanisms are unrelated to the ability of these molecules to inhibit translation. To investigate these possibilities further, we are examining the mechanism of action of the natural product RA-VII (Figure **9**.).

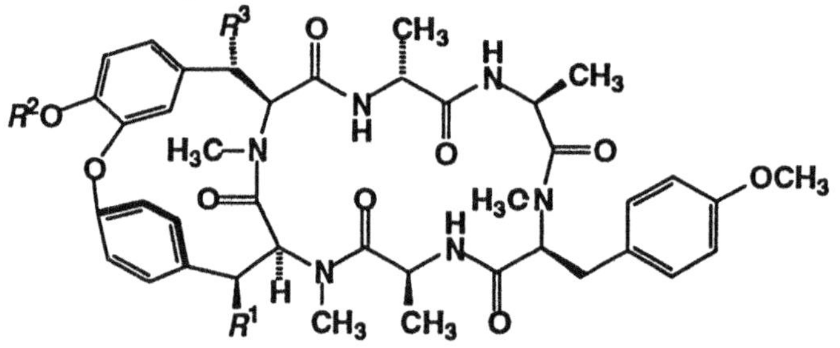

	R^1	R^2	R^3
Bouvardin	OH	H	H
Deoxybouvardin	H	H	H
RA-VII	H	CH_3	H
RA-XII	H	H	1-β-D-Glc

Figure **9**. Some representative RA-peptides.

Mechanism of RA-VII

The RA-peptides (Figure. 9) are named after the *Rubiaceae* plants, from which they are derived.[43] These plants have been used in herbal therapies for many years, and their biological properties led Itokawa and coworkers to identify their active constituents. At least 14 RA-peptides have now been isolated and characterized. Many of the RA-peptides are cytotoxic, and preliminary experiments resulted in their categorization as inhibitors of protein biosynthesis. Vázquez and coworkers studied the mechanism of inhibition by bouvardin in detail and showed that bouvardin prevents EF-1α dependent binding of aminoacyl-tRNA to ribosomes as well as EF-2 dependent translocation.[44] They explained this dual effect by proposing a binding site for bouvardin on eukaryotic (80*S*) ribosomes. Thus, binding of bouvardin to ribosomes could prevent both EF-1α and EF-2 binding since these elongation factors are known to occupy overlapping binding sites.[5] Subsequently, Itokawa and coworkers have provided evidence for RA-XII binding to eukaryotic ribosomes, consistent with the Vázquez model.[45] The other member of this class of peptides that has been shown to inhibit protein synthesis is RA-VII. We were curious to know whether RA-VII displays the same mechanism for inhibition of protein synthesis as bouvardin and consequently we launched an investigation of its effect using *in vitro* protein synthesis assays.

Using similar techniques to the ones used to study didemnin B, we have shown that RA-VII inhibits *in vitro* protein synthesis in a rabbit reticulocyte cell-free translation system with IC_{50} = 80 nM.[46] Furthermore, time-course analyses suggest that inhibition by RA-VII occurs at the elongation stage. Using rabbit reticulocyte ribosomes and yeast elongation factors we have shown that RA-VII does not affect aminoacyl-tRNA binding to ribosomes but instead inhibits both the intrinsic peptidyltransferase activity of 80*S* ribosomes and peptide bond formation between P-site bound Phe-tRNA or Ac-Phe-tRNA, and an A-site bound Phe-tRNA (Fig. 3). Since assays of ribosomal translocation include an obligate peptidyltransferase step to determine the amount of P-site bound peptidyl-tRNA, determination of the effect of a peptidyltransferase inhibitor upon translocation is technically difficult. However, by examining the binding of Phe-tRNA to pre-translocative and post-translocative ribosomes, in conjunction with standard translocation assays, no evidence has been obtained for inhibition of translocation by RA-VII. Thus, in contrast to bouvardin, RA-VII appears to inhibit only one step in protein synthesis, namely the peptidyltransferase step. The reason for these apparently disparate mechanisms is unknown.

As with the didemnins, a correlation between the ability of RA-peptides to inhibit protein synthesis *in vitro* and their cytotoxicity has not been established. This fact is surprising given the level of interest in these compounds as clinical agents (both RA-V and RA-VII are currently undergoing phase-II clinical trials). However, one piece of evidence from our labs strongly indicates that this correlation should be closely examined. The synthetic analog of RA-VII shown in Figure **10** was prepared by Boger *et al.* during the course of synthetic investigations of bouvardin and related peptides.[47] These workers identified the *iso*-dityrosine moiety as the RA-peptide pharmacophore and showed that cytotoxicity is dependent upon the adoption of a cis-amide linkage between these two residues.

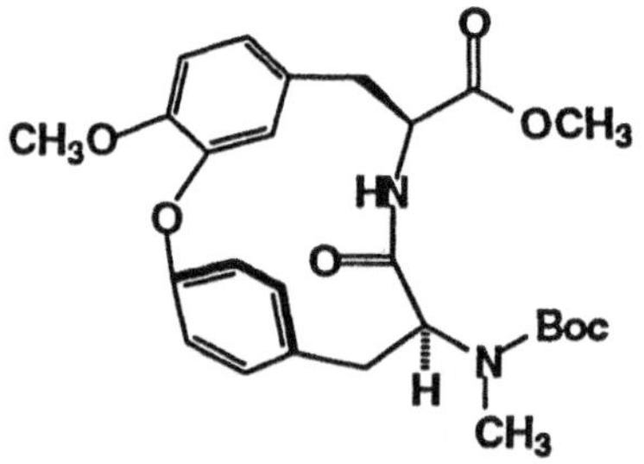

Figure **10**. A synthetic analog of RA-VII. Boc = *tert*-Butyloxycarbonyl.

The analog is approximately 30-fold less potent than RA-VII as an inhibitor of L1210 cell growth, with IC_{50} = 78 nM. However, in direct comparison with RA-VII, the analog is ineffective as an inhibitor of protein synthesis *in vitro* at concentrations up to 1 mM.[48] This result could be interpreted as an indication that the mechanism by which RA-peptides inhibit cell growth is independent of their effect upon the translational machinery, although this conclusion is premature in the absence of supporting data. Further studies of this interesting class of natural products are anticipated to provide new insight into the mode of action of RA-peptides *in vivo*, and possibly may identify a novel mechanism of cytotoxicity.

Conclusion

To summarize, we have provided mechanistic hypotheses for the mode of action of didemnins and RA-peptides through *in vitro* studies of the effects of these compounds upon protein synthesis. Further work is required to establish whether the mechanisms that we propose are responsible for the biological effects observed for these compounds *in vivo*. Some lines of evidence point to the existence of independent mechanisms for both classes of peptide that do not require direct interactions with the translational machinery. However, alternative mechanisms that satisfactorily explain the observed effects of these compounds upon protein synthesis are not currently available.

References

1. Merrick, W. C. (1992) Mechanism and regulation of eukaryotic protein synthesis, *Microbiol. Rev.* **56**, 291-315.
2. Moldave, K. (1985) Eukaryotic protein synthesis, *Ann. Rev. Biochem.* **54**, 1109-1149.
3. Hershey, J. W. B. (1991) Translational control in mammalian cells, *Ann. Rev. Biochem.* **60**, 717-755.
4. Redpath, N. T. and Proud, C. G. (1994) Molecular mechanisms in the control of translation by hormones and growth factors, *Biochim. Biophys. Acta* **1220**, 147-162.
5. Nygard, O. and Nilsson, L. (1990) Translational Dynamics. Interactions between the translational factors, tRNA and ribosomes during eukaryotic protein synthesis. *Eur. J. Biochem.* **191**, 1-17.
6. Vázquez, D. (1979) Inhibitors of protein synthesis, Springer-Verlag, New York.

7. Cannon, M. (1990) Agents that interact with ribosomal RNA and interfere with its functions, in Hansch, C., Sammes, P. G. and Taylor, J. B. (eds) *Comprehensive Medicinal Chemistry*, Pergamon Press, New York, pp. 813-838.

8. Franklin, T. J. and Snow, G. A. (1989) *Biochemistry of Antimicrobial Action*, 4th Edition, Chapman and Hall, New York, pp. 112-135.

9. Rinehart, K. L. Jr., Gloer, J. B., Hughes, R. G., Renis, H. E., McGovern, P. J., Swynenberg, E. B., Stringfellow, D. A., Kuentzel, S. L., and Li, L. H. (1981) Didemnins: antiviral and antitumor depsipeptides from a carribean tunicate, *Science*, **212**, 933-935.

10. Li, W. R. and Joullié, M. M. (1992) The didemnins: biological properties, chemistry and total synthesis, in Attar-ur-Rahman (ed), *Studies in Natural Products Chemistry* **10**, Elsevier, Amsterdam, pp. 241-302.

11. Rinehart, K. L. Jr., Kishore, V., Nagarajan, S., Lake, R. L., Gloer, J. B., Bozich, F. A., Li, K., Maleczka, R. E. J., Todsen, W. L., Munroe, M. H. G., Sullins, D. W., and Sakai, R. (1988) Total synthesis of didemnins A, B and C, *J. Am. Chem. Soc.* **109**, 6846-6848.

12. Schmidt, U., Kroner, M., and Griesser, H. (1988) Total synthesis of the didemnins-2. Synthesis of didemnin A, B, C and prolyldidemnin A, *Tetrahedron Lett.* **29**, 4407-4408.

13. Hamada, Y., Kondo, Y., Shibata, M., and Shioiri, T. (1989) Efficient total synthesis of didemnins A and B, *J. Am. Chem. Soc.* **111**, 669-673.

14. Li, W., Ewing, W. R., Harris, B. D., and Joullié, M. M. (1990) Total synthesis and structural investigations of didemnins A, B and C, *J. Am. Chem. Soc.* **112**, 7659-7672.

15. Jouin, P., Poncet, J., Dufour, A., and Castro, B. (1989) Synthesis of the cyclodepsipeptide nordidemnin B, a cytotoxic minor product isolated from the sea tunicate *Trididemnun cyanophorum*, *J. Org. Chem.* **54**, 617-627.

16. Searle, M. S., Hall, J. G., Kyratzis, I., and Wakelin, L. P. G. (1989) Didemnin B. Conformation and dynamics of an antitumor and antiviral depsipeptide studied in solution by ^{1}H and ^{13}C N.M.R. spectroscopy, *Int. J. Peptide Protein Res.* **34**, 445-454.

17. Kessler, H., Mronga, S., Will, M., and Schmidt, U. (1990) Solution structure of [Me-L-Leu^{7}]didemnin B determined by NMR spectroscopy and refined by MD calculation *Helv. Chem. Acta* **73**, 25-47.

18. Hossain, M. B., van der Helm, D., Antel, J., Sheldrick, G. M., and Sanduja, S. K. (1988) Crystal and molecular structure of didemnin B, an antiviral and cytotoxic depsipeptide, *Proc. Natl. Acad. Sci. (USA)* **85**, 4118-4122.

19. Li, L. H., Timmins, L. G., Wallace, T. L., Krueger, W. C., Prairie, M. D, and Im, W. B. (1984) Mechanism of action of didemnin B, a depsipeptide from the sea, *Cancer Lett.* **23**, 279-288.

20. Crampton, S. L., Adams, E. G., Kuetzel, S. L., Li, L. H., Badiner, G., and Bhuyan, B. K. (1984) Biochemical and cellular effects of didemnins A and B, *Cancer Res.* **44**, 1796-1801.

21. Legrue, S. J., Sheu, T., Carson, D. C., Laidlaw, J. L., and Sanduja, S. K. (1988) Inhibition of T-lymphocyte proliferation by the cyclic polypeptide didemnin B: no inhibition of lymphokine stimulation, *Lymphokine Res.* **7**, 21-29.

22. Shen, G. K., Zukoski, C. F., and Montgomery, D. W. (1992) A specific binding site in Nb2 node lymphoma cells mediates the effects of didemnin B, an immunosuppressive cyclic peptide, *Int. J. Immunopharmacol.* **14**, 63-73.

23. Rinehart, K. L. Jr., Gloer, J. B., Wilson, G. R., Hughes, R. G. Jr., Li, L. H., Renis, H. R., and McGovern, J. P. (1983) Antiviral and antitumor compounds from tunicates, *Fed. Proc.* **42**, 87-90.

24. Dominice, C., Dufour, M-N., Patino, N., Manzoni, O., Grazzini, E., Jouin, P.,and Guillon, G. (1994) Dual effects of nordidemnin on WRK_1 cells: inhibition of phosphoinositide metabolism and cell proliferation, *J. Pharmacol. Exper. Therapeutics* **271**, 107-117.

25. Crews, C. M., Collins, J. L., Lane, W. S., Snapper, M. L., and Schreiber, S. L. (1994) GTP-Dependent binding of the antiproliferative agent didemnin to elongation factor 1α, *J. Biol. Chem.* **269**, 15411-15414.

26. These data were kindly provided to us by Dr. Jill Johnson of the National Cancer Institute.

27. SirDeshpande, B. V. and Toogood, P. L. (1995) Mechanism of protein synthesis inhibition by didemnin B *in vitro*, *Biochemistry*, in press.

28. Merrick, W. C. (1979) Assays for eukaryotic protein synthesis, *Methods Enzymol.* **60**, 108-123.

29. Parmeggiani, A. and Swart, G. W. M. (1985) Mechanism of action of kirromycin-like antibiotics, *Ann. Rev. Microbiol.* **39**, 557-577.

30. van den Broek, L. A. G. M., Liskamp, R. M. J., Colstee, J. H., Lelieveld, P., Remacha, M., Vázquez, D., Ballesta, J. P. G., and Ottenheijm, H. C. J. (1987) Structure-activity relationships of Sparsomycin and its analogues. Inhibition of peptide bond formation in cell-free systems and of L1210 and bacterial cell growth, *J. Med. Chem.* **30**, 325-222.

31. Jouin, P., Poncet, J., Dufour, M., Aumelas, A., and Pantaloni, A. (1991) Antineoplastic activity of didemnin congeners: nordidemnin and modified chain analogues, *J. Med. Chem.* **34**, 486-491.

32. SirDeshpande B. V. and Toogood, P. L. unpublished results.

33. Orlowski, S., Belehradek, J. Jr., Paoletti, C., and Mir, L. M. (1988) Transient electropermeabilization of cells in culture. Increase in the cytotoxicity of anticancer drugs. *Biochem. Pharmacol.* **37**, 4727-4733.

34. Negrutskii, B. S., Stapulionis, R., and Deutscher, M. P. (1994) Supramolecular Organization of the Mammalian Translation System, *Proc. Natl. Acad. Sci. (USA)* **91**, 964-968.

35. Murray Deutscher, personal communication.

36. Pelham, H. R. B. and Jackson, R. J. (1976) An efficient mRNA-dependent translation system from reticulocyte lysates, *Eur. J. Biochem.* **67**, 247-256.

37. Montgomery, D. W. and Zukoski, C. F. (1985) Didemnin B: A new immunosuppressive peptide with potent activity *in vitro* and *in vivo*, *Transplantation* **40**, 49-56.

38. Gschwendt, M., Kittstein, W., and Marks, F. (1987) Didemnin B inhibits biological effects of tumor promoting phorbol esters on mouse skin, as well as phosphorylation of a 100 kD protein in mouse epidermis, *Cancer Lett.* **43**, 187-191.

39. Gschwendt, M., Kittstein, W., and Marks, F.(1989) The immunosuppressant FK506, like cyclosporins and didemnin B, inhibits calmodulin-dependent phosphorylation of the elongation Factor 2 *in vitro* and biological effects of the phorbol ester TPA on mouse skin *in vivo*, *Immunobiol.* **179**, 1-7.

40. Mayer, S. C., Ramanjulu, J., Vera, M. D., Pfezenmayer, A. J., Joullié, M. M. (1994) Synthesis of new didemnin B analogues for investigations of structure/biological activity relationships (SAR), *J. Org. Chem.* **59**, 5192-5205.

41. Furukubo, S., Hamada, Y., and Shioiri, T. (1992) Synthesis of lower analogs of didemnins: cytotoxic cyclic depsipeptides, *Proceedings of the 2nd Japan Symposium on Peptide Chemistry* **11**, 9-13.

42. Willingham, W. Jr., Stafford, E. A., Reynolds, S. H., Chaney, S. G., Lee, K-H., Okano, M. and Hall, I. (1981) Mechanism of eukaryotic protein synthesis inhibition by brusatol, *Biochim. Biophys. Acta* **654**, 169-174.

43. Itokawa, H. and Takeya, K. (1993) Antitumor substances from higher plants, *Heterocycles* **35**, 1467-1501.

44. Zalacain, M., Zaera, E., Vázquez, D., and Jiménez, A. (1982) The mode of action of the antitumor drug bouvardin, an inhibitor of protein synthesis in eukaryotic cells, *FEBS Lett.* **51**, 869-900.

45. Morita, H., Yamamiya, T., Takeya, K., Itokawa, H., Sakuma, C., Yamada, J., and Suga, T. (1993) Conformational recognition of RA-XII by 80S ribosomes: A differential line broadening study in ^{1}H NMR spectroscopy, *Chem. Pharm. Bull.* **41**, 781-783.

46. SirDeshpande B. V. and Toogood, P. L. (1995) Inhibition of Protein synthesis by RA-VII, submitted to *Bioorganic Chemistry*.

47. Boger, D. L., Patane, M. A., Jin, Q., and Kitos, P. A. (1994) Design, synthesis and evaluation of bouvardin, deoxybouvardin and RA-I-XIV pharmacophore analogs, *Bioorg. Med. Chem. Lett.* **2**, 85-100.

48. SirDeshpande B. V. and Toogood, P. L. unpublished results.

Author Index
(main author)

Almaraz, M. 87
Anslyn, Eric V. 1
Berger, Michael 191
Bien, Jeffrey T. 227
Caballero, M. C. 87
Cacciapaglia, Roberta 71
Crego, M. 87
Dowd, Paul 15
Eichhorn, U. 53
Gerisch, S. 53
Ham, Seung Wook 15
Hänsler, M. 53
Hershline, Roger 15
Hughes, Martin Patrick 227
Imperiali, Barbara 35
Jakubke, H.-D. 53
Kerns, Jeffrey 15
Lee, In Quen 111
Liras, Jennifer L. 1
Mackin, Gilbert 111
Mandolini, Luigi 71
Martín, M. 87
Metzger, Axel 191
Morán, J. R. 87
Mussons, M. L. 87
Naganathan, Sriram 15
Nowick, James S. 111
Pairish, Mason 111
Raposo, C. 87
Reinhoudt, David N. 137
Roy, R. Sinha 35
Rudkevich, Dmitry M. 137
Russell, Alan J. 163
Salchert, K. 53
Scheerder, Jurgen 137
Schmidtchen, Franz P. 191
Scrimin, Pablo 211
Shaka, A. J. 111
Smith, Eric M. 111
Smith, Bradley D. 227
Stephen, Holger 191
Tecilla, P. 211
Tonellato, U. 211
Toogood, Peter L. 243
Ullmann, G. 53
Ullmann, D. 53
Veronese, A. 211
Walkup, G. K. 35
Wang, L. 35
Weinrich, V. 87
Yang, Fangxiao 163
Zheng, Barbara Zhizhen 15
Ziller, Joseph W. 111

Subject Index

α-helix 11 228
    dipole 228
acetal hydrolysis 1
acetate binding 232
active site 113
adenine nucleotides 197
adsorption isotherms 163
alcohol dehydrogenase 163
amino acid extraction 201
anions
    halides 230
    recognition 137
    transport 137
Antione's equation 171
artificial
    biomimetic catalysis 35
    catalysts 230
    enzymes 228
    proteins 221
    receptor 113
    transacylase mimics 71
1,9 asymmetric induction 208

β-sheet 111
B-N bond 237
B-N dative bonds 240
blood-clotting 15
boron containing receptors 227
boronic acid diethanolamine esters 237
Bouvardin 256

C-H≡O hydrogen bonding 230, 237
calixarenes 77
carboxylase 23
carboxylation 16
catalytic properties 113
CH≡O hydrogen 236
charge-transfer complexes 236
coenzyme-amino acids 35
cofactor 16
cofactor turnover number 179
complementary 229, 236
co-operativity 213
crown ethers 81
cyclic depsipeptides 244

dative B-N bond 235
decarboxylases 36
deuterium-scrambled 9
diarylurea 230
didemnins 244
differing symmetry 9
dioxetane 18
dipoles 227, 235
    B-N dipoles 236-237
    dipole directionality 229
    oriented B-N dipoles 240
    oriented dipoles 227
    macrodipole, helix 228
    molecular dipole 234

electrostatics 229, 234, 240
    calculated surface electric potentials 231,234
    electrostatic effects 227-228
    helix dipole 228
elongation factors 243
enamine trapping 9
endo vs exocyclic cleavage 1
enthalpy-entropy compensation 240

enzymatic synthesis in frozen
  aqueous systems 58
enzyme 113
  alcohol dehydrogenase
    163
  artificial 228
  carboxylase 23
  lysozyme 2
  protease 53
  peptidyltransferase 248
  serine proteases 227
  zymogen 64
enzyme catalyzed
  peptide synthesis 53,54
  gas phase 163
exo-anomeric 5

flexible hosts 194
folding process 194

γ-carboxyglutamate 16
gas phase 163
  humidity 169
guanidinium salts 192

H-bonded triad 228
^{1}H NMR titrations 238
halides 230
Hartree-Frock methods 234
host
  artificial 113
  bifunctional 137
  electroneutral 198
  ditopic 240
  flexible 194
  neutral, for cations 235
  neutral, for anions
    137, 230
  zwitter-ionic 197
  -guest binding 192
  -guest complex 234
humidity 169
Huttig isotherm 175
hydrogen bonding 137
hydrolysis
  acetal 1
  ester 211
  rates 7

hysteresis 171

inhibitiion studies 4
  didemnins 244
interactions
  Lewis acid
    234, 235, 240
  ion-dipole 233, 236
  ion-ion 233
  ion-pairs 229
ions
  binding cations 235
  binding anions 137, 230
  zwitter-ionic 197
  alkaline-earth metal ions
    73
isotope effect, ^{18}O 1

Job plot 233, 239

kinetics
  inhibitiion studies 4
  maximum catalytic effect
    229
  multiple steady states
    178
  ^{18}O kinetic isotope effect
    1
  turnover catalysis 76

leaving groups 56

Lewis acids 231
lysozyme 2

macrocycle 213
maximum catalytic effect 229
membrane permeation 211
multiple steady states 178

nicotinamide adenine dinucleotide
(NADH) 35
NMR titrations 231
nonaqueous media 163

organoboron receptors 227
oxoanion
binding 192
extraction 197
hole 227

π-interactions 236
aromatic face-to-face orientation 239
π-stacking interaction 237
aromatic face-to-face orientation 239
paraquat 236
peptidyl -helix 227
peptidyltransferase 248
peroxide 18
phase transfer of anions 197
phosphate anions 195
preorganisation 201, 227, 239
flexible hosts 194
host flexibility 198
protease-catalyzed acyl transfer 55
protease-catalyzed peptide synthesis 53
protein synthesis 243
pyridoxal 204
pyridoxal phosphate
(PLP) 35

RA-VII 256
receptor
artificial 113
bifunctional 137
electroneutral 198
ditopic 240
flexible 194
neutral, for cations 235
neutral, for anions 137, 230
tripodal 218
tweezers-like 216
zwitter-ionic 197

S' subsite mapping 54
serine proteases 227
small molecule design 230
solid-gas bioreactor 177
stereoelectronic factors
theory of stereoelectronic control 1
antiperiplanar lone pair hypothesis 5
subsite specificity 55
supramolecular catalysts 211
synthesis in high-density media 62

thiamine pyrophosphate (TPP) 35
transacylase mimics 71
transamination 35, 204
turnover catalysis 76

unnatural amino acids 35

vitamin B_6 204

vitamin E 15
vitamin K 15
vitamin K hydroquinone 17
vitamin K oxide 16
vitamin KH_2 16

zymogen-catalyzed peptide synthesis 64

GPSR Compliance

The European Union's (EU) General Product Safety Regulation (GPSR) is a set of rules that requires consumer products to be safe and our obligations to ensure this.

If you have any concerns about our products, you can contact us on ProductSafety@springernature.com

In case Publisher is established outside the EU, the EU authorized representative is:

Springer Nature Customer Service Center GmbH
Europaplatz 3
69115 Heidelberg, Germany

Batch number: 10761697

Printed by Printforce, the Netherlands